# Industrial Electricity & Motor Controls

## ABOUT THE AUTHORS

**Rex Miller** is Professor Emeritus of Industrial Technology at the State University College at Buffalo. He has taught technical courses on all levels, from high school through graduate school, for more than 40 years. Dr. Miller is the author or coauthor of over 100 textbooks, including McGraw-Hill's best-selling *Carpentry and Construction, Electrician's Pocket Manual,* and *Electricity and Electronics for HVAC.*

**Mark R. Miller** is Professor and Coordinator of Industrial Technology at the University of Texas at Tyler. Dr. Miller is the author or coauthor of over 30 textbooks for trade and technical programs, including Audel's *Refrigeration: Home and Commercial,* Audel's *Air Conditioning: Home and Commercial,* and McGraw-Hill's *Electricity and Electronics for HVAC.*

# Industrial Electricity & Motor Controls

**REX MILLER**
Professor Emeritus
State University College at Buffalo

**MARK R. MILLER**
Professor
University of Texas at Tyler

New York   Chicago   San Francisco   Lisbon   London
Madrid   Mexico City   Milan   New Delhi   San Juan
Seoul   Singapore   Sydney   Toronto

Cataloging-in-Publication Data is on file with the Library of Congress.

McGraw-Hill books are available at special quantity discounts to use as premiums and sales promotions, or for use in corporate training programs. For more information, please write to the Director of Special Sales, Professional Publishing, McGraw-Hill, Two Penn Plaza, New York, NY 10121-2298. Or contact your local bookstore.

Industrial Electricity & Motor Controls

Copyright © 2008 by The McGraw-Hill Companies, Inc. All rights reserved. Printed in the United States of America. Except as permitted under the United States Copyright Act of 1976, no part of this publication may be reproduced or distributed in any form or by any means, or stored in a database or retrieval system, without the prior written permission of the publisher.

1 2 3 4 5 6 7 8 9 0    QPD/QPD    0 1 2 1 0 9 8 7

ISBN 978-0-07-154476-4
MHID 0-07-154476-3

This book is printed on acid-free paper.

**Sponsoring Editor**
  Larry S. Hager

**Acquisitions Coordinator**
  Alexis Richard

**Editorial Supervisor**
  David E. Fogarty

**Project Manager**
  Preeti Longia Sinha

**Copy Editor**
  Ragini Pandey
  Surendra Nath Shivam

**Proofreader**
  Divya Kapoor

**Indexer**
  Dr. Rex Miller

**Production Supervisor**
  Pamela A. Pelton

**Composition**
  International Typesetting and Composition

**Art Director, Cover**
  Jeff Weeks

---

Information contained in this work has been obtained by The McGraw-Hill Companies, Inc. ("McGraw-Hill") from sources believed to be reliable. However, neither McGraw-Hill nor its authors guarantee the accuracy or completeness of any information published herein, and neither McGraw-Hill nor its authors shall be responsible for any errors, omissions, or damages arising out of use of this information. This work is published with the understanding that McGraw-Hill and its authors are supplying information but are not attempting to render engineering or other professional services. If such services are required, the assistance of an appropriate professional should be sought.

# Contents

*Preface xiii*

## Introduction

**Performance Objectives** 2
**What Is Electricity?** 2
**Elements and Atoms** 2
   Free Electrons 3
   Current Flow 3
   Energy 3
**Electrical Materials** 3
   Conductors 3
   Insulators 3
   Semiconductors 3
**Generating Electricity** 3
**Voltage and Current** 6
   Resistance 8
   Wire Size 9
   Copper versus Aluminum Wire 9
   Circuits 9
**Ohm's Law** 12
   Using Ohm's Law 12
**Power** 13
   Kilowatts 13
**Measuring Electricity** 13
   Meters 13
   Other Types of Meters 19
**Controlling Electricity** 20
   Switches 20
   Solenoids 23
   Relays 25
   Diodes 25
**Resistors** 25
   Resistor Color Code 27
**Review Questions** 27

## 1 Tools and Equipment

**Performance Objectives** 30
**Hand Tools** 30
   Screwdrivers 30
   Pliers 31
   Hammers 32
   Hacksaws 33
   Wrenches 33
   Nut Drivers 35
   Tools for Bearings and Bushings 35
   Solder-Less Connector Crimper 36
   Soldering Iron 36
**Soldering Gun** 36
   Wire Gages 36
   Fuse Puller 36
   Tachometer 37
   Knockout Punch 37
   Fish Tape 37
   Cable Stripper 37
   Cable Cutter 38
   Electrician's Knives 38
   Other Tools 38
**Electrical Tools** 38
   Portable Ammeter and Voltmeter 39
   Megohmmeter 39
   Voltage Tester 39
   Volt-Ohm-Milliammeter (VOM) 40
   Digital Logic Probe 40
   Continuity Tester 40
   Polarized Plug Tester 40
   Recorder 42
   Oscilloscope 42
   Phase Sequence Adapter 43
   Balance Analyzer 43
**Special Tools for Specific Jobs** 43
   Printed Circuit Board Puller 43
   Wire Wrap/Unwrap Tool 44
   Conduit Benders 44
   PVC Cutters 44
   Cable Benders 45
   Conduit Reamer 45
**Review Questions** 45
**Review Problems** 45

## 2 Safety in the Workplace

**Performance Objectives** 48
**Electrical Shock** 48
**Electrical Safety Measures** 48
   Using Ground-Fault Circuit Interrupters 48
   Safety Devices 49
   Circuit Protection 49
   Over currents 49
   Overloads 49
   Short Circuits 50
   Fuses 50
**Electrical Codes** 51
   National Electrical Codes 51
   Underwriters Laboratories 51
   Canadian Standards Association 52
   Standards in Other Countries 52
**OSHA** 53
   Equipment Standards 53
   OSHA Color Code 53
**Fire Extinguishers** 53
**Safe Working Practices** 53
**Tool Safety** 53
   Using Hammers Properly 56
   Working Clothes 56
   Disconnects 56
   Using Motors and Generators Safely 57
   Grounding 58
**Review Questions** 59
**Review Problems** 60

## 3 Symbols

**Performance Objectives** 62
**Electrical Symbols** 62
   Common Switch (Button) Symbols 62
   Standard Wiring Diagram Symbols 62
**Electronic Symbols** 62
   Resistor Color Code 62
   Electronic Symbols Compared 62
**Relay Contact Symbols** 62
**Line Diagrams, Wiring Diagrams, and Ladder Diagrams** 62
**Review Questions** 66
**Review Problems** 70

## 4 Control Circuits and Diagrams

**Performance Objectives** 72
**Wiring Diagrams** 72
   Start Circuits 72
**Line Diagram or Ladder Diagram** 74
**Undervoltage Release** 74
**Two-Wire Control Circuits** 74
**Thermostat Control** 75
**Undervoltage Protection** 76
**Three-Wire Control Circuits** 76
**Overcurrent Protection for Control Circuits** 78
**Transformers in Control Circuits** 79
**Review Questions** 81
**Review Problems** 81

## 5 Switches

**Performance Objectives** 84
**Controlling Electricity** 84
**Drum Switch** 84
   Easy Conversion 85
   Three-Phase Switching 85
   DC Switching 86
**Float Switches** 86
**Flow Switches** 86
**Foot Switches** 89
**Joystick** 90
**Interlock Switches** 90
   Mechanical Interlocking 90
   Pushbutton Interlocking 91
   Auxiliary Contact Interlocking 92
**Limit Switches** 93
   Types of Limit Switches 93
   Limit Switch Circuits 96
**Pressure Switches** 96
**Pushbutton Switches** 97
**Safety Switches** 98
**Selector Switches** 99
**Single-Pole Switches** 100
**Start-Stop Switches** 100
**Temperature Switches** 100
**Toggle Switches** 100
**Transistor Switching** 100
**Vacuum Switches** 102
**Review Questions** 102
**Review Problems** 102

## 6 Magnetism and Solenoids

**Performance Objectives** 106
**Solenoids** 106
   Solenoid Valves 107
   Automatic Gas Furnace Solenoid 108
**Construction of Solenoids** 108
   Eddy Currents in the Armature and Core 108
   Air Gap and the Armature 108
**Shading Coils** 109
**Solenoid Coils** 109
   Sealed Current Rating 110
   Coil Voltage 110
   Effects of Voltage Variation 110
**Alternating Current Hum** 111
**Review Questions** 111
**Review Problems** 112

## 7 Relays

**Performance Objectives** 114
**Relays** 114
    Relay Solenoids 114
    Uses for Relays 114
    Relay Armature 114
    Relay Contacts 115
**Solid-State Relays** 115
    Transistor 115
    Surge Protection 116
    Triacs 116
    Silicon-Controlled Rectifiers 117
**Phase-Failure Relays** 119
**Solid-State Monitoring Relays** 121
    Voltage Relay 121
    Current Relay 121
    Over/Under Relay 122
**Types of Solid-State Relay Switching** 122
    Switching Relay Loads 122
    Thermistor Relay 123
    Contact Amplifier Relay 123
    Load Detector and Load Converter Relays 123
    Typical Applications 124
**Thermal Overload Relays** 125
    Types of Thermal Overload Relays 125
    Hand-Reset Melting Alloy (NEMA Style) 126
    Bimetallic Overloads (NEMA Style) 126
    Temperature Compensation 126
**Electromagnetic Relays and Motors** 128
    Motor Winding Relays 129
**Electromechanical Relays** 130
**Relay Operating Characteristics** 131
    Contact Life 131
    Contact Construction 131
    Relay Coils 132
    Shock and Vibration 132
    Relays and Altitude 132
**Review Questions** 132
**Review Problems** 133

## 8 Electric Motors

**Performance Objectives** 136
**Motor Classifications** 136
**DC Motors** 136
    Operating Principles 136
    Counter Electromotive Force 137
    Loads 137
    Types of DC Motors 138
    Series DC Motors 138
    Shunt DC Motors 138
    Compound DC Motors 139
    Types of Armatures 139
    Direction of Rotation 139
    Motor Speed 139
    Armature Reaction 140
    Compensating Windings and Interpoles 140
    DC Motor Starting Resistance 141
    DC Motor Characteristics and Applications 141
    Troubleshooting DC Motors 141
**AC Motors** 142
    Types of AC Motors 142
    Series AC Motor 142
    Magnetic Fields in AC Motors 142
    Rotating Magnetic Field: Two-Phase 143
    Two-Phase Waveforms 143
    Rotating Magnetic Field: Three-Phase 144
    Synchronous Motors 144
    Squirrel-Cage Motor 145
    Starting the Synchronous Motor 145
    Induction Motors 146
    Stator Construction 146
    Slip 147
    Single-Phase Induction Motors 148
    Split-Phase Induction Types 148
    Shaded-Pole Induction Motors 149
    Speed and Slip in Squirrel-Cage Motors 150
    Rotation 150
    Torque and Horsepower 151
    Locked-Rotor Torque 151
    Breakdown Torque 151
    Locked-Rotor kVA/hp 152
    Starting Squirrel-Cage Motors 152
**Review Questions** 152
**Review Problems** 153

## 9 Timers and Sensors

**Performance Objectives** 156
**Time and Timers** 156
**Dashpot or Pneumatic Time-Delay Timers** 156
**Synchronous Clock Timers** 156
**Solid-State Timers** 156
**Time-Delay Relays** 157
**General Purpose Timing Relays** 158
**Control of Programmable Timers** 158
**Digital Solid-State Timer/Counter** 159
    Single-Stage Version 159
    Time/Count Modes 159
    Dual-Stage Version 159
    Thumbwheel Switches 160
**DIP Switches** 161
**Pneumatic Timing Relays** 161
**Motor-Driven Timers** 163
**Sequence Control** 164
**Programmable Timers** 165
**Sensors** 165

Solid State Level Controls   165
Capacitance-Type Sensor   166
Temperature Sensing   166
Thermocouple   167
Thermistor   167
Resistance Temperature Devices   167
Semiconductor Temperature Sensor   169
Pressure Sensors   169
Strain Gage Transducers   169
Piezo-resistive Transducers   169
**Review Questions   170**
**Review Problems   170**

## 10 Sensors and Sensing

**Performance Objectives   172**
**Classes of Sensors   172**
Contact Sensors   172
Noncontact Sensors   172
**Presence Sensing   172**
**Limit Switches   172**
**Speed Switches   174**
Operation of the Speed Switch   175
**Pressure Controls   176**
**Temperature Controls   177**
Wells and Packing Glands   177
**Float Switches   179**
**Encoders   179**
Optical Programmable Controller Encoders   179
Multiplexing   182
**Proximity Switches   182**
Operation   182
Solid-State Switches   182
Inductive Cylindrical Switch   182
Extended-Sensing-Range Inductive Proximity Switch   183
Compact Inductive Proximity Switch   183
Self-Contained Low Profile Proximity Switch   183
**Photoelectric Switches   183**
Operation   184
Features   184
Photoelectric Light Sources   185
Mark Sensors   185
Clear Material Detection   186
**Switch Applications   186**
**Automatic Identification   186**
**Radio-Frequency Identification Systems   187**
The Antenna   188
The RF Tag   188
**Bar Code Readers and Decoders   188**
Hand-Held Scanners   189
Moving-Beam Scanners   189
Multiplexing   189
Vision System   189
**Review Questions   190**

## 11 Solenoids and Valves

**Performance Objectives   192**
**Solenoids   192**
**Industrial Solenoids   192**
Tubular Solenoids   192
Frame Solenoids   192
**Applications   195**
Solenoids as Electromagnets   195
Solenoid Coils   195
Servicing Coils   196
**Solenoid Valves in Circuits   196**
Refrigeration Valve   197
**Review Questions   198**

## 12 Motor Starting Methods

**Performance Objectives   200**
**Electric Motors   200**
**Starting the Motor   200**
Reversibility   201
Uses   201
**Repulsion-Induction Motor   201**
Uses   202
**Capacitor-Start Motor   202**
**Reversibility   202**
Uses   203
**Permanent Split-Capacitor Motor   204**
**Shaded-Pole Motor   204**
Reversibility   205
Uses   205
**Split-Phase Motor   205**
**Polyphase Motor Starters   206**
**Reduced-Voltage Starting Methods   206**
**Primary Resistor Starting   208**
Operation   209
**Autotransformer Starting   209**
Operation   210
**Part-Winding Starting   211**
Operation   212
Advantages and Disadvantages   212
**Wye-Delta or Star-Delta Starters   212**
Operation   213
Advantages and Disadvantages   213
Star-Delta (Wye-Delta) Connections   213
**Multispeed Starters   214**
Low-Speed Compelling Relay   214
Automatic Sequence Accelerating Relay   214
Automatic Sequence Decelerating Relay   214

Consequent-Pole Motor Controller  215
Full-Voltage Controllers  217
Starting Sequence  217
Protection against Low Voltage  218
Time-Delay Protection  218
Review Questions  221

## 13 Solid-State Reduced-Voltage Starters

Performance Objectives  224
Electromechanical Devices  224
Reduced-Voltage Starting  224
Silicon-Controlled Rectifiers  224
Solid-State Stepless Acceleration  226
Logic Rack  226
    B-2 Module  226
    B-3 Module  226
    B-4 Module  227
    B-5 Module  228
    B-6 Module  228
    B-7 Voltage Monitor Module  228
    B-8 Energy-Saving Module  228
Shorted SCR Switch  228
Elementary Wiring Diagrams for Solid State  228
Diac  229
Triac  230
Light-Emitting Diodes  231
Using Solid-State Control and Electromagnetic Devices  231
Surge Suppressors  231
Lightning Protection  232
Review Questions  232

## 14 Speed Control and Monitoring

Performance Objectives  234
Motor Speed Control  234
Squirrel-Cage Motors  234
Synchronous Motors  234
    Excitation  235
    Speed  235
    Starting  236
    Starting Methods  236
    Uses for Synchronous Motors  238
Wound-Rotor Motors  238
    Speed Regulation by Resistance  238
    Types of Speed Control  239
    Multiswitch Starters  239
    Drum Controllers  240
    Magnetic Starters  240
    Resistors  240
    Solid-State Adjustable-Speed Controllers  241
Frequency Speed Control  241
Multispeed Starters  242
Speed Monitoring  242
Review Questions  243

## 15 Motor Control and Protection

Performance Objectives  246
Motor Control  246
Manual Starters  246
Solid-State Motor Controller  248
Sequence Control  249
When Starting any One Requires Another  250
    Automatic Sequence Control  250
Jogging  252
    Forward or Reverse Jogging  253
Plugging  254
    Plugging a Motor to Stop from Either Direction  254
    Antiplugging  254
Braking  255
    Electronic Motor Brake  255
    Mechanical Braking  257
    Thruster Brakes  258
    Magnetic Brakes  258
    Hydraulic Brakes  259
Motor Protection  259
    Surge Protection and Backspin  259
    Overload Protection  260
    Automatic Reset Overload Relay  260
    Inverse Time Current Relay  261
    Solid-State, Line-Voltage and Line-Current Monitor Relays  261
    Programmable Motor Protection  262
    Protective Module  262
    Remote Temperature Detector Module  262
Review Questions  263

## 16 Three-Phase Controllers

Performance Objectives  266
The Three-Phase Motor  266
Starters  266
    Simplest Type  266
    Full-Voltage Reversing Starter  267
    Forward-Reverse-Stop  267
    Start-Stop Jog  267
    Reversing Starter  268
    Two-Speed Starter  268
Duplex Motor Controllers  269
Medium-Voltage Controllers  269
Solid-State Motor Controller  274
    Modules  275
    Solid-State Advantages  275
Motor Control Centers  275
Review Questions  276

## 17 Drives

Performance Objectives   278
Fans, Blowers, and Pumps   278
Adjustable-Speed Drives   278
Adjustable-Frequency AC Drives   278
CSI Inverters   279
Variable-Voltage Inverters   279
Pulse-Width-Modulated Inverters   280
Variable-Pitch Drives   281
Wound-Rotor AC Motor Drives   282
DC Drives   282
Regulated Speed Drives   282
Open-Loop Controls (Traditional Approach)   282
Closed-Loop Controls   284
DC Drives: Advantages and Disadvantages   284
DC Drives and Systems   284
Solid-State Digital AC Drives   285
Review Questions   286

## 18 Transformers

Performance Objectives   288
Transformers   288
Iron-Core Transformer   288
  Turns Ratio   288
Autotransformers   289
  Turns Ratio   289
Transformer Losses   289
  Eddy Current Losses   289
  Hysteresis Losses   290
  Copper Losses   290
Core Construction   290
Transformer Oil   291
  EPA Highlights   292
Autotransformers: Three- and Single-Phase   292
  Impedances   293
  Third Harmonics   293
  Grounding   294
Buck and Boost Transformers   294
  Calculating the Load   294
  Calculating Single-Phase Kilovolt-Amperes   295
  Three-Phase Buck and Boost Transformers   296
  Calculating Three-Phase Kilovolt-Amperes   296
Dry-Type Transformers   296
  Cooling   297
  Insulation and Temperature Rise   297
  Taps   297
Special Transformers   297
Liquid-Filled Transformers   299
  Safety Notice   299
Submersible Transformer   299
  Maintenance   299
Transformer Applications   301
  Distributing Power at High Voltages   301
  Double-Wiring Elimination   301
  Operating 120/240-V Equipment from Power Circuits   301
  Isolating Circuits   302
  Changing from a Four-Wire to Three-Wire Circuit   302
  Stepping-Up or Stepping-Down Voltage   302
Motor Transformers   302
Review Questions   306
Review Problems   306

## 19 Power Generation

Performance Objectives   308
Basic Principle   308
  Sine Wave   309
Components of the Alternator   309
  Rotors   309
  Stators   309
  Exciters   310
Types of Alternators   311
  Single-Phase   311
  Two-Phase   311
  Three-Phase   312
Wye and Delta Configurations   312
  Wye Connection   312
  Delta Connection   312
  Power in a Balanced Wye   313
  Power in a Balanced Delta   313
Frequency   313
Load Changes   314
  Armature Resistance   314
  Armature Reactance   314
  Armature Reaction   314
Voltage Regulation   314
Standby or Emergency Power Sources   314
  Transfer Switch   314
  Types of Standby Generators   315
  Automatic Transfer Switches   316
  Electronic Loads   317
  Uninterruptible Power Systems   317
  Multiple-Engine Generator Sets   317
  Systems for Paralleling Multiple Power Sources   318
  Selective Load Transfer Systems   318
Prime Mover Systems   319
  Two-Source Systems   319
  Multiple-Source Systems   319

Power Management Systems **320**
  Peak Load-Sharing Systems 320
  Utility-Isolated Peaking Transfer System 320
  Utility-Paralleled Peaking Systems 321
**Review Questions 321**
**Review Problems 322**

# 20 Power Distribution Systems

**Performance Objectives 324**
**Sources of Commercial AC 324**
  Falling-Water Generators 324
  Switchyard 325
  Nuclear Power Generators 325
  Fossil-Fuel Power Generators 325
**Distribution 328**
  Types of Distribution Systems 328
**Systems Analysis 333**
  Short-Circuit Currents 334
  Fault Currents 334
  Three-Phase Power 334
**Measurement of Power 335**
**Transformers 336**
  Transformer Construction 336
  Three-Phase Transformer Connections 337
**Switching 341**
  Automatic Emergency Power Switching 342
  Switchboards 344
  Panel Boards 345
  Raceways 345
  Aluminum Lay-In Wall Duct and Floor Trench Duct 346
  Cable Tray 346
**Review Questions 348**
**Review Problems 348**

# 21 Programmable Controllers

**Performance Objectives 352**
  NEMA PLC Definition 352
**Solid-State Electronics 352**
  Standard Electronic Features in Controls 352
**The Programmable Controller 353**
  Input-Output 353
  Interfacing 353
  ASCII Code 353
  Parallel Ports 354
  Serial Ports 354
**The Environment Affects Performance 354**
**Electrical Noise 354**
  Installation Practices 355
  Grounding 355
**Solid-State Reliability 355**
  Backup Operations 355
**Controllers 356**
  System Hardware and Programming Equipment 356
  Processors 356
  Input-Output Modules 358
  Intelligent I/O Modules 358
  Display Systems 359
  Cell Controllers 360
  Micro-Cell Controller 360
  Local Area Network 361
  Future of PLCs 361
**Review Questions 363**

# 22 Troubleshooting and Maintenance

**Performance Objectives 366**
**Troubleshooting and the Electrician 366**
**Preventive Maintenance 366**
  Damp and Wet Areas 366
  Prevention of Accidental Shock 367
  Ground-Fault Receptacles 367
  Wiring Devices 367
**Maintenance of Small Electric Motors 370**
  Adequate Wiring 370
  Check Internal Switches 370
  Check Load Condition 370
  Extra Care in Lubrication 370
  Keep Commutators Clean 370
  Motors Must Have Proper Service Rating 370
  Replace Worn Brushes 370
**Motor Problems 370**
  Ball Bearing Motors 370
  Sleeve Bearing Motors 371
**Common Motor Problems and Their Causes 371**
  Problem Diagnosis 372
  Centrifugal Switches 372
  Commutator-Type Motors 372
**Troubleshooting Aids 372**
  Connection Diagrams 372
  Small Three-Phase Motor Rating Data 372
**Power Supply Disturbances 375**
  Voltage Fluctuations 376
  Transients 376
  Power Outages 377
  Looking for Shorts 377
**Motors with Squirrel-Cage Rotors 377**
**Testing the Centrifugal Switch in a Single-Phase Motor 377**

Testing for Short Circuits between Run
  and Start Windings   378
**Capacitor Testing**   378
  Measuring Capacity of a Capacitor   378
**Using Meters to Check for Problems**   379
  Using a Volt-Ammeter for Troubleshooting
    Electric Motors   379
  Clamp-On Volt-Ammeter   379
  Looking for Grounds   380
  Looking for Opens   380
**Troubleshooting Guide**   380
**Motor Life**   384
  Ventilation   384
  Ambient Temperature   384
**Performance Characteristics**   385
  Voltage Unbalance   385
**Motor Protection**   386
**DC Motor Problems**   386
**Solid-State Equipment
  Troubleshooting**   391
**Review Questions**   391

## 23  Careers in Electricity

**Performance Objectives**   394
  Industrial Electrician   394
**Industries That Employ Electricians**   394
**Careers in Electricity**   394
  Electrical Engineer   395
  Construction Electrician   395
  Maintenance Electrician   396
  Limited Energy Electrician   396
**General Information**   396
  Information Sources   397

**Starting Your Own Business**   397
**The Development of Leadership**   398
  Employment of Electricians   398
  Earnings   398
  Related Occupations   398
  Sources of Additional Information   399

## Glossary

## Appendices

  I. DC Motor Trouble Chart   405
  II. Wound-Rotor Motor Trouble
      Chart   407
  III. Fractional-Horsepower Motor
       Trouble Chart   409
  IV. Selection of Dual-Element Fuses
      for Motor-Running Overload
      Protection   415
  V. Tables and Formulas   417
  VI. Full-Load Currents of AC and
      DC Motors   423
  VII. Power Factor Correcting
       Capacitors   425
  VIII. Switch Symbols   429
  IX. Wiring Diagram Symbols   431
  X. Unit Prefixes   433
  XI. Conversion Factors   435
  XII. Decibel Table   437

*Index   439*

# Preface

*Industrial Electricity & Motor Controls* has been designed for use by industrial electricians and apprentices who install and operate electrical systems as well as for others concerned with motor controls. These include electricians, technicians, engineers, electrical contractors and such as drafters and designers of electrical systems. We assume that the reader has a working knowledge of basic electricity and electric motor theory and operation, but a quick review of basics is provided for those who can use it. The book builds on that base to provide a working knowledge of the many aspects of maintenance systems.

All the information needed for a thorough understanding of motor controllers and their theory, operation, installation, and maintenance of various electrical systems is provided. The book is designed primarily for use by apprentice training programs, journeyman training, vocational-technical schools and two-year colleges where a need for a better grasp of the terminology and processes are present.

A glossary is provided for use by the technician and apprentice to aid in keeping current with the rapidly evolving terminology and equipment.

There are ample illustrations for showing the equipment long established in this field. However, the field is rapidly progressing and this material can serve as a valuable resource for those who are attempting to keep up with the latest developments in motor control. Obviously, not all problems can be presented here; a great deal of on-the-job ingenuity is required.

As you know, it is not possible to learn to swim without first getting a feel for the water. Neither is it possible to learn all there is to know about electrical controls by reading about them. You must be willing to get practical experience and devote time to the development of skills related to the job.

It is the authors' hope that the book will serve as an appetizer to those really interested in going into this exciting, yet demanding field.

REX MILLER
MARK R. MILLER

# Introduction

## A Quick Review of the Basics

## PERFORMANCE OBJECTIVES

After studying this chapter, you will be able to:

- Give a simple definition of current electricity.
- List seven ways in which electricity is generated.
- Identify the four factors that determine resistance.
- Identify the various types of circuits.
- State Ohm's law.
- Discuss the ways in which electricity is measured.
- Identify the switches used to control electricity.
- Read a resistor color code.

## WHAT IS ELECTRICITY?

Though you cannot see electricity, you are aware of it everyday. You see it used in countless ways. You cannot taste or smell electricity, but you can feel it. You can taste food cooked with its energy. You can smell the gas (ozone) formed when lightning passes through the air.

Basically there are two kinds of electricity: static (stationary) and current (moving). This book is chiefly about current electricity because that is the kind commonly put to use. Current electricity can be simply defined as the flow of electrons along a conductor. To understand the definition, you must know something about chemical elements and atoms.

## ELEMENTS AND ATOMS

Elements are the most basic materials in the universe. Ninety-four elements, such as iron, copper, and nitrogen, have been found in nature. Scientists have made 11 others in laboratories. Every known substance—solid, liquid, or gas—is composed of elements.

It is very rare for an element to exist in a pure state. Nearly always the elements are found in combinations called compounds. Even such a common substance as water is a compound rather than an element (Fig. I-1).

An atom is the smallest particle of an element that retains all the properties of that element. Each element has its own kind of atom. That is, all hydrogen atoms are alike, and they are different from the atoms of all other elements. However, all atoms have certain things in common. They all have an inner part, known as the nucleus, which is composed of tiny particles called protons and neutrons. An atom also has an outer part consisting of other tiny particles, called electrons, which orbit around the nucleus (Figs. I-2 and I-3).

**Fig. I-1** *Two or more atoms linked are called a molecule. Here two hydrogen atoms and one oxygen atom form a molecule of the compound water ($H_2O$).*

**Fig. I-2** *Atomic Structure: (A) Hydrogen atom; (B) Aluminum atom. Atoms contain protons, neutrons, and electrons.*

**Fig. I-3** *Molecular structure.*

Neutrons have no electrical charge, but protons are positively charged. Electrons have a negative charge. Because of these charges, protons and electrons are energy particles, that is, these charges form an electric field of force within the atom. Stated very simply, these charges are always pulling and pushing each other, which makes energy in the form of movement.

The atoms of each element have a definite number of electrons and protons. A hydrogen atom has one electron and one proton. An aluminum atom has 13 of each. The opposite charges: negative electrons and positive protons, attract each other and tend to hold electrons in orbit. As long as this arrangement is not changed, an atom is electrically balanced.

However, electrons of some atoms are easily pushed or pulled out of their orbits. This ability of electrons to move or flow is the basis of current electricity.

## Free Electrons

In some materials, heat loosens electrons from their atoms. In other materials such as copper, electrons may be easily forced to drift, even at room temperatures. When electrons leave their orbits, they may move from atom to atom at random, drifting in no particular direction. Electrons that move in such a way are referred to as free electrons. However, a force can be applied to direct them in a definite path.

## Current Flow

If the movement of free electrons is channeled in a given direction, a flow of electrons, commonly referred to as current flow, occurs. Thus you see that the movement of electrons is related to current electricity.

## Energy

Electrons are incredibly small. The diameter of an electron is about 0.00000000000022 in. You may wonder how anything so small can be a source of energy. Much of the answer lies in the fact that electrons move with nearly the speed of light. Also, billions of them can move at once through a wire. The speed and concentration produce great energy.

## ELECTRICAL MATERIALS
### Conductors

A conductor is a material through which electrons move. Actually, all metals and most other materials are conductors to some extent. Some, however, are better than others. Thus the term conductor is usually used to mean a material through which electrons move freely.

What makes one material a better conductor than another? A material that has many free electrons tends to be a good conductor. For practical purposes, however, there are other points that must be considered when choosing a material to use as a conductor. For example, gold, silver, aluminum, and copper are all good conductors. However, the cost of gold and silver limits their use. Copper, because of its superior strength in both hot and cold weather, is preferred over aluminum for many uses.

### Insulators

An insulator is a substance that restricts flow of electrons. Such materials have a very limited number of free electrons. Thus you see that the movement of free electrons classifies a material as either a conductor or an insulator. No material is known to be a perfect insulator, that is, entirely void of free electrons. However, some materials are such poor conductors that for all practical purposes, they are placed in the insulator class.

Wood, glass, mica, and polystyrene are insulators (Fig. I-4). They have varying degrees of resistance to the movement of their electrons. The higher the line on the chart in Fig. I-4, the better are the insulating qualities of the material.

### Semiconductors

You have heard the word "semiconductor" in relation to transistors and diodes used in electronic equipment. Materials used in the manufacture of transistors and diodes have a conductivity between that of a good conductor and a good insulator. Therefore, the name semiconductor is given to them. Germanium and silicon are the two most commonly known semiconductors. Through the introduction of small amounts of other elements, these nearly pure (99.999999%) elements become limited conductors. The manufacture of semiconductors is a fascinating process. However, it would take too long to go into details at this time. You may wish to research the topic on your own by checking out a book from your library or using the Internet.

## GENERATING ELECTRICITY

There are several ways to produce electricity. Remember that electricity is the flow of electrons along a conductor. Friction, pressure, heat, light, chemical action, and magnetism are among the more practical methods

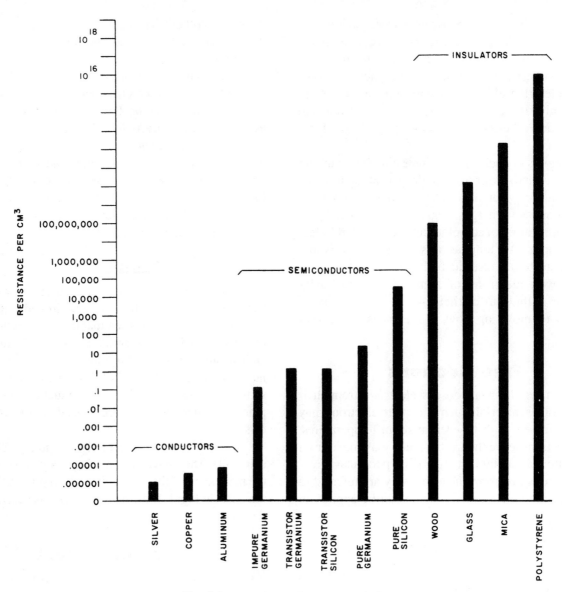

***Fig. I-4*** *Resistances of various materials.*

used to make electrons move along a conductor. Other methods (sometimes called "exotic") are used to generate electricity for special purposes. For instance, experimental cells developed for the space program are termed "exotic."

- **Friction**. Electricity is produced when two materials are rubbed together. The movement of your shoes against the carpet can cause static electricity. Some practical applications of static electricity are manufacture of sandpaper and cleaning of polluted air (Fig. I-5).
- **Pressure**. Electricity is produced when pressure is applied to certain crystals, these are usually Rochelle salts or quartz. Special properties of the crystals are utilized in crystal microphone (Fig. I-6). Here, bending of the crystal produces a small electrical output. This phenomenon is known as the piezoelectric effect. The small voltage thus produced can be amplified to drive a speaker. In fact, crystal pickups are used in inexpensive record players and for some industrial jobs.
- **Heat**. Electricity is produced when heat is applied to the junction between two dissimilar metals. This junction is usually referred to as a thermocouple. Thermocouples are used to measure temperatures in industrial applications. This is especially true in checking the temperature of kilns for ceramic work (Fig. I-7).

**4   Introduction**

**Fig. I-5** *Electrostatic precipitator uses a two-stage method of cleansing air by collecting ionized particles on charge plates. Ionizing wires of tungsten are charged with 12,000-V dc (+). All particles are then electrically charged by ionization (+). The positively charged particles are attracted to the negatively charged plates with 6000 V (−). The negatively charged plate is the collecting cell. Clean air is exhausted from the precipitator.*

**Fig. I-6** *(A) Basic principle of crystal microphone operation; (B) crystal microphones: directly actuated type and diaphragm type.*

**Fig. I-7** *A thermocouple.*

**Generating Electricity**

- **Light**. Electricity is produced when light strikes a photosensitive material. (The word "photo" means light.) Photoelectric cells are used in cameras, spacecraft, and radios (Fig. I-8).

**Fig. I-8** Photoelectric cell.

- **Chemical action**. Electricity is produced when a chemical action takes place between two metals in a cell. A single unit is called a cell. Connecting two or more cells together produces a battery. Batteries are used in flashlights, radios, hearing aids, toys, cameras and calculators. Automobiles use a lead-acid cell combination. You cannot start cars today without a battery. Many types of cells are available today (Fig. I-9).
- **Magnetism**. Electricity is produced when a magnet is moved past a piece of wire or a piece of wire is moved through a magnetic field. The result is the same. Motion, a magnetic field, and a piece of wire are needed to produce electricity. To date, magnetism is the most inexpensive way of producing electrical power. We use magnetism to produce electricity for homes and cars. An electric generator is found under the hood of every automobile. This device can produce great amounts of electrical energy. The electric generator is called an alternator because it generates alternating current (ac). AC flows first in one direction and then in the other. Direct current (dc) flows in one direction only (Fig. I-10).
- **Exotic generators**. The fuel cell is one of the latest developments for production of electricity. The oxygen-concentration cell includes an electrolyte. The electrolyte conducts electric charge in the form of oxygen ions, but acts as an insulator to electrons. The electrolyte is located between two electrodes. (The electrolyte is wet, and the electrodes are usually metal rods or sheets.) By causing oxygen of different concentrations to pass by the electrodes, it is possible to produce electricity.

The hydrogen-oxygen cell produces water and electricity. Such a cell was used on one of the spaceflights to supply both drinking water and electricity in a very small space. Other exotic cells—not all of them perfected yet—are the redox fuel cell, hydrocarbon fuel cell, ion-exchange membrane, and magnetohydrodynamic (MHD) generator (Fig. I-11). In the MHD generator, hot plasma is generated and seeded in a burner, similar to a rocket engine. It then travels through a magnetic field applied at right angles to the flow and past electrodes exposed to this stream of gas. Electrons in the gas are deflected by the field. Between collisions with the particles in the gas, they make their way to one of the electrodes. Electricity flows as the electrons move from the cathode, through the load, to the anode, and back again to the gas stream. There are thousands of other methods of producing electricity.

## VOLTAGE AND CURRENT

So far you have become aware of what electricity is. You have learned some of the ways it is produced. Now it is time to learn how electrical energy is measured. The units of measurement most frequently used are volt (V) for voltage and ampere (A) for current.

- **Volts**. We measure the difference in potential between two plates in a battery in terms of volts. It is actually the electrical pressure exerted on electrons in a circuit. (A circuit is a pathway for the movement of electrons.) An external force exerted on electrons to make them flow through a conductor is known as electromotive force (EMF), which is measured in volts. Electrical pressure, potential difference, and EMF mean the same thing. The terms "voltage drop" and "potential drop" can be interchanged.
- **Current**. For electrons to move in a particular direction, there must be a potential difference between two points of the EMF source. If $6.28 \times 10^{18}$ electrons pass a given point in 1 s, a current of 1 A flows through the point. The same number of electrons stored on an object (static charge) and not moving is called a coulomb (symbol C). Current is assumed to flow from the negative ($-$) to positive ($+$) terminal of a battery or generator.

Current is measured in amperes. In electronics, it is sometimes necessary to use smaller units of measurement. The milliampere (mA) is used to indicate 1/1000 of an ampere (0.001 A). If an even smaller unit is needed, it is usually the microampere ($\mu$A). The microampere is one-millionth of an ampere, which may also be written as 0.000001 A. The Greek letter

*Fig. 1-9* (A) A secondary cell is one that can be recharged. Here the charge and discharge cycles are shown. (B) Cutaway view of a mercury cell.

**Voltage and Current 7**

**Fig. I-10** Automobile alternator.

**Fig. I-11** One exotic power source is the MHD generator.

"mu" ($\mu$) is used to indicate micro. (Table I-A lists the Greek alphabet and the terms they designate.)

A voltmeter is used to measure voltage. An ammeter is used to measure current in amperes. A microammeter or a milliammeter may be used to measure smaller units of current.

## Resistance

The movement of electrons along a conductor meets with some opposition. This opposition is resistance. Resistance is useful in electrical and electronics work. Resistance makes it possible to generate heat, control electron flow, and supply the correct voltage to a device.

Resistance in a conductor depends on four factors: material, length, cross-section area, and temperature.

- **Material.** Some materials offer more resistance than others. The resistance depends upon the number of free electrons present in the material.
- **Length.** The longer the wire or conductor, the more resistance it has. The resistance varies directly with the length of the wire.
- **Cross-sectional area.** Resistance varies inversely with the cross-sectional size of the conductor. In

**8** Introduction

**Table I-A** Greek Alphabet

| Name | Capital | Small | Used to Designate |
|---|---|---|---|
| alpha | A | α | Angles, area, coefficients, and attenuation constant |
| beta | B | β | Angles and coefficients |
| gamma | Γ | γ | Electrical conductivity and propagation constant |
| delta | Δ | δ | Angles, increment, decrement, and determinants |
| epsilon | E | ε | Dielectric constant, permittivity, and base of natural logarithms |
| zeta | Z | ζ | Coordinates |
| eta | H | η | Efficiency, hysteresis, and coordinates |
| theta | Θ | ϑ θ | Angles and angular phase displacement |
| iota | I | ι | Coupling coefficient |
| kappa | K | κ | |
| lambda | Λ | λ | Wavelength |
| mu | M | μ | Permeability, amplification factor, and prefix micro |
| nu | N | ν | |
| xi | Ξ | ξ | |
| omicron | O | o | |
| pi | Π | π | Pi = 3.1416 . . . |
| rho | P | ρ | Restivity and volume charge density |
| sigma | Σ | σ ς | Summation |
| tau | T | τ | Time constant and time-phase displacement |
| upsilon | Υ | υ | |
| phi | Φ | ϕ φ | Magnetic flux and angles |
| chi | X | χ | Angles |
| psi | Ψ | ψ | Dielectric flux |
| omega | Ω | ω | Resistance in ohms and angular velocity |

other words, the larger the wire, the smaller the resistance per foot.

- **Temperature**. For most materials, the higher the temperature, the higher the resistance. However, there are some exceptions to this in devices known as thermistors. Thermistors change resistance with temperature. The resistance decreases as the temperature increases. Thermistors are used in meters and as temperature indicators.

Resistance is measured in the unit called *ohm*, which is denoted by the Greek letter "omega" ($\Omega$).

## Wire Size

As you become more familiar with electricity and circuits and with some of the requirements for wiring a house or building, you will become more aware of the current carrying abilities of wire. Size of the wire is given in numbers. This size usually ranges from 0000 (referred to as 4 noughts) to No. 40. The larger the wire, the smaller its number.

For instance, No. 32 wire is smaller than No. 14. Table I-B shows the resistance (in ohms per 1000 ft) in relation to the cross-sectional area. Note how the temperature affects the resistance at 77°F and 149°F (25°C and 65°C). Temperature can make quite a difference in resistance for long wires. Long wires pick up heat and expand when exposed to summer weather.

## Copper versus Aluminum Wire

Although silver is the best conductor, its use is limited because of high cost. Two commonly used conductors are aluminum and copper. Each has advantages and disadvantages. For instance, copper has high conductivity and is more ductile (can be drawn out thinner). It is relatively high in tensile strength and can be soldered easily. But it is more expensive than aluminum.

Aluminum has only about 60% of the conductivity of copper. It is used in high-voltage transmission lines and sometimes in home wiring. Its use has increased in recent years. However, many electricians will not use it to wire a house today. There are a number or reasons for this, which will become apparent as we progress through the book.

If copper and aluminum are twisted together, as in a wire nut connection, it is possible for moisture to get to the open metals over a period of time. Corrosion will take place, causing a high-resistance joint. This can result in a dimmer light or a malfunctioning motor.

## Circuits

- **Complete circuit**. A complete circuit is necessary for the controlled flow or movement of electrons along a conductor (Fig. I-12A). A complete circuit is made up of a source of electricity, a conductor,

*Table I-B   Standard Annealed Solid Copper Wire*

| Gage Number | Diameter (Mils) | Cross Section | | Ohms Per 1000 Feet | | Ohms Per Mile 25°C (= 77°F) | Pounds Per 1000 Feet |
|---|---|---|---|---|---|---|---|
| | | Circular Mils | Square Inches | 25°C (= 77°F) | 65°C (= 149°F) | | |
| 0000 | 460.0 | 212 000.0 | 0.166 | 0.0500 | 0.0577 | 0.264 | 641.0 |
| 000 | 410.0 | 168 000.0 | 0.132 | 0.0630 | 0.0727 | 0.333 | 508.0 |
| 00 | 365.0 | 133 000.0 | 0.105 | 0.0795 | 0.0917 | 0.420 | 403.0 |
| 0 | 325.0 | 106 000.0 | 0.829 | 0.100 | 0.116 | 0.528 | 319.0 |
| 1 | 289.0 | 83 700.0 | 0.0657 | 0.126 | 0.146 | 0.665 | 253.0 |
| 2 | 258.0 | 66 400.0 | 0.0521 | 0.159 | 0.184 | 0.839 | 201.0 |
| 3 | 229.0 | 52 600.0 | 0.0413 | 0.201 | 0.232 | 1.061 | 159.0 |
| 4 | 204.0 | 41 700.0 | 0.0328 | 0.253 | 0.292 | 1.335 | 126.0 |
| 5 | 182.0 | 33 100.0 | 0.0260 | 0.319 | 0.369 | 1.685 | 100.0 |
| 6 | 162.0 | 26 300.0 | 0.0206 | 0.403 | 0.465 | 2.13 | 79.5 |
| 7 | 144.0 | 20 800.0 | 0.0164 | 0.508 | 0.586 | 2.68 | 63.0 |
| 8 | 128.0 | 16 500.0 | 0.0130 | 0.641 | 0.739 | 3.38 | 50.0 |
| 9 | 114.0 | 13 100.0 | 0.0103 | 0.808 | 0.932 | 4.27 | 39.6 |
| 10 | 102.0 | 10 400.0 | 0.00815 | 1.02 | 1.18 | 5.38 | 31.4 |
| 11 | 91.0 | 8 230.0 | 0.00647 | 1.28 | 1.48 | 6.75 | 24.9 |
| 12 | 81.0 | 6 530.0 | 0.00513 | 1.62 | 1.87 | 8.55 | 19.8 |
| 13 | 72.0 | 5 180.0 | 0.00407 | 2.04 | 2.36 | 10.77 | 15.7 |
| 14 | 64.0 | 4 110.0 | 0.00323 | 2.58 | 2.97 | 13.62 | 12.4 |
| 15 | 57.0 | 3 260.0 | 0.00256 | 3.25 | 3.75 | 17.16 | 9.86 |
| 16 | 51.0 | 2 580.0 | 0.00203 | 4.09 | 4.73 | 21.6 | 7.82 |
| 17 | 45.0 | 2 050.0 | 0.00161 | 5.16 | 5.96 | 27.2 | 6.20 |
| 18 | 40.0 | 1 620.0 | 0.00128 | 6.51 | 7.51 | 34.4 | 4.92 |
| 19 | 36.0 | 1 290.0 | 0.00101 | 8.21 | 9.48 | 43.3 | 3.90 |
| 20 | 32.0 | 1 020.0 | 0.000802 | 10.4 | 11.9 | 54.9 | 3.09 |
| 21 | 28.5 | 810.0 | 0.000636 | 13.1 | 15.1 | 69.1 | 2.45 |
| 22 | 25.3 | 642.0 | 0.000505 | 16.5 | 19.0 | 87.1 | 1.94 |
| 23 | 22.6 | 509.0 | 0.000400 | 20.8 | 24.0 | 109.8 | 1.54 |
| 24 | 20.1 | 404.0 | 0.000317 | 26.2 | 30.2 | 138.3 | 1.22 |
| 25 | 17.9 | 320.0 | 0.000252 | 33.0 | 38.1 | 174.1 | 0.970 |
| 26 | 15.9 | 254.0 | 0.000200 | 41.6 | 48.0 | 220.0 | 0.769 |
| 27 | 14.2 | 202.0 | 0.000158 | 52.5 | 60.6 | 277.0 | 0.610 |
| 28 | 12.6 | 160.0 | 0.000126 | 66.2 | 76.4 | 350.0 | 0.484 |
| 29 | 11.3 | 127.0 | 0.0000995 | 83.4 | 96.3 | 440.0 | 0.384 |
| 30 | 10.0 | 101.0 | 0.0000789 | 105.0 | 121.0 | 554.0 | 0.304 |
| 31 | 8.9 | 79.7 | 0.0000626 | 133.0 | 153.0 | 702.0 | 0.241 |
| 32 | 8.0 | 63.2 | 0.0000496 | 167.0 | 193.0 | 882.0 | 0.191 |
| 33 | 7.1 | 50.1 | 0.0000394 | 211.0 | 243.0 | 1,114.0 | 0.152 |
| 34 | 6.3 | 39.8 | 0.0000312 | 266.0 | 307.0 | 1,404.0 | 0.120 |
| 35 | 5.6 | 31.5 | 0.0000248 | 335.0 | 387.0 | 1,769.0 | 0.0954 |
| 36 | 5.0 | 25.0 | 0.0000196 | 423.0 | 488.0 | 2,230.0 | 0.0757 |
| 37 | 4.5 | 19.8 | 0.0000156 | 533.0 | 616.0 | 2,810.0 | 0.0600 |
| 38 | 4.0 | 15.7 | 0.0000123 | 673.0 | 776.0 | 3,550.0 | 0.0476 |
| 39 | 3.5 | 12.5 | 0.0000098 | 848.0 | 979.0 | 4,480.0 | 0.0377 |
| 40 | 3.1 | 9.9 | 0.0000078 | 1,070.0 | 1,230.0 | 5,650.0 | 0.0299 |

(American wire gage—B & S)

and a consuming device. The flow of electrons through the consuming device produces heat, light, or work.

In order to form a complete circuit, these rules must be followed:

1. Connect one side of the power source to one side of the consuming device (A to B).
2. Connect the other side of the power source to one side of the control device, usually a switch (C to D).
3. Connect the other side of the control device to the consuming device it is supposed to control (E to F).

This method is used to make a complete path for electrons to flow from one terminal of the battery or power source containing excess of electrons to the terminal which has a deficiency of electrons. The movement of the electrons along the completed path provides energy. Of course, in order for the path to be complete, the switch must be closed.

If the circuit is so arranged that the electrons have only one path, the circuit is called a series circuit. If there are two or more paths for electrons, the circuit is called a parallel circuit.

- **Series circuit.** Figure I-12B shows three resistors connected in series. The current flows through each of

**Fig. I-12** *(A) A simple circuit; (B) series circuit; (C) parallel circuit; (D) series-parallel circuit.*

them before returning to the positive terminal of the battery.

*Kirchoff's Law of Voltages* states that the sum of all voltages across resistors or loads is equal to the applied voltage. Voltage drop is considered across the resistor. In Fig. I-12B, the current flows through three resistors. The voltage drop across $R_1$ is 5 V. Across $R_2$, it is 10 V, and across $R_3$ it is 15 V. The sum of the individual voltage drop is equal to the total or applied voltage, 30 V.

To find the total resistance in a series circuit, just add the individual resistances:

$$R_T = R_1 + R_2 + R_3 + \cdots$$

- **Parallel circuit.** In a parallel circuit, each load (resistance) is connected directly across the voltage source. There are as many separate paths for current flow as there are branches (Fig. I-12C).

The voltage across all branches of a parallel circuit is the same because all branches are connected across the voltage source. Current in a parallel circuit depends on the resistance of the branch. Ohm's law (discussed later) can be used to determine the current in each branch. You can find the total current for a parallel circuit by adding the individual currents. As a formula, this reads:

$$I_{RT} = I_{R1} + I_{R2} + I_{R3} + \cdots$$

The total resistance of a parallel circuit cannot be found by adding the resistor values. Two formulas are used for finding parallel resistances. If there are only two resistors in parallel, a simple formula can be used:

$$R_T = \frac{R_1 \times R_2}{R_1 + R_2}$$

If there are more than two resistors in parallel, you can use the following formula: (You can also use this formula if there are only two resistors.)

$$\frac{1}{R_T} = \frac{1}{R_1} + \frac{1}{R_2} + \frac{1}{R_3} + \cdots$$

One thing should be kept in mind in parallel resistances: The total resistance is always less than the smallest resistance.

- **Series-parallel circuits.** Series-parallel circuits are a combination of the two circuits. Fig. I-12D shows a series-parallel resistance circuit.

- **Open circuit.** An open circuit is one which does not have a complete path for electrons to follow. Such an incomplete path is usually brought about by a loose connection or the opening of a switch (Fig. I-13).

**Fig. I-13** *Open circuit caused by an open switch.*

- **Short circuit.** A short circuit is one which has a path of low resistance to electron flow. It is usually created when a low-resistance wire is placed across a consuming device. A greater number of electrons will flow through the path of least resistance rather than the consuming device. A short circuit usually generates an excess current flow which results in overheating, possibly causing a fire or other damage (Fig. I-14).

**Voltage and Current**

**Fig. I-14** *A short circuit. The wire has less resistance than the lamp.*

It is easy to compute the amount of current flowing in a circuit if the voltage and the resistance are known. The relationship between voltage, current, and resistance in any circuit is given by Ohm's law.

## OHM'S LAW

There are three basic quantities of electricity, and each has a relationship to the other two. A physicist named Georg S. Ohm discovered the relationship in 1827. He found that in any circuit where the only opposition to the flow of electrons is resistance, there is a relationship among the values of voltage, current, and resistance. The strength or intensity of the current is directly proportional to the voltage and inversely proportional to the resistance.

It is easier to work with Ohm's law when it is expressed in a formula. In the formula, $E$ represents EMF, or voltage. $I$ is the current, or the intensity of electron flow. $R$ stands for resistance. The formula is $E = I \times R$ and is used in finding the EMF (voltage) when the current and resistance are known.

To find the current when the voltage and resistance are known, the formula to use is

$$I = \frac{E}{R}$$

To find the resistance when the voltage and current are known, the formula to use is

$$R = \frac{E}{I}$$

### Using Ohm's Law

There are many times in electrical work when you will need to know Ohm's law: for example, to determine wire size in a particular circuit or to find the resistance in a circuit.

The best way to become accustomed to solving problems is to start with something simple, such as:

1. If the voltage is given as 100 V and the resistance is known to be 50 Ω, it is a simple problem and a practical application of Ohm's law to find the current in the circuit.

$$I = \frac{E}{R}$$
$$I = \frac{100 \text{ V}}{50 \text{ Ω}}$$
$$I = 2 \text{ A}$$

2. If the current is given as 4 A (shown on an ammeter) and the voltage (read from a voltmeter) is 100 V, it is easy to find the resistance.

$$R = \frac{E}{I}$$
$$I = \frac{100 \text{ V}}{4 \text{ A}}$$
$$R = 25 \text{ Ω}$$

3. If the current is known to be 5 A, and the resistance is measured (before current is applied to the circuit) and found to be 75 Ω, it is then possible to determine how much voltage is needed to cause the circuit to function properly.

$$E = I \times R$$
$$E = 5 \text{ A} \times 75 \text{ Ω}$$
$$E = 375 \text{ V}$$

Figure I-15 illustrates the way the formula works.

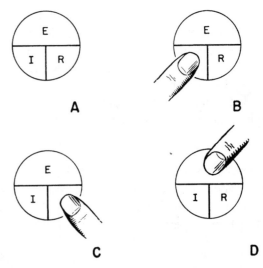

**Fig. I-15** *Ohm's law. This illustrates the way the formula works. Place finger on the unknown value. The remaining two letters will produce the formula to use for finding that unknown value.*

# POWER

Power is defined as the rate at which work is done. In metric measurement it is expressed in watts for power and joules (J) for energy or work. (For some common metric conversions, see Appendix A.) One watt is the power that gives rise to the production of energy at the rate of 1 J in 1 s (1 W = 1 J/s). One joule is the work done when the point of application of a force of 1 N (newton) is displaced a distance of 1 m in the direction of the force (1 J = 1 N × 1 m).

It has long been the practice in United States to measure work in terms of horsepower (hp). Electric motors are still rated in horsepower, and probably will be for some time. The U.S. did not change to metric when Canada and England did and probably never will. Power can be electrical or mechanical. When a mechanical force is used to lift a weight, work is done. The rate at which the weight is moved is called *power*. Horsepower is defined in terms of moving a certain weight over a certain distance in 1 min. Energy is consumed in moving a weight, or work is done in the process. The findings in this field have been equated with the same amount of work done by electrical energy. It takes 746 W of electrical power to equal 1 hp (Table I-C).

*Table I–C* Horsepower

---

One horsepower is usually defined as the amount of work required to move a 550-pound weight a distance of one foot in one second.

In most cases the modern way to measure power is in kilowatts rather than horsepower. In case a motor is specified in terms of horsepower, but is rated in watts or kilowatts, the conversion is simple:

1 horsepower = 746 watts

Divide the number of watts or kilowatts by 746 or 0.746, respectively, to find the horsepower rating.

---

The horsepower rating of electric motors is arrived at by multiplying the voltage by the current drawn under full load. This power is measured in watts. In other words, 1 V times 1 A equals 1 W. When put into a formula,

Power = Volts × Amperes   or   $P = E \times I$

## Kilowatts

The prefix "kilo" means one thousand. Thus 1000 W equals 1 kilowatt (kW). One kilowatt-hour (kWh) is equivalent to 1000 W used for 1 h. Electric bills are figured in kWh. Usage for an entire month is computed on an hourly basis and then read in the kWh unit.

Power formulas are sometimes needed to figure the wattage of a circuit. The three most commonly used formulas are

$$P = E \times I$$
$$P = \frac{E^2}{R}$$
$$P = I^2 R$$

This means that if any one of the three—voltage, current, or resistance—is missing, it is possible to find the missing quantity by using the relationship between the two known quantities. In later chapters, you will encounter the problem of the $I^2R$ losses and some other terminology related to the formulas just shown.

The milliwatt (mW) is sometimes used in referring to electrical equipment. For instance, the rating of a speaker in a portable transistor radio may be given as 100 or 300 mW, which means that a 0.1- or 0.3-W rating, since the prefix "milli" means one-thousandth. Transistor circuits are designated in milliwatts, but power-line electrical power is usually in kilowatts.

# MEASURING ELECTRICITY

Electricity must be measured if it is to be sold, or if it is to be fully utilized. There are a number of ways to measure electricity. It can be measured in volts, amperes, or watts. The kWh meter is the most commonly used device to measure power.

## Meters

In order to measure anything, there must be a basic unit in which to measure. In electricity, current (flow of electrons) is measured in a basic unit called *ampere*. The current is usually measured with a permanent magnet and an electromagnet arranged to indicate the amperes. Such a device is necessary since we are unable to see an electron—even with the most powerful microscopes. Obviously, counting the number of electrons passing a given point in a second is impossible when there are no visible particles to count. Therefore, a magnetic field is used to measure the effect of electrons.

The D'Arsonval meter movement uses a permanent magnet as a base over which a wire or electromagnet is pivoted and allowed to move freely. When current flows through the coil, a magnetic field is set up (Figs. I-16A and I-16B). The strength of the magnetic field determines how far the coil will be deflected. The polarity of the moving coil is the same as that of the permanent magnet. A repelling action results, in proportion to the strength of the magnetic field generated by the current flowing through the coil. The number of

**Fig. I-16** (A) D'Arsonval meter movement; (B) D'Arsonval meter movement showing completed unit in diagram form; (C) Assembled arrangement of the D'Arsonval meter movement.

turns in the coil, times the current through the coil determines the strength of the magnetic field. Since the meter coil is pivoted on jeweled bearings to reduce any friction, the movement is calibrated against a known source of current or against another meter. The scale on a new device is calibrated to read in amperes, milliamperes, or microamperes (Fig. I-16C).

- **AC ammeter**. If ac is to be measured by a dc meter movement, a rectifier is inserted in the circuit (meter circuit) to change the ac to dc. Otherwise, the ac will make the needle on the meter vibrate rapidly. This vibration means there is little or no movement from zero (Fig. I-17).
- **Shunts**. A shunt is a means of bypassing current around a meter movement. A resistor of proper size is inserted across the meter movement to bypass the current around the movement. Most of the current is bypassed with only the necessary amount left to

**Fig. I-17** ac meter movement is made by adding a diode to the dc meter movement.

cause the meter to deflect at its designed limit. The meter is calibrated to read on its scale the full amount of current flowing in the circuit (Fig. I-18).

A number of shunts can be placed in a meter case and switched. A different resistor (shunt) can be switched in for each range needed. A meter with more than one range is called a *multimeter*.

**Fig. I-18** *Meter movement with a shunt to increase its range to 10 mA.*

A multimeter (one which can measure volts and ohms as well) is shown in Fig. I-19. It can measure from 0 to 1 mA, 0 to 10 mA, and 0.1 to 1 A. This is a 1-mA meter movement and needs no shunt when used to measure as high as 1 mA. A shunt is switched in, however, to measure a range of 0 to 10 mA and 0.1 to 1 A. A switch on the meter can also add a diode to help with measuring ac.

**Fig. I-19** *A multimeter.*

COMMON and POSITIVE holes have test leads inserted to attach to the circuit being measured. COMMON is negative (−), or black, and POSITIVE is positive (+), or red. However, polarity isn't necessary to measure ac, either lead can be used at any terminal in an ac circuit. Ammeters are always connected in series in a circuit. This usually means the circuit has to be broken and the meter inserted in the line.

- **Voltmeter.** The voltmeter measures electrical pressure, or volts. It is nothing more than an ammeter with a resistor added in the meter circuit. The high resistance of the voltmeter makes it possible to connect in parallel across a power source (Fig. I-20). A number of resistors, called multipliers, can be switched into a meter circuit to increase its range or make it capable of measuring higher voltages. The voltmeter in Fig. I-21 is capable of measuring voltages ranging from 0 to 150, 0 to 300, and 0 to 750 V by connecting the proper multiplier into the meter circuit. Note how the terminals on top of the meter allow connection of test probes in a number of different positions for varying the range of measurement.

**Fig. I-20** *Voltmeter's internal circuit with resistor (multiplier) added.*

**Fig. I-21** *DC voltmeter.*

- **Ohmmeter**. The basic unit for measuring resistance is *ohm*. An ohmmeter is a device for measuring resistance in ohms (Ω). It is an ammeter (or milliammeter or micrometer) movement modified to measure resistance (Fig. I-22).

**Fig. I-22** *Multimeter used to measure ohms, volts, and milliamperes as well as microamperes.*

**Fig. I-23** *(A) Basic circuit of an ohmmeter. (B) Test leads crossed (shorted) so ohms adjust resistor can be adjusted to make the meter read zero. (C) test leads touching the leads of a resistor to measure it.*

Figure I-22 shows a multimeter capable of measuring ohms with three different ranges: R × 1, R × 100, and R × 10K (R × 10,000). This means the meter can measure from 0 to 200 Ω on the R × 1 scale and 0 to 200 kΩ on the R × 100 range. Within the R × 10K range, it is capable of measuring from 0 to 20 MΩ (megohms; mega means 1 million). The meter scale has to be multiplied by 100 or 10,000 in order to have it read the proper value. By changing resistors, it is possible to vary the resistance measuring range of an ohmmeter. Figure I-23 shows a basic ohmmeter using a 1-mA meter movement and its necessary parts. Note how the battery serves as the power source. This makes it necessary to turn off power whenever you read the resistance of a circuit. THE OHMMETER HAS ITS OWN POWER SOURCE. Do not connect it to a LIVE CIRCUIT or one with the power on, because that will result in destruction of the meter movement.

Adjust the ohmmeter so that the meter reads zero before starting resistance measurement, which means you have adjusted the meter circuit to compensate for the battery voltage changes. Battery voltages decrease with shelf life. It doesn't matter whether or not the battery is used. It will, in time, lose its voltage.

**16  Introduction**

Some meters are not portable. They need external sources of power. Electronic circuits using vacuum tubes or transistors are used to improve the meter capabilities. Figure I-24 shows a digital voltmeter set to read dc voltages.

- **Digital meters**. The digital meter is entirely electronic. It uses printed circuits and integrated circuit chips to measure and calculate voltage, resistance, and current. There are no coils or magnets.

Figure I-24 shows a portable and self-contained digital meter. The meter indicates the reading on a liquid crystal display (LCD). This type of digital readout is found on many instruments, clocks, and watches.

*Fig. I-24* A digital voltmeter set to read dc voltages.

Most of these meters have a number of voltage ranges. The ranges must be selected each time you measure a circuit. Some meters are made with auto-ranging. They select the proper range and measure the voltage without the need for preselection. The operator must indicate if volts, ohms, or amperes are to be measured.

Digital meters sample the circuit about 5 times per second and then display the average of the 5 samples for very accurate readings. In most instances, the electrician is not concerned with the 0.01-V accuracy that these meters are capable of reaching.

The meter is highly accurate and easy to use. You simply turn on the meter to the needed function, then select ohms, and use the probes across the resistor to measure resistance. If you use this unit as a milliammeter, you have to insert it properly into the circuit being measured. If a circuit is being measured for resistance, make sure there is no power turned on when the meter is connected.

This type of meter is rugged but can be damaged, however, if it is set for measuring ohms when instead, volts are being measured. The display can be damaged permanently when left on for too long or when the meter is dropped rather hard. To extend the battery life, turn the meter off when not in use.

Prices of digital meters are dropping rapidly. Gradually, the digital meter can be expected to replace all other types. However, for some purposes, the digital meter can be outperformed by the D'Arsonval movement. For example, some meter indicators simply need a deflection of the needle to show proper operation. In digital meters, you have to wait for the numbers to be counted up or down, which takes concentration on the part of the user. The D'Arsonval movement simply shows a deflection of the needle.

- **AC clamp-on meter**. Figures I-25 and I-26 show two types of ac clamp-on meters. The meters are inserted over a wire carrying ac. The magnetic field around the wire induces a small current in the meters. The scale is calibrated to read amperes or volts. Because the wire is run through the large loop extending past the meter movement, it is possible to read the ac voltage, or current, without removing the insulation from the wire. These meters are very useful when working with ac motors.

*Fig. I-25* Clamp-on type of portable ac volt-ammeter.

**Fig. I-26** *Clamp-on type of ac voltmeter. (Weston)*

- **Wattmeter.** A wattmeter measures electrical power. Electrical power is figured by multiplying the voltage times the current. A wattmeter has two electromagnetic coils (a coil with many turns of fine wire for voltage and a coil with a few turns of heavy wire for current). The voltage coil is connected across the incoming line, and the current coil is connected in series with one of the incoming wires. The two coils are stationary and in series with a moving coil. The strength of the magnetic fields determines how much the moving coil is deflected. The deflection of the needle is read on a scale calibrated in watts. In this way, the wattmeter measures the power consumed in 1 s (Fig. I-27).

**Fig. I-27** *(A) Wattmeter hookup; (B) kilowatt-hour meter.*

18   Introduction

For measuring the electrical power used over a longer period of time, the kWh meter was designed. The kWh meter is often seen on the side of a house or building. It measures power used over a certain period, for example, a month. A kWh meter measures power consumed in terms of thousands of watts. Electric power is priced at a certain rate per kWh.

The kWh meter is a small induction motor. Meter torque is produced by an electromagnet called a stator, which has two sets of windings. One winding, called a potential coil, produces a magnetic field representing circuit voltage. Another winding, called a current coil, produces a magnetic field that represents the load current. These two coils are arranged so that their magnetic fields create a force on the meter disc, which is directly proportional to the power, or watts, drawn by the connected load.

Permanent magnets are used to introduce a retarding, or braking, force that is proportional to the speed of the disc. The magnetic strength of these retarding magnets regulates the disc speed for any given load so that each revolution of the disc always measures the same quantity of energy or watt-hours. Disc revolutions are converted to kWh on the meter register.

Most meters are inserted into a socket on the wall of a structure. Removing the meter interrupts or terminates power, without handling of dangerous high-voltage wires. Three-phase and single-phase power, to be discussed in later chapters 16, 19, and 20, each require different watt-hour meters. Kilowatt-hour meters are tested by computers in the service centers of power companies. Fig. I-28.

*Fig. I-28* Computer checking kilowatt-hour meters.

## Other Types of Meters

There are other types of meters used to measure voltage and current. The D'Arsonval movement is only one of many types used today. The taut band type is basically the same as the D'Arsonval movement except that a tightly stretched and twisted hand is used to hold the coil and needle in place between the permanent magnet poles. In addition, no moving points touch the meter case; so jeweled bearings are unnecessary. The band is twisted when it is inserted into the meter frame so that it will cause the coil to spring back to its original resting place upon interruption of current through the coil.

- **Electrodynamometer**. The electrodynamometer type of meter uses no permanent magnet. Two fixed coils produce the magnetic field. The meter also uses two moving coils. This meter can be used as a voltmeter or an ammeter (Fig. I-29). It is not as sensitive as the D'Arsonval meter movement.

*Fig. I-29* Electrodynamometer: (A) Circuit when used as a voltmeter; (B) circuit when used as ammeter; (C) internal construction.

**Measuring Electricity**

**Fig. I-30** *Inclined-coil iron-vane meter.*

- **Inclined-coil iron-vane meter.** The inclined-coil iron-vane meter is used for measuring ac or dc, where large amounts of current are present. This meter can be used as a voltmeter or ammeter (Fig. I-30).

## CONTROLLING ELECTRICITY

In order to make electricity useful, it is necessary to control it. You want electricity in the proper place at the proper time. Otherwise, it can do great damage—even kill. Electricity can be controlled by using switches, relays, or diodes. These devices are used to direct the current where it will work for you. Each device is carefully chosen to do a specific job. For example, the relay is used for remote-control work, and a diode is used to control large and small currents in electrical as well as electronic equipment. A diode is a device which allows current to flow in one direction only. It can be used to change ac to dc.

## Switches

There are a number of switches used for controlling electricity. Each switch has a different name, which helps designate it according to the job it performs. For instance, the single-pole, single-throw (SPST) is just that, a single pole which is moved either to make connection between two points or to not make connection. In the off position the contacts are not touching, and the flow of electrons is interrupted (Fig. I-31).

The double-pole, double-throw (DPDT) switch can be used to control more than one circuit at the same time. It can be used to reverse the direction of rotation of a dc motor by reversing polarity (Fig. I-32).

**Fig. I-31** *Knife switch. Single pole-single throw.*

20  Introduction

*Fig. I-32*  *Knife switch. Double pole, double throw.*

Figure I-33 shows a single-pole, double-throw (SPDT) switch. These switches: SPST, DPDT, and SPDT, are open switches and should not be used with more than 12 V. They are usually referred to as radio or battery switches and are used here to show simple operation of a switch. Switches used for higher voltages are totally enclosed and protected from body contact. This prevents shocks or injuries from electrical shorts or contact with high voltages.

*Fig. I-33*  *Knife switch. Single pole, double throw.*

The DPST switch is used to control two circuits at the same time. It can be used as a simple on-off switch for two circuits. Opening the switch interrupts the current in the two circuits. When it is closed it completes the circuits for proper operation. This switch is meant to be used only on low voltages where the danger of shock is very much reduced. Figure I-34 illustrates this type of switch.

*Fig. I-34*  *Knife switch. Double pole, single throw.*

The doorbell, or door chime, switch is extremely simple. It controls a circuit from the low-voltage transformer to the chime or bell (Fig. I-35). When the button is pressed, the switch completes the circuit to the bell or chime, causing current to flow from the transformer to the chime (Fig. I-36).

*Fig. I-35*  *Door chime circuit. Switch open.*

*Fig. I-36*  *Door chime circuit. Switch closed.*

- **Toggle switches**. Toggle switches are used to turn various devices on and off, or to switch from one device to another. They are made in a number of configurations to aid in selection for a particular job (Fig. I-37). These switches usually have a metal handle and are mounted through a round hole. Screw terminals are usually provided for attaching wires. However, some may have wire leads furnished. A wire nut is used to attach the switch leads to the circuit wires. A wire nut is a device that makes a connection between two pieces of wire by twisting ends together. It insulates the connection with a plastic coating.

**Controlling Electricity**

**METAL HANDLE TOGGLE SWITCH**
SINGLE POLE — SINGLE THROW
With 6" No. 18 gauge 105°C type AWM plastic wire leads.

**WITH WIRE LEADS**

**CIRCUIT CONTROL A.C. ROCKER SWITCHES**

**Fig. I-37** Toggle switches.

Note the ac rocker switch in Fig. I-37. It is mounted with screws to attack to the switch's steel bracket. Holes are tapped for a No. 6-32 screw on "1" centers. These switches are usually rated for use on 120-V ac, with special attention to current. The switches are capable of switching 240 V. Current is usually doubled in a 120-V rating as compared to a 240-V rating. This means a switch rated at 120 V and 6 A must be de-rated to 3 A for 240-V circuits.

- **Residential toggle switches**. Various shapes are encountered when switches are needed for use in business, industry, or the home. For instance in Fig. I-38, you will find examples of some of those used to switch 120-V ac and 240-V ac in common circuits used in lighting and small motors. Figure I-38A shows a residential toggle switch, rated 10 A on a 125-V line. Note the absence of "plaster ears" near the long screws. Switch B is like A, except that it has wide plaster ears (the extensions around the screw holes) that can be removed easily if not required to hold the switch rigid in its box. The plaster ears are scored (marked) so that they can be easily bent and removed. Switch C is more expensive than either A or B. It has specification grade (top of the line) quality with wide plaster ears. Note that the screws for attaching circuit wiring are located topside, instead of along the side of the switch.

22  Introduction

**Fig. I-38** *Switches for home and office use.*

Switch D is a high-capacity, heavy-duty, industrial type, rated 20 A at 125 V. Switch E is an extra-heavy-duty industrial type. It is a more expensive switch which minimizes the arcing of contacts when turned on and off. The arc, which occurs each time a switch is turned on or off, creates high heat. The heat can cause the contacts of the switch to become pitted and make a high-resistance contact. The contacts in switch E have an extended life, made possible by the use of arc snuffers.

Switch F is a "no-klik," or quiet, heavy-duty switch. It has eliminated the noise associated with the on-off operation of a switch. Switch G is also a heavy-duty, quiet, high-capacity-specification-grade switch, and is good for 15 A at 120 or 277 V. Switch H is a quiet 15-A ac, side- or back-wired with binding screw, and pressure, or screw-less, terminals. Some switches have a grounding strap designed for use with nonmetallic systems that use Bakelite boxes and for bonding between the device strap and steel boxes.

These newer switches feature a green grounding-screw terminal. They are available in either ivory or brown.

- **Three-way switches.** Three-way switches are used where you need to control a light or device from more than one location. These switches have three terminals instead of two and do not have the labels ON and OFF on the handles. The three-way switch looks like a regular switch with the exception of the three terminals for wiring into a circuit.
- **Four-way switches.** Four-way switches have double poles, and are used where a light or device needs to be controlled from three or more locations. If three controls are preferred, you need two three-way switches and one four-way switch. The four-way switch resembles the three-way, but has four terminals for connection into a circuit. Other types of switches are available for different jobs of current control. They will be shown as they are introduced in connection with a specific job.

Switches are used to turn the flow of electricity on or off, thereby causing a device to operate or cease operation. Switches can be used to reverse polarity, and as in the case of electric motors, the direction of rotation can be reversed by this action. Switches, as you have already seen, come in many shapes and sizes. The important thing to remember is to use a switch with proper voltage and current rating, for the job to be done. A careful study of the types presented in chapter 5 will help in proper selection of a switch for a particular job.

## Solenoids

Solenoids are devices that turn electricity, gas, oil, or water on and off. Solenoids can be used, for example, to turn the cold water on and the hot water off in a washing machine, to get a proper mix of warm water. To control the hot water solenoid, a thermostat is inserted in the circuit.

Figure I-39 shows a solenoid for controlling natural-gas flow in a hot-air furnace. Note how the coil is wound around the plunger. The plunger is the core of the solenoid. It has a tendency to be sucked into the coil whenever the coil is energized by current flowing through it. The electromagnetic effect causes the plunger to be attracted upward into the coil area. When the plunger is moved upward by the pull of the electromagnet, the soft disc (10) is also pulled upward, allowing gas to flow through the valve. This basic technique is used to control water, oil, gasoline, or any other liquid or gas.

**Controlling Electricity** 23

**Fig. I-39** *Solenoid for controlling natural gas flow to a hot-air furnace.*

The starter solenoid on an automobile uses a similar procedure, except that the plunger has electrical contacts on the end that complete the circuit from the battery to the starter. The solenoid uses low voltage (12 V) and low current to energize the coil. The coil in turn sucks the plunger upward. The plunger then touches heavy-duty contacts designed to handle the 300-A current needed to start a cold engine. In this way, low voltage and low current are used, from a remote location, to control low voltage and high current.

Solenoids are electromagnets. An electromagnet is composed of a coil of wire wound around a core of soft iron. When current flows through the coil, the core will become magnetized. The magnetized core can be used to attract an armature and act as a magnetic circuit breaker (Fig. I-40). (A circuit breaker, like a fuse, protects a circuit against short circuits and overloads.) In Fig. I-40, the magnetic circuit breaker is connected in series with both the load circuit to be protected and with the switch contact points. When excessive current flows in the circuit, a strong magnetic field in the electromagnet causes the armature to be attracted to the core. A spring attached to the armatures causes the switch contacts to open and break the circuit. The circuit breaker must be reset by hand to allow the circuit to operate again. If the overload is still present, the circuit breaker will "trip" again. It will continue to do so until the cause of the short circuit or overload is found and corrected.

**Fig. I-40** *Magnetic circuit breaker.*

## Relays

A relay is a device that can control current from a remote position through the use of a separate circuit for its own power. Figure I-41 shows a simple relay circuit.

**Fig. I-41** *Simple relay circuit.*

**Fig. I-42** *(A) Diode control of current; (B) diode control for two lamps.*

When the switch is closed, current flows through the electromagnet, or coil, and energizes it. The pull of the electromagnet causes the soft iron armature to be attracted toward the electromagnet core. As the armature moves toward the coil, it touches the contacts of other circuits, thereby completing the circuit for the load. When a switch opens, the relay coil de-energizes, and the spring pulls the armature back. This action breaks the contact and removes the load from the 12-V battery. Relays are remote switches that can be controlled from almost any distance if the coil is properly wired to its power source.

Many types of relays are available. They are used in telephone circuits and in almost all automated, electrical machinery.

## Diodes

Diodes are semiconductor devices that allow current to flow only in one direction. By properly connecting a diode, or diodes, in a circuit, it is possible to control current flow by controlling the direction of the current (Fig. I-42A). AC flows first in one direction and then in the reverse/opposite direction 120 times per second in a 60-hertz (Hz) circuit. The diode is a rectifier which allows current to flow in only one direction. It changes ac to dc which flows in only one direction.

It is possible to use four diodes, two switches, and two wires to control two lamps located some distance away from the switches. For example, Fig. I-42B shows how switch A is depressed, allowing current to flow through it and diode No. 1. The current then passes through the wire to diode No. 2, through lamp A, and back to the 120-V source. Lamp A will light.

When switch B is pressed, current flows through the bottom wire to lamp B, diode No. 3, and along the top wire to diode No. 4. The current then flows to switch B and, back through the wire, to the 120-V ac source, causing lamp B to light. Whenever lamp B is lit, lamp A is out because current cannot flow through diode No. 2. However, if both switches are de-pressed at the same time, both lamps will light. This is due to the ac taking turns, flowing first in one direction, for instance in lamp A, and then the other, in lamp B. There are pulses of current, or pulsating direct current (PDC), through each lamp. At 60 Hz, it is not possible for the human eye to see the on-off condition of the lamps. The pulses are so fast that the eye cannot respond, thus causing the lamps to appear glowing continuously.

You will find applications of diodes in computers and in other less-complicated electrical devices. Just remember that the diode allows current to flow in only one direction.

## RESISTORS

A resistor is a device used to provide a definite, required amount of opposition to current in a circuit. Resistance is the basis for generation of heat. It is used in circuits to control the flow of electrons and to ensure that the proper voltage reaches a particular device. Resistors are usually classified as either wire-wound or carbon-composition.

- **Wire-wound resistors.** These resistors are used to provide sufficient opposition to current flow to dissipate power of 5 W or more. They are made of resistance wire (Fig. I-43). Variable wire-wound resistors

*Fig. I-43*  *Wire-wound resistor.*

*Fig. I-44*  *Variable wire-wound resistor.*

*Fig. I-45*  *Adjustable wire-wound resistor.*

are also available for use in circuits where voltage is changed at various times (Fig. I-44). Some variable resistors have the ability to be varied but also adjusted for a particular setting (Fig. I-45).

High-wattage, wire-wound resistors are available in many sizes and shapes (Fig. I-46).

**Carbon-composition resistors.** These resistors are usually found in electronic circuits of low wattage,

*Fig. I-46*  *Various sizes and shapes of wire wound resistors.*

since they are not made in sizes beyond 2 W. They can be identified by three- or four-color bands around them. Their resistance can be determined by reading the color bands and checking the resistor color code.

The wattage ratings of carbon-composition resistors are determined by physical size. They come in $1/4$-, $1/2$-, 1-, and 2-W sizes. By examining them and becoming familiar with them through use, you should be able to identify wattage rating by sight. The larger the physical size of the resistor, the larger the wattage rating (Fig. I-47).

**Fig. I-47** *Carbon-composition resistors.*

## Resistor Color Code

Take a close look at a carbon-composition resistor. The bands should be to your left (Fig. I-48). Read from left to right. The first band gives the first number according to the color code. In this case, it is yellow, or 4. The second band gives the next number, which is violet, or 7. The third band represents the multiplier or divisor. If the third band is a color in the range 0 to 9 in the color code, it states the number of zeros to be added to the first two numbers. Orange is 3; so the resistor in Fig. I-48 has a value of 47,000-Ω resistance.

If there is no fourth band, the resistor has a tolerance rating of ±20% (± means "plus or minus"). If the fourth band is silver, the resistor has a tolerance of ±10%. If the fourth band is gold, the tolerance is ±5%.

Silver and gold may also be used for the third band, where according to the color code, the first two numbers (obtained from the first two color bands) must be divided by 10 or 100. Silver means divide

**Fig. I-48** *47,000-ohm resistor. See color code.*

by 100; gold means divide by 10. For example, if the bands on a resistor are red, yellow, gold, and silver, the resistance would be 24 divided by 10, or 2.4 (±10%) (Fig. I-49).

**Fig. I-48** *2.4-ohm resistor with color code.*

Resistors are available in hundreds of sizes and shapes. Once familiar with electronics and electrical circuits, you will be able to identify various components by their shape, size, or markings. Products for such circuits are constantly changing with new items being marketed almost every day. To stay informed about these products, it is necessary to read the literature written about the industry. Each area of electrical energy has its own magazines to keep those on the job informed and up-to-date in their special fields of interest.

## REVIEW QUESTIONS

1. What is electricity?
2. What is an atom?
3. What are elements?
4. What is the difference between static and current electricity?
5. What are free electrons?
6. What is a conductor?
7. What is an insulator?
8. What is a semiconductor?

9. Name six methods used to generate electricity.
10. What is an exotic generator?
11. In what unit of measurement is electrical current measured?
12. In what unit of measurement is voltage measured?
13. In what unit of measurement is electrical resistance measured?
14. What is the relationship between wire size and its numbering system?
15. What is a complete circuit?
16. What is an open circuit?
17. What is a short-circuit?
18. State Ohm's law.
19. Define a kilowatt.
20. What is a kilowatt-hour?
21. How is a meter shunt used?
22. What has to be done to a dc meter to convert it to an ac meter?
23. What is the difference between an ohmmeter and a voltmeter?
24. Where is an inclined-coil, iron-vane meter used?
25. What do SPST, DPDT, SPDT, and DPST represent?
26. What is a toggle switch?
27. What is the difference between a three-way switch and a four-way switch?
28. What is the definition of a solenoid? A relay?
29, What is a diode?
30  What is a resistor?

# CHAPTER 1

# Tools and Equipment

## PERFORMANCE OBJECTIVES

After studying this chapter, you will be able to:

- Identify the hand tools needed for motor control work.
- Operate a volt-ohm-milliammeter (VOM), scope, voltage tester, and digital probe.
- Identify wire wrap/unwrap tools, conduit benders, polyvinyl chloride (PVC) cutters, cable benders, and conduit reamers.

A quick review of the tools and equipment used in this trade will provide you with a refresher and better overall view of them. You will be handling these tools as well as others. Identification and utilization of tools and equipment is necessary for your everyday work.

Tools are an extension of a person. They allow a human being to do things that are impossible using the human hands alone. Tools are also expensive and represent a sizable investment in money and in the time necessary to keep them in working condition. Two types of tools are needed by electricians: hand tools, and electrical instruments for measurement. We discuss both types in this chapter.

## HAND TOOLS

Tools may be placed in an electrician's pouch (Fig. 1-1) or in a toolbox (Fig. 1-2). Keeping tools organized saves time and money. If tools do not have a place to call their own, you will spend too much time looking for them. Not only do you need to have a definite place for a tool, but you should also develop the habit of returning it to that place after you have used it. Whenever

**Fig. 1-1** *Electrician's pouch.* (Klein)

**Fig. 1-2** *Portable toolbox.* (Klein)

your tools are used on a workbench or around a piece of equipment, you need a toolbox.

Some tools can be mounted on a silhouetted outline on a board in front of the workbench. That way, if a tool is missing, you know which one it is, and when using it, you know where to put it when you are finished with it. Pegboard or a piece of plywood with cutouts attached and pegs on which to hang the tools can be used.

## Screwdrivers

These tools are often incorrectly used. They may be used for opening paint cans, punching holes, and prying as well as their intended use. Keep in mind that the right tool should be used for the job.

Screwdrivers are most often misused. Two types of screwdrivers are encountered most common: the Phillips-head and the standard or slot type. Screwdrivers come in thousands of variations.

Generally, screwdrivers are available with either wooden or plastic handles, but plastic is most common. Plastic handles are very helpful when working around electricity. Plastic is supposed to be shockproof if the handle is kept clean and dry. The blade tip may vary in size from $1/8$ to $1/4$ in. The shaft is 4 to 8 in. long and is usually made of nickel-plated chrome-vanadium steel. The tips or points must withstand force applied when a screw sticks or is difficult to remove.

The main thing to remember when using a screwdriver is to get a good fit between the tip of the screwdriver and the slot in the screw. This will prevent damage to both the screw head and the screwdriver.

Other types of screwdrivers are also available. They are usually adapted for a particular purpose. The main reason for changing the screw head from the slot to some other type is to increase efficiency in getting the driver into the screw head for positive mating and more positive driving force. See Fig. 1-3 for other types of screw-head configurations.

**Fig. 1-3** *Screw-head configurations.* (Klein)

## Pliers

There are a number of pliers available for special jobs. The pliers shown in Fig. 1-4 are indicative of the variety available for work in the electrical and electronics field. Each of the pliers is designated a particular job.

1. 4-in. midget for close work.
2. 4-in. plier for fast, clean tip cutting. It has a tapered nose and nearly flush cutting edges and will cut to the tip, producing burr-free cuts.
3. 7-in. diagonal pliers for heavy-duty cutting.
4. $4^{1}/_{2}$-in. thin needle-nose plier, with cutter at the tip.
6. 5-in. thin chain-nose plier, whose smooth jaws are slightly beveled on the inside edges.

Numbering is not in sequential order

7. 5-in. plier with fine serrated jaws for firm gripping or looping wire.
8. Slim serrated jaws (6 in. long); permit entry in areas inaccessible to regular long-nose pliers.
9. Long-nose plier ($6^{1}/_{2}$ in. long) with side cutter.
10. Long-nose plier ($6^{1}/_{2}$ in. long) without side cutter.
11. Thin 60°-bent nose plier (5 in.) with fine serrated jaws for thin wire applications.
12. and 13 were removed from consideration in this type of work.
14. 8-in. serrated upper- and lower-jaw plier with side cutter.
15. 8-in. chrome-plated combination plier for general use.
16. Four-position 10-in. utility plier with forged rib and lock design with serrated jaws.

Hand Tools 31

**Fig. 1-4** *Various types of pliers.*

## Hammers

As with other hand tools, there are a number of hammer types. The three most common are best suited for electricians or technicians. The claw hammer (Fig. 1-5) is used to drive nails and to work mainly with wood; it has claws with which to remove nails. It is the most common type of hammer, but is used only occasionally in most work where a ball-peen hammer is not available.

**Fig. 1-5** *Claw hammer.* (Klein)

The ball-peen hammer (Fig. 1-6) has a rounded top and a flat, large-diameter bottom surface. It is used in work around machineries.

**Fig. 1-6** *Ball-peen hammer.* (Klein)

In some cases, a mallet is needed to force a connection or to make a slight adjustment. Mallets can be obtained with either a soft face or a hard face. Leather, plastic, rubber, wood, and lead mallets are used for various jobs, as the need arises (Fig. 1-7). In most instances, a rubber mallet is used to enable you to do the job without marring the surface of the metal being hit. Keep in mind that you should be careful with your hammer or mallet blows. Use eye protection.

**Fig. 1-7** *Plastic-tip mallet.* (Klein)

CAUTION Hammers are perhaps the most abused and misused of all hand tools. Improper use of hammers can cause injury. Use of damaged hammers can cause injury. Use of hammers for jobs other than those for which they have been specifically designed can cause injury.

The following rules for proper use apply to all hammers:

1. Strike square blows only. Avoid glancing blows.
2. Be sure striking face of the hammer is at least $3/8$ in. larger than tools to be struck (chisels, punches wedges, etc.). The hammer face should be slightly crowned with edges beveled.
3. Never use one hammer to strike another hammer.
4. Always use a hammer of the right size and weight for the job.
5. Never strike with the side or "cheek" of a hammer.
6. Never use a hammer with a loose or damaged handle. Replace the handle.
7. Discard any hammer having chips, cracks, dents, mushrooming, or excessive wear. Replace the hammer. DO NOT TRY TO REPAIR IT.
8. Always wear safety goggles to protect your eyes.

For instructions on safe use, see label on each hammer.

## Hacksaws

Hacksaws are very useful for cutting metal. Figure 1-8 shows a hacksaw blade being installed. The blade will break if it is not properly inserted and tightened; so once the blade is pointing in the right direction—away from the handle—tighten it. Note that the cutting takes place when the hacksaw is pushed away from the operator. Lift up on the saw when bringing it back to the starting position. Riding the blade as it is drawn back through the metal ruins the blade.

***Fig. 1-8*** *Inserting a hacksaw blade.*

Blades come in 8-, 10-, and 12-in. lengths. The hacksaw is usually adjustable to fit any of the three blade lengths. Hacksaw blades also come in a number of tooth sizes. Use 14-teeth-per-inch blades for cutting 1-in. or thicker sections of cast iron, machine steel, brass, copper, aluminum, bronze, or slate. Use 18-tooth-per-inch blades for cutting $1/4$- to 1-in.-thick sections of annealed tool steel, high-speed steel, rail, bronze, aluminum, light structural shapes, and copper. The 24-tooth-per-inch blade is used for cutting $1/8$- to $1/4$-in.-thick section of materials. It is usually best for iron, steel, brass, and copper tubing, wrought iron pipe, drill rod, conduit, light structural shapes, and metal trim. The 32-tooth-per-inch blade is used for cutting material similar to that recommended for 24-tooth-per-inch blades.

## Wrenches

Wrenches are used to tighten and loosen nuts and bolts. Two general types of wrenches are adjustable and nonadjustable wrenches. Adjustable wrenches have one jaw that can be adjusted to accommodate different nut and bolt sizes and may range from 4 to 18 in. in length for different types of work (Fig. 1-9). When adjustable wrenches are used, there are two rules to remember:

1. Place the wrench on the nut or bolt so that the force will be placed on the fixed jaw.
2. Tighten the adjustable jaw so that the wrench fits the nut or bolt snugly (Fig. 1-10).

***Fig. 1-9*** *Adjustable wrench.* (Klein)

***Fig. 1-10*** *Note the direction of applied force.*

Nonadjustable wrenches have fixed openings to fit nuts and bolt heads. Figure 1-11 shows a nonadjustable open wrench, and Fig. 1-12 shows a nonadjustable box-end wrench. These wrenches are available in sets and also in metric sizes. Openings are usually 0.005 to 0.015 in. larger than the size marked on the wrench to allow the wrench to be slipped easily over the nut or bolt head. Make sure, however, that the wrench fits the nut or bolt

***Fig. 1-11*** *Combination open-end, box-end wrench.* (Klein)

***Fig. 1-12*** *Box-end wrench.* (Klein)

Hand Tools  33

head properly (Fig. 1-13); otherwise, the nut or bolt head may get damaged. It is safer to pull than to push a wrench. If you exert pressure on a wrench and the nut or bolt head suddenly breaks loose, there is a chance that you will injure your hand. If the wrench must be pushed rather than pulled, use the palm of the hand so that the knuckles will not be injured if there is a slip.

**Fig. 1-13** *Proper fit of wrench.*

**Allen Wrenches** Allen wrenches are designed to be used with headless screws. These screws are used in many devices such as setscrews (Fig. 1-14). Allen wrenches come in a variety of sizes to fit any number of setscrews (Fig. 1-15). A complete set is especially helpful in motor work.

**Fig. 1-14** *Allen wrench and setscrew.*

**Fig. 1-15** *Allen wrench set.*

Another type of recessed setscrew head has an Allen-type hole, but ridges in each flat side make it difficult for an Allen wrench to fit. Newer hex-socket key sets are made in $3/32$ to $1/4$-in. sizes with 8 blades in a set (Fig. 1-16).

**Fig. 1-16** *Hex-socket key set.* (Klein)

**Socket Wrenches** Socket wrenches may be used in locations that are not easily accessible to the box-end or open-end wrench. Sockets are easily taken off the ratchet and replaced with another size. Sockets come in both a 12-point and a 6-point arrangement. Make sure that you use the proper size for each nut or bolt head. Socket wrenches are also available in metric sizes. Figure 1-17 shows some sockets, extensions, flexible handles, and universal joints that make sockets effective in almost any location.

**Fig. 1-17** *Socket wrench set.* (Klein)

**Torque Wrenches** Torque wrenches, which are designed to apply proper torque to various bolts and nuts, are made to fit various socket drives. Two popular sizes of torque wrenches are the $3/8$-in. drive and the $1/2$-in. drive. Torque wrenches are made to measure in pound inches (lb-in.) and in pound feet (lb-ft).* Use the proper wrench for the torque that has to be applied. The wrenches come in various handle lengths. Normally, the longer the wrench, the greater the torque the wrench will measure. A typical torque wrench is shown in Fig. 1-18.

**Fig. 1-18** *Torque wrench.* (Klein and Allen-Bradley)

## Nut Drivers

The nut driver is nothing more than a socket attached to a screwdriver handle. It is an excellent tool for most on-the-bench work. Nut drivers come in a variety of sizes and usually have the size stamped on the handle. Sometimes they are color coded according to size. Figure 1-19 shows a set of nut drivers.

**Fig. 1-19** *Nut driver set.*

*The term in.-lb and ft-lb were formally used for torque. Recently they have been more accurately stated as lb-in. and lb-ft.

## Tools for Bearings and Bushings

In motor repair, it is sometimes necessary to remove a bearing from the end bell of a motor. A bearing tool (Fig. 1-20) makes this task somewhat easier. The set illustrated has nine adapters that easily and quickly remove or insert any sleeve motor bearing or bushing with a $1/2$- to 1-in. inside diameter. A bearing tool eliminates the chance of broken bearings or end bells.

**Fig. 1-20** *Bearing tool with adapters.*

Some bearing removals need a different approach. The pulley or gear puller can, in some instances, be used to remove a bearing that has stuck on the motor shaft. These pullers come in a variety of sizes and styles (Fig. 1-21).

**Fig. 1-21** *Gear Pullers.*

Bushing tools have been designed for removing or inserting bushings in motors. They are handy time-savers. A complete set usually consists of 20 pieces: the box, 3 drivers, and 16 adapters of lengths varying from $3/8$ to 1 in. (Fig. 1-22).

**Fig. 1-22** *Bushing tool set.*

## Solder-Less Connector Crimper

The solder-less connector crimper (Fig. 1-23) is very useful in motor work. It takes a good connection to withstand the vibration of a motor. A number of contactors have been designed for electrical work. The tool and kit of connectors and lugs of various sizes are available at most electrical supply houses.

**Fig. 1-23** *Solder-less connector crimper.* (Klein)

## Soldering Iron

The soldering iron (Fig. 1-24) comes in handy when it is necessary to make a solder connection that will take vibration and withstand corrosion. Soldering irons of about 15 watts (W) to over 600 W are available in the market. The best all-purpose soldering iron for use in the shop is about 100 W. This will do the job in most cases where larger wires are involved. The small 15-W irons are very useful in electronics work on printed circuit boards.

**Fig. 1-24** *Soldering iron.*

## SOLDERING GUN

The soldering gun (Fig. 1-25) is very handy for making quick disconnects of soldered joints. Cold solder joints result when the operator heats the tip, places solder on it, and then lets the solder cool on the joint. The wires being soldered on the metal surface and the wire being connected to it must be heated up to the temperature sufficient to melt the solder. This means that the gun must be left in one spot long enough to cause the joint to be heated to the temperature needed to melt the solder. The secret is to heat the material, not the solder.

Do not hold the soldering gun tip on the joint too long. You do not want the soldering gun tip to overheat the printed circuit board, or the copper strip will lift up from the board.

**Fig. 1-25** *Soldering gun.*

## Wire Gages

Wire gages are needed to measure wire size. The numbers on the gage (Fig. 1-26) tell you the size of the wire. Keep in mind, however, that wire with Formvar insulation will read one size larger. Keep in mind also that the wire is moved through the slit in the gage. The hole is there to pass the wire through. The slot does the measuring. Pull the wire free of the slot and through the hole. Decimal equivalents are usually stamped on the metal disk on the opposite side of the gage numbers. Every toolbox needs a wire gage.

**Fig. 1-26** *Wire gage.*

## Fuse Puller

Fuse pullers are made of phenolic material that has been shaped so that you can pull at least two sizes of fuse with them. A fuse puller catches the round body of the fuse and allows you to extract the fuse from its holder, without coming in contact with the live circuit (Fig. 1-27).

**Fig. 1-27** *Fuse puller.*

## Tachometer

To check the speed of rotation of an electric motor, it is best to use a handheld tachometer, which is available in both analog and digital forms representing speed in revolutions per minute (rpm) (Fig. 1-28). The tachometer is a very useful device for measuring motor speed. It can help locate possible problems and can indicate if a motor is operating as it should after it has been repaired.

**Fig. 1-29** *Knockout punch in operation.*

**Fig. 1-28** *Tachometer. (Biddle)*

The shaft speed will read out in digital form in the tachometer as shown in Fig. 1-28. The tachometer can be placed on the open end of a motor shaft, or it can be used on motors, saws, compressors, fans, pumps, grinders, and other equipment. A cone-shaped tip is used for shafts with center holes; a cup-shaped tip is used for flat-end shafts. Tachometers that use a strobe light to detect the number of revolutions per minute are available, but they are somewhat more expensive.

## Knockout Punch

Knockout punches come in many sizes. They are used to make holes in metal boxes for enclosures of various control devices. They range from a simple punch that uses wrenches to large hydraulic units utilized to punch holes in very heavy gage metals (Fig. 1-29).

## Fish Tape

For pulling wires through conduit and walls, or into junction boxes, fish tape is essential (Fig. 1-30).

**Fig. 1-30** *Fish tape and reel. (Ideal Industries)*

## Cable Stripper

A larger-gage wire (cable) has to be stripped for a T-tap or mid-span strip. In most instances, a cable stripper is used for stripping insulation so that placing lugs on the end of the wire or cable is much easier. Figure 1-31 shows how the stripper can be used to strip insulation from a conductor without scoring.

**Soldering Gun**    37

1. Make two cuts around circumference of cable, rotating jaws around cable for a clean, accurate insulation cut.

2. Slit insulation between cuts, parallel to the cable. Integral steel slitter, guided by tool, makes it safe and easy. No knife needed.* Won't damage cable.

3. Pinch edge of slitted insulation between the built-in grippers at tip of tool and peel it away.

**Fig. 1-31** Cable stripper. (Klein)

## Cable Cutter

Hand-operated cable cutters are used to make a shear-type cut for large-size wire. Long fiberglass handles give leverage. A cable cutter can cut cables up to 750 MCM (750,000 circular mils)† (Fig. 1-32). The clean cut that results makes it easier to fit cable ends into lugs.

**Fig. 1-32** Cable cutter. (Klein)

†M is the Roman numeral for 1000. CM stands for circular mil. 1 mil = 0.001 in. 1 circular mil = 0.001 in. in cross-sectional area. 1 MCM = 1000 CM.

## Electrician's Knives

The jack knife (Fig. 1-33) is used to cut insulation. Every toolbox should have a handy jack knife with very sharp blades. Another knife that comes in handy for the electrician or anyone working around electrical equipment is the skinning knife (Fig. 1-34). This knife fits in the toolbox or in the tool pouch with the proper cover.

**Fig. 1-33** Jack knife. (Klein)

**Fig. 1-34** Skinning knife. (Klein)

## Other Tools

Other tools that may be useful on the job are a flashlight, digital camera, polarized receptacle tester, wire markers, and various wire-pulling apparatus and threading tools for rigid conduit. A heavy-duty electric drill or one with hammer action to drill through concrete is also handy. Various tools will come to mind as you develop your workshop or toolbox. As different situations arise on the job, you will be better equipped to know which tools to invest in. Keep in mind that your tools are an investment. Mark them with your name or the name of your company.

## ELECTRICAL TOOLS

There are a number of electrical instruments that are needed for the motor repair technician to install and maintain motors and their controls. Everything from a simple VOM to an oscilloscope is needed to perform various tests and to increase the efficiency of the trouble shooter.

## Portable Ammeter and Voltmeter

The clamp-on type of ammeter and voltmeter has extended leads with replaceable probe tips that make voltage reading easy and fast. The clamp-on jaw goes around a conductor of 1-in. diameter, 500-MCM or 2-in. diameter, or 2000-MCM diameter to take the current reading without interrupting the service. One hand operation is possible: Open the jaws, change ranges, and read the current. The meter comes in a handy, sturdy case to protect it while not in use (Fig. 1-35).

*Fig. 1-36* Megger. (Biddle)

*Fig. 1-35* (A) Analog clamp-on ammeter; (B) digital clamp-on ammeter. (Simpson)

## Voltage Tester

The voltage tester is a handy device that checks 10 ac/dc voltage levels. It fits in a shirt pocket. The lighted windows indicate the voltage level, which enables it to be easily read in dimly lighted areas. The coiled lead cord extends to 50 in. A test button distinguishes normal readings from those due to distributed capacitance or high-resistance leakage. The voltage tester is also helpful in checking out 115-V ac grounded convenience outlets, and will operate on 25 to 800 Hz (Fig. 1-37).

## Megohmmeter

The megohmmeter is usually referred to as a *megger*. The megger shown in Fig. 1-36 has ranges for measuring insulation resistances of motors, compressors, conductors, or anything else that needs measuring. It can also read continuity and can measure low resistances of motor windings.

Test voltage is generated by a hand crank. This means that the megger does not require any other power source and is therefore always ready for use. The ohmmeter range of the unit illustrated is especially well suited for measuring the resistance of motor windings and other low resistances. The guard terminal eliminates the effect of any surface leakage that may influence a reading. The megger comes with a manual that shows how to use it for various purposes.

*Fig. 1-37* Voltage tester. (Amprobe Instruments)

Electrical Tools  39

## Volt-Ohm-Milliammeter (VOM)

The volt-ohm-milliammeter, usually called a VOM, is available in many sizes designed to read many ranges. VOMs check out voltages, currents, and resistances. They are portable and contain a battery to power the resistance checking ranges. Most are also capable of testing capacitors and inductances as well as decibels (Figs. 1-38 and 1-39). A flip-open carrying case makes it easy to store and provides a place for lead storage so that you know where they are the next time you need the meter.

*Fig. 1-38*  Analog VOM. (Simpson)

*Fig. 1-39*  Digital VOM.

## Digital Logic Probe

The digital logic probe has become a necessity in troubleshooting programmable controllers and other equipment that uses computer logic to switching or sequencing. It is a quick way to peek inside TTL, LSI, and CMOS digital circuits.[‡] The probe shown in Fig. 1-40 has color-coded light-emitting diodes (LEDs) to indicate high, low, or pulsed logic states (up to 10 MHz). It puts out a tone that really speeds up the testing.

The next order of business involves the operation of the probe and what it indicates at various points in the test procedure. Typical signals and corresponding LED indications are shown in Fig. 1-40. To operate the probe, apply power to it by connecting the black clip to ground or as the English say, earth $(-)$. Connect the red clip to the $V_{cc}$, $(+)$. Make sure that the voltage is less than 20 V.

The TTL/CMOS switch can be switched to TTL mode for use in TTL circuits. The TTL logic 1 threshold is $2.3 \pm 0.2$ V; the logic 0 threshold is $0.8 \pm 0.2$ V. When switching to CMOS mode, the CMOS logic 1 threshold is 70% $V_{cc}$; the logic 0 threshold is 30% $V_{cc}$. The logic probe can also detect and memorize the level transition. Either positive or negative level transition can be detected or memorized, depending on the mode (pulse/memory) selected. On the PULSE position, the memory function is inoperative. Input state transition is "0" to "1"; or "1" to "0" will activate the pulse indicator (flicker for 500 ms) and will generate a chop sound for the beeper. On the MEMORY position, the memory function is activated. The pulse indicator lights and a beeper generates sound until reset if any pulse or level transition occurs.

The beeper generates sound when the red or green LED lights. Some typical combinations of indications from the probe are shown in the table of typical signals in Fig. 1-40.

## Continuity Tester

Whenever an ohmmeter is not available or handy, it is convenient to use the continuity tester, since the tester is usually carried in the toolbox or on the tool belt. The tester is a simple device consisting of a battery, lamp, and probe, with another lead to complete the circuit and light the lamp to indicate a complete path for current flow. It can be used for testing for opens, shorts, or continuity (Fig. 1-41).

## Polarized Plug Tester

There are a number of plug-in devices to indicate if a circuit has the proper ground and if the hot side of the receptacle is where it is supposed to be. Figure 1-42 shows one of these plug-in devices and how the readout in the LEDs indicates the various conditions of the circuit.

---

[‡] TTL stands for transistor logic, LSI for large scale integrated, and CMOS for complementary metal-oxide semiconductor.

**TYPICAL SIGNALS AND CORRESPONDING LED INDICATION:**

| ITEMS | WAVEFORM | LED INDICATIONS LEVEL PULSE RED GREEN YELLOW | | | BEEPER |
|---|---|---|---|---|---|
| Logic "1" no pulse activity | 1 ―――<br>0 ------ | ● | ○ | ○ | High tone |
| Logic "0" no pulse activity | 1 ------<br>0 ――― | ○ | ● | ○ | Low tone |
| Signal level between "1" & "0" | 1 ------<br>0 ------ | ○ | ○ | ○ | |
| Logic "1" with pulse | 1 ⎍⎍⎍⎍⎍<br>0 | ● | ○ | ☆ | Intermittent high tone. |
| Logic "0" with pulse | 1 ⎍⎍⎍⎍⎍<br>0 | ○ | ● | ☆ | Intermittent low tone. |
| Pulse train with freq. < 200KHZ | 1 ⊓⊔⊓⊔<br>0 | ● | ● | ☆ | 1. Alternate and Intermittent sound.<br>2. Mixed and Intermittent sound. |
| Pulse train with freq. > 200KHZ | 1 ⊓⊔⊓⊔<br>0 | ○ | ○ | ☆ | |

● LED On   ○ LED Off   ☆ LED blinks   ------ ref. level   ――― signal

**Fig. 1-40** *Digital logic probe.* (B&K Instruments)

**Electrical Tools**

**Fig. 1-41** Continuity tester. (Klein)

**Fig. 1-42** Polarized receptacle tester.

Strip Chart

(A)

(B)

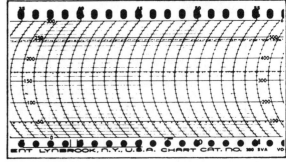

Strip Chart

**Fig. 1-43** (A) Voltage recorder; (B) ammeter recorder.

## Recorder

The strip recorder comes in handy to check out line voltages that may vary under a number of conditions that are not easily checked. The recorder can be placed on the line and will record variations in line voltages over a long period of time. Knowing when the variations occur aids in locating a number of problems. The recorder shown in Fig. 1-43A monitors voltage conditions over a wide variety of ranges. It helps pinpoint over-voltage conditions, which shorten lamp and motor life, and under-voltage conditions, which affect oven, tool, and lighting output. The ammeter-type recorder (Fig. 1-43B), which uses the split-core transformer principle for clamp-on transducers, simply clamps around the conductor in which the current is to be measured. There is no need to interrupt the service. With this ammeter-type device, you can monitor all three phases; in conjunction with a voltmeter-type recorder, it can be utilized to check out when various power surges occur and can aid in regulating placement of various machines on the line at the most economical time for power consumption.

## Oscilloscope

The oscilloscope (Fig. 1-44A) is also a voltage-indicating device. It shows the shape of power being used. It can aid in tracing pulses on the line that may cause timing problems with programmable controllers or other computer-operated machines. By using a function reference signal generator, such as that shown in Fig. 1-44B, with an oscilloscope, it is possible to properly adjust and tune the stability circuit of a high-gain motor drive and regulators. The appropriate stability

**Fig. 1-44** (A) Oscilloscope; (B) function reference signal generator.

circuits can be optimized by using a step function into the regulator and observing the feedback loop output with an oscilloscope or chart recorder.

## Phase Sequence Adapter

The phase sequence adapter (Fig. 1-45) is used in conjunction with any Amprobe volt/ammeter or other voltmeter with appropriate ac range. It permits you to determine the phase sequence of any electrical equipment using three-phase lines up to 550 V, 25 to 60 Hz. The carrying case and manual are included.

**Fig. 1-45** Phase sequence adapter.

## Balance Analyzer

The balance analyzer (Fig. 1-46) has a digital meter that provides readings of displacement, velocity, and acceleration in the vibration mode. The strobe mode allows the operator to select the strobe and find the rate for stop-motion analysis. The balance/analyze mode allows the model to be tuned to specific frequencies for signature analysis and balancing purposes. Although this analyzer is a very expensive instrument, it is usually worth its price in locating and eliminating problems caused by vibration and off-balance loads.

**Fig. 1-46** (A) Balance analyzer; (B) vibration meter and bearing tester. (Vitec)

## SPECIAL TOOLS FOR SPECIFIC JOBS

### Printed Circuit Board Puller

Motor controls are sometimes incorporated into printed circuit boards and mounted into racks. Extreme caution must be taken when installing or removing boards because some boards are extremely sensitive to static electricity discharges. The printed circuit board puller shown in Fig. 1-47 clamps firmly onto the circuit board. It is covered with plastic to prevent physical damage to the board, and its use prevents you from touching the board while removing it.

**Fig. 1-47** *Printed circuit board puller.* (Reliance)

**Fig. 1-49** *Conduit bender, handle, and bending formula.* (Klein)

## Wire Wrap/Unwrap Tool

Soldering is not the only way that wire terminations are made at present. Wire wrapping provides fast, secure electrical connections without soldering. Wire wrapping and unwrapping tools (Fig. 1-48) are made of case-hardened steel for long life. They are hand operated, so there is no need for air or electrical supply. This method of termination eliminates wire crystallization due to soldering heat and subsequent fracturing from on-the-job vibration. The tool shown can handle 18- to 22-gage wire.

**Fig. 1-48** *Manual wire wrapper and unwrapper.* (Reliance)

## PVC Cutters

Cutting PVC plastic has traditionally been done using a backsaw. This saw cut can be rather ragged when finished, because the kerf of the saw blade is designed for cutting wood. Making clean, burr-free cuts through PVC is a quick, simple task with specially designed cutters. They use ratchet action for maximum cutting power with very little effort (Fig. 1-50). The one shown in the figure is only 9 in. long, so it can fit in the toolbox or can be hung on the tool pouch. It will cut $1/4$- to $1 1/4$-in. PVC. Just open the handles completely, place the conduit in the lower jaw hook, and alternately squeeze and release the handles, letting the ratchet action do the work until the shear is complete. By using a cutter with longer handles, it is possible to cut $1/2$- to 2-in. PVC as easily as the short-handled cutter cuts smaller diameters. The advantage of this type of cutter is reduction in pipe deformation and the possibility of PVC conduit cracking during cutting. A clean cut is also obtained. The blades can easily be replaced when they become dull.

## Conduit Benders

Conduit benders take the guesswork out of this job. Benders have built-in benchmark symbols, degree scales, and multiplier scales. An arrow points to the beginning of a bend; a star indicates back-of-bend locations. A teardrop symbol indicates the exact center of a 45° bend.

Benders for $1/2$-in. EMT also have a cast-in offset formula and multiplier, providing instant, on-the-spot information for making accurate offset-angle bends (Fig. 1-49).

**Fig. 1-50** *PVC cutters.* (Klein)

## Cable Benders

Some cables require a great deal of effort to bend to fit their intended placement. The forged one-piece bender has been made of steel to meet the demands for such a tool designed to aid in the bending operation (Fig. 1-51). The head is bent at 22° for ease in bending the cable in tight, hard-to-access places. A $7/8$-in. opening will allow inserting cables up to 300-MCM capacity. A $1\,1/64$-in. opening is used for cables up to 500-MCM capacity. The handle is plastic coated and contoured for good gripping action. For easier bends, two benders should be used.

*Fig. 1-51*  Cable bender. (Klein)

## Conduit Reamer

The conduit reamer (Fig. 1-52) locks onto a screwdriver, and reams and de-burrs $1/2$-; $3/4$-, and 1-in. thinwall conduit ends. The smooth ends protect wire being pulled through the conduit, which permits correct installation of fittings. The conduit reamer reams inside and outside at the same time. Two setscrews hold it tightly on round or square screwdriver shanks. The reamer can be left on the screwdriver for normal use. The steel cutting blade is replaceable to keep it cutting easily.

*Fig. 1-52*  Conduit reamer fits onto standard screwdriver. (Klein)

## REVIEW QUESTIONS

1. List at least five types of pliers that may be put to good use by an electrician.
2. What is the difference between a ball-peen hammer and a claw hammer?
3. Why do hacksaw blades have different amounts of teeth?
4. Describe an Allen wrench.
5. Why are torque wrenches necessary?
6. What is a nut driver, and how is it used?
7. What is the wattage rating of the best all-purpose soldering iron?
8. What are fuse pullers made of? Why?
9. What are the two types of electrician's knives?
10. What does a megger do?
11. Why do you need a digital logic probe?
12. What does CMOS stand for?
13. How is a continuity tester used?
14. What function does an oscilloscope serve?
15. What is a balance analyzer?
16. How is PVC conduit cut?
17. What does MCM stand for?
18. Why is a conduit reamer needed?

## REVIEW PROBLEMS

One of the most important tools that an electrician can have is a good grasp of Ohm's law:

$$E = I \times R \quad I = \frac{E}{R} \quad R = \frac{E}{I}$$

1. What drop in voltage exists across a 5000-Ω resistor, if it carries 100 mA?
2. What voltage is necessary to cause a current of 5 A to flow through a 150-Ω resistor?
3. How much electromotive force (EMF) is needed to cause 10 mA to flow through a resistance of 1000 Ω?
4. A resistor has a resistance of 60. What is the voltage drop across it when 100 mA flows through it?
5. A magnetic brake coil requires 6.5 A to load a motor. If the coil has a resistance of 0.5 Ω, what voltage is required?
6. An automobile headlight pulls 7.5 A from a 12.6-V car battery. What is the resistance of the lamp?
7. A solenoid coil pulls a current of 4.5 A when connected to a 12-V supply. What is the resistance of the coil?

# CHAPTER 2

# Safety in the Workplace

# PERFORMANCE OBJECTIVES

After studying this chapter, you will be able to:

1. List safety measures for working safely on the job.
2. List values associated with the use of ground-fault circuit interrupters (GFCIs).
3. Identify conditions that lead to overloads, short circuits, and blown fuses.
4. Identify fuses by their voltage rating and ampere rating.
5. Describe how electrical codes promote safety.
6. Describe the role of Underwriters Laboratories, Inc. (UL) and Canadian Standards Association (CSA) in electrical safety.
7. Identify the Occupational Safety and Health Act (OSHA) safety color code colors.
8. Choose the proper fire extinguisher for electrical fires.
9. Use hammers properly.
10. Wear the correct working clothes.
11. Use motors and generators safely.
12. Select proper equipment for doing a job safely.
13. Explain the need for proper grounding of electrical equipment.

Safety concerns everyone working with you as well as yourself. Working safely is of great concern for all who work around electricity and electrical equipment.

Most fatal electrical shocks happen to people who should know better. Below we provide some electromedical facts that should make you think twice before taking chances. For a safe workplace you have to think before you act. Working safely involves a number of considerations. Before working around motors that require large currents and usually operate on very high voltages, there are a number of things of which you should be aware.

# ELECTRICAL SHOCK

It is not the voltage but the current that kills. People have been killed by 110 V in the home and also with as little as 42-V direct current (dc). The real measure of a shock's intensity lies in the amount of current (in milliamperes) forced through the body, not in the voltage. Any electrical device used on a house wiring circuit can, under certain conditions, transmit a fatal current.

Since you do not know how much current went through the body in an accident, it is necessary to perform artificial respiration to try to get the person breathing again, or if the heart is not beating, perform CPR.

NOTE: A heart that is in fibrillation cannot be restricted by closed-chest cardiac massage. A special device called a defibrillator is available in some medical facilities and ambulance services.

Muscular contractions are so severe with 200 mA and above that the heart is forcibly clamped during the shock. Clamping prevents the heart from going into ventricular fibrillation, making the victim's chances for survival better.

# ELECTRICAL SAFETY MEASURES

Working with electricity can be dangerous. However, electricity can be safe if properly respected.

## Using Ground-Fault Circuit Interrupters

Some dangerous situations have been minimized by using ground-fault circuit interrupters (GFCIs; see Fig. 2-1). Since 1975 the *National Electrical Code*

***Fig. 2-1*** *Ground-fault protection system. (Westinghouse)*

(NEC) has required installation of GFCIs in outdoor and bathroom outlets in new construction, but most homes built before then have no GFCI protection.

Ground-fault protection is also available for installation in motor control centers and high-voltage starters.

The ground-fault protection shown in Figs. 2-1 and 2-2 consists of a ground-fault relay and a current sensor. Two types of relays are available from Westinghouse. Type GR operates instantaneously and fulfills the basic need for fast, sensitive ground-fault detection. Type GRT incorporates a time delay and is adjustable from instantaneous to 36 cycles (0.6 second at 60 Hz). Both relays employ reliable solid-state circuits and are self-powered by fault current.

**Fig. 2-2** *(A) Typical Groundgard® application. (Westinghouse) (B) Typical ground-fault circuit interrupter most often used in residential buildings—bathrooms, kitchens, laundry rooms, and garages as well as outdoor outlets.*

Current sensors are window current transformer-type devices through which the cable or bus of all phases is run. Various "window" sizes are available for a wide variety of applications. Relay contacts are available on relays both tripped open for application in standard contactor holding coil circuits and tripped closed for mechanically latched contactors.

## Safety Devices

Electricity can create conditions almost certain to result in bodily harm, property damage, or both. It is important for those who work with or around electricity to understand the hazards involved when they are working around electrical power tools, maintaining electrical equipment, or installing equipment for electrical operation. Available safety devices are essentially over current devices: specifically, fuses and circuit breakers as well as GFCIs.

## Circuit Protection

Electrical distribution systems are often quite complicated. They cannot be absolutely fail-safe. Circuits are subject to destructive over currents. Harsh environments, general deterioration, accidental damage or damage from natural causes, excessive expansion, or overloading of the electrical distribution system are factors that contribute to the occurrence of over currents. Reliable protective devices prevent or minimize costly damage to transformers, conductors, motors, and the many other components and loads that make up the complete distribution system.

Reliable circuit protection is essential to avoid the severe monetary losses that can result from power blackouts and prolonged downtime of facilities. It is the need for reliable protection, safety, and freedom, from fire hazards that have made the fuse a widely used protective device.

## Over currents

An over current is either an overload current or a short-circuit current. The overload current is an excessive current relative to normal operating current but one that is confined to the normal conductive paths provided by the conductors and other components and loads of the distribution system. As the name implies, a short-circuit current is one that flows outside the normal conducting paths.

## Overloads

Overloads are most often between one and six times the normal current level. Usually, overloads are caused by harmless temporary surge-currents. They may occur when motors are started or transformers are energized. Such overload currents or transients are normal occurrences. Since they are of brief duration, any temperature rise is trivial and has no harmful effect on the circuit components. It is important to ensure that protective devices do not react to them.

Continuous overloads can result from defective motors (such as worn motor bearings), overloaded equipment, or too many loads on one circuit. Such sustained overloads are destructive and must be cut off by protective devices before they damage the distribution system or system loads. However, since they are of relatively low magnitude compared to short-circuit currents, removal of the overload current within a few seconds will generally prevent equipment damage. A sustained overload current results in overheating of conductors and other components and will cause deterioration of

**Electrical Safety Measures**

insulation, which may eventually result in severe damage and short circuits if not interrupted.

## Short Circuits

Overload currents usually occur at rather modest levels; the short circuit or fault current can be many hundreds of times larger than the normal operating current. A high-level fault may be 50,000 A or larger. If not cut off within a matter of a few milliseconds, damage and destruction can become rampant. There can be

- severe insulation damage
- melting of conductors
- metal vaporization
- ionization of gases
- arcing
- fire

Simultaneous high-level, short-circuit currents can develop huge magnetic field stresses. The magnetic forces between bus bars and other conductors can be many hundreds of pounds per lineal foot; even heavy bracing may not be adequate to keep them from being warped or distorted beyond repair.

## Fuses

The fuse is a reliable over current protective device. The fundamental element of the basic fuse is a fusible link or links encapsulated in a tube and connected to contact terminals. The link's electrical resistance is so low that it simply acts as a conductor. However, when destructive currents occur, the link very quickly melts and opens the circuit to protect the conductor and other circuit components and loads. Fuse characteristics are stable. Fuses do not require periodic maintenance or testing. Fuses have three unique performance characteristics:

1. They are safe. Fuses have a high interrupting rating, which means that they can withstand very high fault currents without rupturing.
2. Properly applied fuses prevent blackouts. Only the fuse nearest to a fault opens without up-stream fuses (feeders or mains) being affected—fuses thus provide selective coordination.
3. Fuses provide optimum component protection by keeping fault currents to a low value. They limit current.

**Voltage Rating** Most low-voltage power distribution fuses have 250- or 600-V ratings (others have ratings of 125 V and 300 V). The voltage rating of a fuse must at least equal the circuit voltage. It can be higher but can never be lower. The voltage rating determines the ability of the fuse to suppress the internal arcing that occurs after a fuse link melts and an arc is produced. If a fuse is used with a voltage rating lower than the circuit voltage, arc suppression will be impaired, and under some fault current conditions, the fuse may not safely clear the over current.

**Ampere Rating** Fuses have ampere ratings. In selecting the ampacity of a fuse, consideration must be given to the type of load and code requirements. The ampere rating of a fuse should normally not exceed the current-carrying capacity of the circuit. There are some specific circumstances where the ampere rating is permitted to be greater than the current-carrying capacity of the fuse. A typical example is a motor circuit. The dual-element fuse can be sized up to 175% and non–time-delay fuses up to 300% of the motor full-load amperes (Fig. 2-3). Generally, the ampere rating of a fuse and switch combination should be selected at 125% of the load current (this usually corresponds to the circuit capacity, which is also selected at 125% of the load current). There are exceptions, such as when the fuse-switch combination is approved for continuous operation at 100% of its rating.

The true dual-element fuse has distinct and separate overload and short-circuit elements.

Under sustained overload conditions, the trigger spring fractures the calibrated fusing alloy and releases the "connector."

The "open" Dual Element fuse after opening under an overload condition.

Like the single element fuse, a short-circuit current causes the restricted portions of the short-circuit elements to melt and arcing to burn back the resulting gaps until the arcs are suppressed by the arc quenching material and increased arc resistance.

The "open" Dual Element fuse after opening under a short circuit condition.

**Fig. 2-3** *Dual-element fuse operation.* (Bussmann Division, Cooper Industries, Inc.)

# ELECTRICAL CODES

Thomas Edison's first electric station could transmit electricity only 5000 ft. Modern power pools enable a customer to use electricity produced in a power plant many states away. Early generators had a power-producing capacity of only 100 kW, but generators with the ability to produce millions of kilowatts are now in operation. Since 1900 transmission voltages have increased from 30,000 V to 500,000 V, and lines of up to 765,000 V are now in operation.

In the early 1880s the New York Board of Fire Underwriters were concerned with the new method of electric lighting proposed by Thomas Edison. Although he did not realize the danger of the giant force that he was helping to create, the New York Board recognized that unless proper precautions were followed, the new method of lighting could prove to be as hazardous as the open flame that it was replacing. In 1881, one person was appointed to inspect every electrical installation before power was turned on. This was the beginning of the Electrical Department of the New York Board. It was necessary for the inspector to check not only the installation within the building but to carry the investigation back to the power station, then only one or two blocks distant. At that time, the board investigated the safety of the entire power system and required the power companies to make weekly tests for grounds and open circuits and report to the board the results of their tests. In 1882, the Committee of Surveys drew up a *set of safe-guards* for arc and incandescent lighting, which was the forerunner of the present National Electrical Code.

## National Electrical Codes

The *National Electrical Code* is the most widely adopted code in the world. The combined sales of its handbooks are over 1 million copies each time it is published, which is once every three years. The *National Electrical Code* is a nationally accepted guide for the safe installation of electrical conductors and equipment and is, in fact, the basis for all electrical codes used in the United States. It is also used extensively outside the United States, particularly where American-made equipment is installed. No electrician should be caught without a copy in the toolbox.

The *National Electrical Code Handbook* is published by the National Fire Protection Association to assist those concerned with electrical safety in understanding the intent of the new edition of the Code. A verbatim reproduction of the new National Electrical Code is included, and added where necessary are comments, diagrams, and illustrations that are intended to clarify further some of the intricate requirements of the Code.

The Code is purely advisory as far as the NFPA (National Fire Protection Association) and ANSI (American National Standards Institute) are concerned but is offered for use in law and for regulatory purposes in the interest of life and property protection. Anyone noticing any errors is asked to notify the NFPA executive office, the chairman, and the secretary of the committee.

## Underwriters Laboratories

Underwriters Laboratories, Inc. (UL) was founded in 1894. William Henry Merrill came to Chicago to test the installation of Thomas Edison's new incandescent electric light at the Columbian Exposition. He later started UL as a laboratory where insurance companies could test products for electrical and fire hazards. It continued as a testing laboratory for insurance underwriters until 1917. It then became an independent, self-supporting, safety-testing laboratory. The National Board of Fire Underwriters (now the American Insurance Association) continued as sponsors of UL until 1968. At that time, sponsorship and membership were broadened to include representatives of consumer interests, governmental bodies or agencies, education, public safety bodies, public utilities, and the insurance industry, in addition to safety and standardization experts. UL has expanded its testing services to more than 13,000 manufacturers throughout the world. Over 1 billion UL labels are used each year on products listed by Underwriters laboratories.

UL is chartered as a not-for-profit organization without capital stock, under the laws of the state of Delaware, to establish, maintain, and operate laboratories for the examination and testing of devices, systems, and materials. Its stated objectives are "By scientific investigation, study, experiments and tests, to determine the relation of various materials, devices, products, equipment, constructions, methods and systems to hazards appurtenant thereto or to the use thereof, affecting life and property, and to ascertain, define and publish standards, classifications and specifications for materials, devices, equipment, construction, methods and systems affecting such hazards, and other information tending to reduce and prevent loss of life and property from such hazards."

The corporate headquarters, together with one of the testing laboratories, is located on West Street in Northbrook, Illinois; Melville, New York; Santa Clara, California; and Tampa, Florida.

Underwriters Laboratories has a total staff of over 2000 employees. More than 700 persons are engaged in engineering work, and of this number, approximately 425 are graduate engineers. Supplementing the engineering staff are more than 500 factory inspectors. The engineering functions are divided among these six departments:

Burglary protection and signaling

Casualty and chemical hazards

Electricity

Fire protection

Heating, air conditioning, and refrigeration

Marine equipment

The electrical department is the largest of the six engineering departments. Safety evaluations are made on hundreds of different types of appliances for use in homes, commercial buildings, schools, and factories. The scope of the work in this department includes electrical construction materials that are used in buildings to distribute electrical power from the meter location to the electrical load.

Underwriters Laboratories publishes annual lists of manufacturers whose products have met UL safety requirements. These lists are kept up-to-date by quarterly supplements. Eleven lists are published each year. UL presently publishes more than 300 *Standards for Safety* for materials, devices, constructions, and methods. Copies of these are available to interested persons, and a free catalog is available.

For your own safety, the products you are using to wire a house, building, or installation of any kind should be marked "UL" (Fig. 2-4).

**Fig. 2-4**  *UL stickers and label.*

## Canadian Standards Association

The Canadian organization that parallels UL is the *Canadian Standards Association (CSA)*. However, CSA has more authority to remove from the market products that do not meet standards; the UL program is strictly voluntary. If an electrical product (or in some cases, another type of product) used in Canada is connected in any way with the consumption of power from the electrical sources owned by the provinces, that product must have CSA approval. This is a measure in the interest of public safety.

Products certified by CSA are eligible to bear the CSA certification mark. Misuse of the mark may result

**Fig. 2-5**  *Canadian Standards Association trademark.*

in suspension or cancellation of certification. CSA may resort to legal action to protect its registered trademark in the event of abuse. Figure 2-5 shows the CSA mark. In addition, CSA information tags and other markings are made available for certified products and their containers, to supplement the CSA mark.

## Standards in Other Countries

Other countries also have standards and testing laboratories. Figure 2-6 shows some of the marking used by other countries.

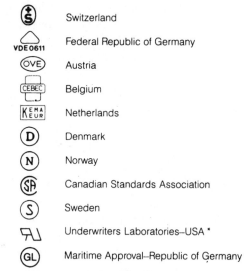

*Products recognized under the component program are identified by the recognized marking ꞋU. Recognized components are suitable for factory installation on other equipment where their use and limitations are determined by UL.

**Fig. 2-6**  *Approval monograms with country of origin.*

52   Safety in the Workplace

## OSHA

The Occupational Safety and Health Act (OSHA) of 1970 sets uniform national requirements for safety in the workplace—anywhere that people are employed. Originally, OSHA adopted the 1971 NEC as rules for electrical safety. The involved process for modifying a federal law such as OSHA made it impossible for the act to adopt each new NEC revision, as the code is amended every three years. To avoid this problem, the OSHA administration in 1981 adopted its own code, a condensed version of the NEC containing only those provisions considered related to occupational safety. OSHA was amended to adopt this code, based on the NFPA standard 70E, Part 1, which is now federal law.

## Equipment Standards

Equipment standards have been established by the National Electrical Manufacturers Association (NEMA). The American National Standards Institute (ANSI) is another standards source. Underwriters Laboratory (UL) has standards that equipment must meet before UL will list or label it. Most jurisdictions and OSHA require that where equipment listed as safe by a recognized laboratory is available, unlisted equipment may not be used. UL is by far the most widely accepted national laboratory, although Factory Mutual Insurance Company lists some equipment, and a number of other testing laboratories have been recognized and accepted. The Institute of Electrical and Electronics Engineers (IEEE) publishes a number of books (the "color book" series) on recommended practices for the design of industrial buildings, commercial buildings, emergency power systems, grounding, and the like. Some of these IEEE standards have been adopted as ANSI standards. They are excellent guides, although they are not in any way mandatory.

## OSHA Color Code

In order to establish a safe environment for everyone, OSHA has standardized some colors for specific applications. Red, yellow, orange, purple, and green have been chosen to indicate various locations, machines, and devices.

- Red
  - Fire protection equipment and apparatus
  - Portable containers of flammable liquids
  - Emergency stop buttons and switches
- Yellow
  - For caution and for marking physical hazards
  - Waste containers for explosive or combustible materials
  - Caution against starting, using, or moving equipment under repair
  - Identification of starting point or power source of machinery
- Orange
  - Dangerous parts of machines
  - Safety starter buttons
  - The exposed parts (edges) of pulleys, gears, rollers, cutting devices, and power jaws
- Purple
  - Radiation hazards.
- Green
  - Safety.
  - Location of first-aid equipment (other than fire-fighting equipment).

## FIRE EXTINGUISHERS

Fire extinguishers are limited in their application. They are used to control small fires that are identified properly by the person selecting the extinguishers. Extinguishers are made for various uses. Needless to say, the water type is not useful for electrical fires. Electrical fires are classified as class C. Class C fires are described as those dealing with energized electrical equipment, where the electrical nonconductivity of the extinguishing medium is of importance. Electrical fires call for carbon dioxide, dry chemicals, multipurpose dry-chemical, and Halon-1211 types of extinguishers (Fig. 2-7).

## SAFE WORKING PRACTICES

Safety equipment is available for use by the electrician or anyone working with electric motors. In most instances a person working in a commercial or industrial location will need steel-toed shoes, a hard hat, goggles, or a face shield. Fuse pullers are insulated to protect the user, but must be handled properly. Using pliers and cutting equipment can be harmful if certain common-sense rules are not followed.

## TOOL SAFETY

Some precautions to be observed when using pliers:

1. Never cut any wire or metal unless your eyes are protected. Safety goggles or other protective equipment are an absolute necessity. It is easy to forget to

| Extinguisher Classifications† | A | | AB | BC | | | ABC | | | D |
|---|---|---|---|---|---|---|---|---|---|---|
| Extinguishing Agent | Water Types (includes antifreeze) | | AFFF Foam and FFFP | Carbon Dioxide | Dry Chemical Types Purple K, Super K, Monnex, Potassium Bicarb. Urea based | | Halogenated Types 1211 1301 1211/1301 | Multipurpose Dry Chemical | | Halogenated Types 1211 1211/1301 | Dry Powder |
| Discharge Method | Stored Pressure | Pump Tank | Stored Pressure | Self Expelling | Stored Pressure | Cartridge Operated | Stored Pressure | Stored Pressure | Cartridge Operated | Stored Pressure | Cartridge Operated |
| Sizes Available | 2 ½ Gal. | 2 ½-5 Gal. | 2 ½ Gal. (33 Gal.) | 5-20 lb. (50-100 lb.) | 2 ½-30 lb. (50-350 lb.) | 4-30 lb. (125-350 lb.) | 1-5 lb. | 2 ½-20 lb. (50-350 lb.) | 5-30 lb. (125-350 lb.) | 5 ½-22 lb. (50-150 lb.) | 30 lb. (150-350 lb.) |
| Horizontal Range (Approx.) | 30-40 ft. | 30-40 ft. | 10-25 ft. (30 ft.) | 3-8 ft. (3-10 ft.) | 10-15 ft. (15-45 ft.) | 10-20 ft. (15-45 ft.) | 10-16 ft. | 10-15 ft. (15-45 ft.) | 10-20 ft. (15-45 ft.) | 9-16 ft. (20-35 ft.) | 5 ft. (15 ft.) |
| Discharge Time (Approx.) | 1 Min. | 1-3 Min. | 50-65 Sec. (1 Min.) | 8-15 Sec. (10-30 Sec.) | 8-25 Sec. (25-60 Sec.) | 8-25 Sec. (25-60 Sec.) | | 8-25 Sec. (20-60 Sec.) | 8-25 Sec. (25-60 Sec.) | 10-18 Sec. (30-45 Sec.) | 20 Sec. (150 lb., 70 Sec.; 350 lb., 1 ¾ Min.) |
| Operating Precautions and Agent Limitations | Conductor of electricity. Needs protection from freezing. (except antifreeze). Use on flammable liquids and grease will spread fire. | | Conductor of electricity. Needs protection from freezing. Not effective on water-soluble flammable liquids such as alcohol, unless otherwise stated on nameplate. AFFF not effective on pressurized flammable liquid/gas fires. | Smothering occurs; in high concentrations. Avoid contact with discharge horn. Limited effectiveness under windy conditions. Severely reduced effectiveness at sub-zero (F) temperatures. | Extensive cleanup, particularly on delicate electronic equipment. Obscures visibility in confined spaces. | | Avoid high concentrations and unnecessary use. | Extensive cleanup. Damages electronic equipment. Obscures visibility in confined spaces. Limited penetrating ability on deep-seated Class A fires. | | Avoid high concentrations and unnecessary use. | Not listed. |
| | NOTE: Protection required below 40°F and above 120°F. | | | | | | | NOTE: Only dry chemical types are effective on pressurized flammable gases and liquids; for deep fat fryers, multipurpose ABC dry chemicals are not acceptable. | | | |

NOTE: These photos are not proportional in relation to one another.

**Fig. 2-7** Fire extinguishers. *(National Association of the Fire Equipment Distributors)*

# How to Select the Proper Fire Extinguisher

Different types of extinguishers are used to put out different kinds of fires. For example, a Halon 1211 extinguisher is one of the types recommended for putting out an electrical fire (halon does not conduct electricity). A water type extinguisher should never be used on such a fire (water does conduct electricity).

To help you select the proper extinguisher for the type of hazard you are most likely to encounter in the home, car, or workplace, the charts here classify both fires and extinguishers into types. Types of fires are classified as: A — ordinary combustible materials, such as wood, cloth, paper, and many plastics; B — flammable liquids, gases, and grease; and C — energized electrical equipment (such as computers). There is a class D — combustible metals, but this hazard is rare and found usually in industrial situations.

Unlike fires, extinguishers have more classifications than simply A, B, C, or D. They are A, AB, BC, ABC, and D; this is because some extinguishers can put out more than just one type of fire. For example, a BC dry chemical fire extinguisher can put out a grease fire (Class B), as well as an electrical fire (Class C).

You might wonder, if that is so, why doesn't everyone just use ABC extinguishers, which put out virtually all types of fires? The answer is, although an ABC extinguisher is capable of putting out Class A, B, and C fires, it is not always as effective as another extinguisher designed solely for putting out only certain kinds of fires.

For example, there are *multipurpose dry chemical* extinguishers, classified as ABC, and simply *dry chemical* extinguishers classified as BC. An ABC multipurpose dry chemical extinguisher is prohibited for use on deep fat fryers in restaurants. Only BC dry chemical types are effective on deep fat fryers. The BC dry chemical reacts with the hot grease to create a kind of soapy foam that suffocates the fire (this process if called saponification). The multipurpose dry chemical of ABC fire extinguishers breaks down this foam once it is formed and allows the fire to re-ignite.

For general home use, however, the ABC extinguishers are fine for all applications. We will, however, show you what other kinds of extinguishers may be used in different areas of the home, car, or office.

All extinguishers are labeled with picture symbols.* These symbols depict the kinds of fires on which the extinguisher is effective. A slash through a picture means the extinguisher is not to be used on that class fire. If it is used, it may actually spread the fire rather than put it out.

*Picture symbols for Class D fires are not practical.

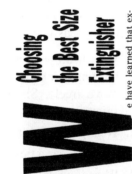

| Classification of Fires | Picture Symbols | Recommended Extinguishers |
|---|---|---|
| **A** Ordinary combustible materials, such as wood, cloth, paper, rubber, and many plastics. | TRASH•WOOD•PAPER | WATER<br>AFFF<br>ABC Dry Chemical<br>Halon 1211 |
| **B** Flammable liquids, oils, grease, tars, oil base paints, lacquers and gases. | LIQUIDS•GREASE | LIQUIDS:<br>AFFF<br>Regular<br>Dry Chemical<br>ABC Dry Chemical<br>Purple K<br>Dry Chemical<br>Halon 1211<br>$CO_2$<br>GASES:<br>Regular & Purple K<br>Dry Chemical |
| **C** Energized electrical equipment where the electrical non-conductivity of the extinguishing agent is essential. (When electrical equipment is de-energized, extinguishers for Class A or B fires may be used safely.) | ELECTRICAL EQUIP | Regular<br>Dry Chemical<br>ABC Dry Chemical<br>Purple K<br>Dry Chemical<br>Halon 1211<br>$CO_2$ |

# Choosing the Best Size Extinguisher

We have learned that extinguishers are classified by the type of fire they put out. They are also *rated* with a number that tells you the *size* of fire they can put out.

For example, a five pound *multipurpose dry chemical* extinguisher (ABC) has a rating of 2-A:10-B:C. That means the extinguisher can put out approximately twice as much fire as a 1-A extinguisher. A 1-A rated extinguisher is required to put out a blazing wood crib consisting of 50 pieces of 20 inch long 2' x 2's and an 8' x 8' wood panel. It can put out five times the size of a Class B fire that consists of 3.25 gallons of liquid fuel burning in a 2.5 square foot pan.

**Recommendations**
- Kitchen — 2½–5 lb. dry chemical, UL rating of 5B:C, and/or halon type
- Car/garage — 1.5 lb. Halon 1211 or multipurpose dry chemical, UL rating of 1A:10B:C
- Personal computer — 1-2½ lb. Halon 121 , UL rating of 2B:C

The National Association of Fire Equipment Distributors

*Fig. 2-7* (*Continued*).

wear them. It is a big bother to put them on for "just one cut." You have heard all the reasons and excuses. But none of them make any sense. They are all part of the lazy person's way, not the professional's way—the safe way.

2. Never cut any wire or metal unless your fellow workers' eyes are also protected. Although you may not have heard that precaution before, you'll see that it makes sense. The wire that does not get you may get someone else. So think about others as well as yourself.

3. Never depend on plastic-dipped handles to insulate you from electricity. Plastic-dipped handles are for comfort and a firmer grip. They are not intended for protection against electric shock.

4. Always wear protective goggles!

## Using Hammers Properly

There are some precautions that should be observed when using hammers on the job. Hammers are perhaps the most abused—and misused—of all hand tools. Improper use of hammers can cause injury. Use of damaged hammers can cause injury, as can use of hammers for jobs other than those for which they were specifically designed.

There are some basic rules for the proper use of hammers:

- Strike square blows only. Avoid glancing blows.
- Make sure that the striking face of the hammer is at least 8 in. larger than tools to be struck (chisels, punches, wedges, and the like). The face should be slightly crowned with the edges beveled.
- Never use a hammer to strike another hammer. Always use a hammer of the right size and weight for the job.
- Never strike with the side or "cheek" of a hammer.
- Never use a hammer with a loose or damaged handle. Replace the handle.
- Discard any hammer having chips, cracks, dents, mushrooming, or excessive wear. Replace the hammer. Do not try to repair it.
- Always wear safety goggles to protect your eyes (Fig. 2-8).

The professional takes both the job and tools seriously. The right tool for a job means saved time and professional results. Using tools safely also means saving time and getting the job done properly. There is only one way to do a job right and that is the safe way.

**Fig. 2-8** *(A) General-purpose cover-type protective goggles are made of soft vinyl plastic. This type protects the eyes from all angles. (B) For more protection you may want to add a hardhat and face shield as a combination safety step while working in hazardous locations.* (Klein)

## Working Clothes

Clothes worn on the job should be selected to protect you while you work. Of course, everyone will not work under the same conditions; therefore, some general safety tips will be mentioned here. Others may present themselves as you work in a specific area.

1. Wear head protection—a hardhat if the job requires it. Long hair should be concealed and not allowed to move freely. Short hair is preferred for personal safety.
2. Wear goggles or tempered glasses or both.
3. Do not wear a tie.
4. Wear long-sleeved shirts with the cuffs tightly buttoned near the wrists.
5. Wear your trousers tucked into your safety shoes or tightly fastened near the ankles.
6. Do not wear rings, earrings, necklaces, or bracelets.
7. Wear safety shoes. These should have an arch support and steel-tipped or safety toes. Soles should be such as not to slip on oily or wet surfaces. Wear rubber boots in wet or damp locations.

## Disconnects

Electrical shock can be prevented if the circuit is not "live" when you are working on it or any equipment connected to the circuit. Remove the power. Remove the fuse or disconnect. If it has a lockout feature, make sure that it is used—that the lock is locked and you have the key.

Report to your supervisor any unsafe conditions observed. Keep in mind both your own safety and that of others.

# Using Motors and Generators Safely

The use of electric motors and generators, like that of all other utilization of concentrated power, is potentially hazardous. The degree of hazard can be greatly reduced by proper design, selection, installation, and use, but hazards cannot be completely eliminated. The reduction of hazards is the joint responsibility of the user, the manufacturer of the driven or driving equipment, and the manufacturer of the motor or generator.*

Most manufacturers make their equipment to meet the NEMA standards for safe operation. However, even well built equipment can be installed or operated in a hazardous manner. It is important that safety considerations be observed. The user must properly select, install, and use the apparatus with respect to load and environment.

**Selection of Apparatus**  Before selecting a piece of equipment you should study the recommendations of the manufacturer of the apparatus, generally available in catalogs. Manufacturers usually maintain engineering departments to assist you in making sure that you have the right equipment for the job. Local sales representatives or factory representatives can usually arrange for more details, if needed.

**Installation of Apparatus**  The equipment manufacturer or the person installing the apparatus must take care in the installation. The National Electrical Code (NEC), sound local electrical and safety codes, and when applicable, the Occupational Safety and Health Act (OSHA) should be followed when installing the apparatus to reduce hazards to persons and property.

**Use of Apparatus**  The chance of electric shocks, fires, or explosions can be reduced by giving proper consideration to the use of grounding, thermal and over-current protection, type of enclosure, and good maintenance procedures.

**Safety Considerations**

1. All motors, gear motors, and controls must be grounded adequately and securely. Failure to ground properly may cause serious injury to personnel.

2. Do not insert objects into ventilation openings, motors, or other apparatus.

3. Sparking of starting switches in ac motors so equipped, and of brushes in commutator type motors, can be expected during normal operation. In addition, enclosures may eject flame in the event of an insulation or component failure. Therefore, avoid, protect from, or prevent the presence of flammable or combustible materials in the environment of motors, gear motors, and controls.

4. When dealing with hazardous locations (flammable or explosive gas, vapor, dust) an explosion-proof or dust-ignition-proof product is the recommended approach. Exceptions are allowed by the National Electrical Code. The NEC and NEMA safety standard should be studied thoroughly before exercising this option.

5. Ventilated products are suitable for clean, dry locations where cooling air is not restricted. Enclosed products are suitable for dirty, damp locations. For outdoor use, wash down, and so on, enclosed products must be protected by a cover while allowing adequate air flow.

6. Moisture will increase the electrical shock hazard of electrical insulation. Therefore, consideration should be given to the avoidance of (or protection from) liquids in the area of motors and controls. Use of totally enclosed motors/gear motors will reduce the hazard if all openings are sealed.

7. Chassis controls should be properly guarded or enclosed to prevent possible human contact with "live" circuitry.

8. Proper consideration should be given to rotating members. Before starting, make sure that keys, pulleys, and so on, are securely fastened. Proper guards should be provided for rotating members to prevent hazards to personnel.

9. Before servicing or working on equipment, disconnect the power source (this applies especially to equipment using automatic restart devices instead of manual restart devices. See Fig. 2-9 where a specially designed switch and panel box has double interlocking hasp of tempered steel with interlocking tabs for extra security. For safety reasons

---

*See Standards Publication No. ANSI C51.1/NEMA MG2 Safety Standard for Construction and Guide for Selection, Installation and Use of Electric Motors and Generators, available from National Electrical Manufacturers Association, 1300 N. 17th Street, Suite 1752, Rosslyn, VA 22290.

*Fig. 2-9* Locked box for safety purposes. (Klein)

the box should be locked whenever someone is working on the circuit or equipment. In this way another person is unable to apply power while someone else is working on equipment.

10. In selecting, installing, and using equipment, the hazard of mechanical failure should be considered in addition to electrical hazards. Mechanical considerations include proper mounting and alignment of apparatus and safe loads on shafting and gearing. Do not depend on gear friction to hold loads.

11. Ambient temperatures around apparatus should not exceed 40°C (104°F) unless the nameplates specifically permit higher values.

12. Power supply to all equipment must be that shown on the nameplate.

## Grounding

In industrial installations, the effect of a shutdown caused by a single ground fault could be disastrous. An interrupted process could cause the loss of all materials involved, often ruin the process equipment itself, and sometimes create extremely dangerous situations for operating personnel.

Grounding encompasses several different but interrelated aspects of electrical distribution system design and construction, all of which are essential to the safety and proper operation of the system and equipment supplied by it. Among these are equipment grounding, system grounding, static and lightning protection, and connection to earth as a reference (zero) potential.

**Equipment Grounding** Equipment grounding is essential to the safety of personnel. Its function is to ensure that all exposed non–current-carrying metallic parts of all structures and equipment in or near the electrical distribution system are at the same potential, and that this is the zero reference potential of the earth. Grounding is required by both the National Electrical Code (Article 250) and the National Electrical Safety Code. Equipment grounding also provides a return path for ground-fault currents, permitting protective devices to operate.

Accidental contact of any energized conductor of the system with an improperly non–current-carrying metallic part of the system (such as a motor frame or panel board enclosure) would raise the potential of the metal object above ground potential. Any person coming in contact with such an object while grounded could be seriously injured or killed. In addition, current flow from the accidental grounding of an energized part of the system could generate sufficient heat (often arcing) to start a fire.

The equipment grounding system must be bonded to the grounding electrode at the source or service. However, it may also be connected to ground at many other points. This will not cause problems with safe operation of the electrical distribution system.

Where computers, data processing, or microprocessor-based industrial process control systems are installed, the equipment grounding system must be designed to minimize interference with their proper operation. Often, isolated grounding of this equipment, or a completely isolated electrical supply system, is required to protect microprocessors from power system "noise" that does not in any way affect motors or other electrical equipment.

**System Grounding** System grounding connects the electrical supply—from the utility, transformer secondary windings, or a generator-to ground. A system can be solidly grounded (no intentional impedance to ground), impedance grounded (through a resistance or reactance), or ungrounded (with no intentional connection to ground).

Solidly grounded three-phase systems are shown in Fig. 2-10. They are usually wye-connected, with the neutral point grounded. Less common is the "red-leg" or high-leg delta, a 240-V system supplied by some utilities with one winding center-tapped to proved 120 V to ground for lighting. This 240-V, three-phase, four-wire system is used where the 120-V lighting load is small compared to the 240-V power load, because the installation is low in cost to the utility.

A corner-grounded three-phase delta system is sometimes used, with one phase grounded to stabilize all voltages to ground. Better solutions are available for new installations. Ungrounded systems (Fig. 2-11) can be either wye or delta, although the ungrounded delta system is far more common.

Resistance-grounded systems (Fig. 2-12) are simplest with a wye connection, grounding the neutral

**Grounded Wye**

**Center-Tapped (Red-Leg) Delta**

**Corner-Grounded Delta**

*Fig. 2-10* Solidly grounded systems. (Westinghouse)

**Ungrounded Delta**

**Ungrounded Wye**

*Fig. 2-11* Ungrounded systems. (Westinghouse)

**Resistance-Grounded Wye**

**Delta With Derived Neutral Resistance-Grounded Using Zig-Zag Transformer**

*Fig. 2-12* Resistance-grounded systems. (Westinghouse)

point directly through the resistor. Delta systems can be grounded by means of a zigzag or other grounding transformer. This derives a neutral point, which can be either solidly or impedance grounded. If the grounding transformer has sufficient capacity, the neutral created can be solidly grounded and used as part of a three-phase, four-wire system. Most transformer-supplied systems are either solidly grounded or resistance grounded. Generator neutrals are often grounded through a reactor, to limit ground-fault (zero sequence) currents to values the generator can withstand. Medium-voltage resistance-grounded neutrals are often grounded through a grounding transformer, although many 2400-V and some 4160-V systems are grounded directly, like low-voltage systems.

**Which System Is Best?** There is no best distribution system for all applications. That is why you may find any one of these systems in any plant. In choosing among solid-grounded, resistance-grounded, or ungrounded power distribution, the characteristics of the system must be weighed against the requirements of power loads, lighting loads, continuity of service, safety, and cost.

# REVIEW QUESTIONS

1. If it is not the voltage that kills, what does?
2. How much current does it take to kill?
3. What is ventricular fibrillation?
4. What does GFCI stand for?
5. What does *NEC* stand for?
6. Give two examples of two over current devices.
7. How is an overload described?
8. What is a fusible link?
9. How is a dual-element fuse sized?
10. What is the highest transmission voltage used today?
11. What is the National Electrical Code?
12. What is the purpose of the Underwriters Laboratories?
13. What does the Canadian Standards Association do?
14. What does OSHA stand for?
15. What do IEEE and ANSI stand for?
16. What does the OSHA color orange denote?
17. Why isn't a water-type fire extinguisher used on electrical fires?
18. How can the chance of electrical shocks be reduced?

19. What is grounding? What is a system ground?
20. Where do you encounter resistance-grounded systems?

## REVIEW PROBLEMS

Electrical energy can be converted to heat energy by the use of resistances. Heat and power are closely related in the electrical field, especially with large resistors. To help you handle and locate such heat-producing devices safely, a review of the power formulas will come in handy.

$$P = E \times I \quad P = E^2/I \quad P = I^2R$$

1. How much power is required to operate a solenoid that draws 4.5A when connected to a 12-V supply?
2. How much power is drawn when a spotlight draws 2.5A at 120V?
3. What is the current needed when a 120-V, 150-W bulb is connected to a power source?
4. A 24-V transformer delivers 1500 W of power. What is the current?
5. If a resistor of 500 is connected across a source of 277 V, how much power is consumed?
6. How much current will a 240-V $\frac{1}{2}$-hp electric motor pull?
7. If a 5-hp motor is connected to a 240-V line, what will the current requirements be?
8. What current do you need for a 2-hp motor on a 120-V line?

# 3
CHAPTER

# Symbols

# PERFORMANCE OBJECTIVES

After studying this chapter, you will be able to:

1. Draw the symbols used in electrical wiring diagrams and ladder diagrams.
2. Use the resistor color code to select the proper size resistors.
3. Identify relay contact symbols.
4. Identify electronic schematic symbols.

Electrical wiring diagrams and ladder diagrams, which are the main means of communicating between engineer and technician in the motor controls field, are made up of nothing more than lines connecting certain symbols.

Symbols are used as a sort of shorthand that allows many electrical functions to be identified and sketched out in a small area. People familiar with the symbols used in the electrical field are able to identify certain components and figure out how they are wired and how they operate.

Symbols are the heart of any wiring diagram. Because they are so important a part of the job, it is best to become acquainted with them at an early stage in the learning of a trade.

# ELECTRICAL SYMBOLS

Electrical symbols are used on blueprints and in wiring diagrams for buildings and plants. Figure 3-1 shows the standardized symbols utilized by drafters who draw wiring diagrams for the placement of electrical equipment and control devices.

## Common Switch (Button) Symbols

Motor controls technicians work with switches (buttons) and their associated components to make a motor do what it is supposed to do when it is supposed to do it. Some of the symbols used for switching are shown in Fig. 3-2. Note the pressure and vacuum switches, liquid-level switch, temperature-actuated switch, and flow switch. These will be utilized by the motor control technician in their many applications. Switches may also be referred to as buttons.

## Standard Wiring Diagram Symbols

Some basic symbols used in wiring diagrams that you will be studying are shown in Fig. 3-3. All of these will be encountered in a normal wiring diagram for the control of electric motors in an industrial or commercial installation.

# ELECTRONIC SYMBOLS

The field of electronics has its own symbols that are used by those who draw wiring diagrams. More and more electronics is being utilized in the control of motors (Figs. 3-4 and 3-5).

## Resistor Color Code

It is a good idea to learn how the resistor color code is read. Many electronic controls have resistors identified by the color code. The color code is used on resistors from 0.1 to 2 W. The small resistors are useful primarily in electronic circuits that draw very little current. They usually have large resistances, as opposed to the larger wire-wound types used directly in series with motor windings (Fig. 3-6).

## Electronic Symbols Compared

Figure 3-7 shows the usual electronic symbol on the left and compares it with industrial control symbols. This will give you an idea as to how they may differ when you are looking at a wiring diagram. In most instances, the capacitor symbol is a straight line with a curved line parallel to it. The older symbol shown is there for those schematics or wiring diagrams you may find hidden away in a closet from bygone days.

# RELAY CONTACT SYMBOLS

One of the things you do work with a great deal in controlling motors is relays. Figure 3-8 shows some of the basic relay contact symbols that may be used. Keep in mind what make and break mean when working with relays. The single pole, single throw is similar to switches symbols. Remember, the relay contacts are doing exactly what the switch contact is doing—making and breaking or closing and opening the circuit.

# LINE DIAGRAMS, WIRING DIAGRAMS, AND LADDER DIAGRAMS

Different types of drawings and diagrams are covered in Chap. 4 in greater detail. However, at this point it may be well to take a closer look at symbols in action.

**1. Lighting Outlets**

*Ceiling*  *Wall*

1.1  Surface or Pendant Incandescent, Mercury-Vapor, or Similar Lamp Fixture

1.2  Recessed Incandescent, Mercury-Vapor, or Similar Lamp Fixture

1.3  Surface or Pendant Individual Fluorescent Fixture

1.4  Recessed Individual Fluorescent Fixture

1.5  Surface or Pendant Continuous-Row Fluorescent Fixture

1.6  Recessed Continuous-Row Fluorescent Fixture

1.7  Bare-Lamp Fluorescent Strip

1.8  Surface or Pendant Exit Light

1.9  Recessed Exit Light

1.10 Blanked Outlet

1.11 Junction Box

1.12 Outlet Controlled by Low-Voltage Switching When Relay Is Installed in Outlet Box

**2. Receptacle Outlets**

*Grounded*  *Ungrounded*

2.1  Single Receptacle Outlet

2.2  Duplex Receptacle Outlet

2.3  Triplex Receptacle Outlet

2.4  Quadruplex Receptacle Outlet

2.5  Duplex Receptacle Outlet—Split Wired

2.6  Triplex Receptacle Outlet—Split Wired

2.7  Single Special-Purpose Receptacle Outlet

2.8  Duplex Special-Purpose Receptacle Outlet

2.9  Range Outlet (typical)

2.10 Special-Purpose Connection or Provision for Connection

2.11 Multioutlet Assembly

2.12 Clock Hanger Receptacle

2.13 Fan Hanger Receptacle

2.14 Floor Single Receptacle Outlet

2.15 Floor Duplex Receptacle Outlet

2.16 Floor Special-Purpose Outlet

**3. Switch Outlets**

3.1  Single-Pole Switch  **S**

3.2  Double-Pole Switch  **S2**

3.3  Three-Way Switch  **S3**

3.4  Four-Way Switch  **S4**

3.5  Key-Operated Switch  **SK**

3.6  Switch and Pilot Lamp  **SP**

3.7  Switch for Low-Voltage Switching System  **SL**

3.8  Master Switch for Low-Voltage Switching System  **SLM**

3.9  Switch and Single Receptacle

3.10 Switch and Double Receptacle

3.11 Door Switch  **SD**

3.12 Time Switch  **ST**

3.13 Circuit Breaker Switch  **SCB**

3.14 Momentary Contact Switch or Pushbutton for Other Than Signaling System  **SMC**

3.15 Ceiling Pull Switch

**5. Residential Occupancies**

5.1  Pushbutton

5.2  Buzzer

5.3  Bell

5.4  Combination Bell-Buzzer

5.5  Chime

5.6  Annunciator

**Fig. 3-1**  *Electrical symbols for architectural plans.*

## COMMON SWITCH SYMBOLS

## OTHER SWITCHES

**Fig. 3-2** *Common switch symbols.*

**Fig. 3-3** Standard wiring diagram symbols. *(Square D)*

**Fig. 3-4** *Electronic schematic symbols.*

Figure 3-9 shows the symbols put to work in the typical ladder diagram. Note that there are both a line 1 (L1) and a line 2 (L2), where the power is brought into the circuit Then the various devices are connected to both sides of the power line. At this time you should focus on how the contacts are drawn in the circuit in reference to what device is being controlled by the normally open (NO) and normally closed (NC) contacts. Also take a close look at the START and STOP switches. As soon as the START switch (button) is closed, M in the circle will energize. This causes the M contacts (located across the START switch), which are NO, to close. That completes the circuit from L1 through the stop switch and the START switch through coil M and the overload contacts (O.L.) to the other side of the line L2. Once the coil is energized, the NO contacts across the START switch stay closed and the START switch can be released. However, when the STOP switch is opened, it causes the circuit to de-energize and the contacts M to open again until the START button is again closed.

Now drop down to the next line in the diagram. Note that the NC contacts have an M above them. This means that they are connected to the M coil or relay. When the M coil is energized, these contacts open and turn off the pilot light connected in series with the contacts.

The pushbutton station wiring diagram is a representation of the physical station, showing relative positions of units, suggested internal wiring, and connections with the starter.

## REVIEW QUESTIONS

1. Where are electrical symbols used?
2. Draw five basic symbols used in wiring diagrams that you will need in studying electric motor controls.
3. What are the resistor color code colors? What do they mean?
4. What are the basic relay symbols used in diagrams?

**Fig. 3-5** *More electronics symbols (old and new).*

| Color | Significant Figure | Multiplying Value |
|---|---|---|
| Black | 0 | 1 |
| Brown | 1 | 10 |
| Red | 2 | 100 |
| Orange | 3 | 1,000 |
| Yellow | 4 | 10,000 |
| Green | 5 | 100,000 |
| Blue | 6 | 1,000,000 |
| Violet | 7 | 10,000,000 |
| Gray | 8 | 100,000,000 |
| White | 9 | 1,000,000,000 |
| Gold | ±5% tolerance | 0.1 |
| Silver | ±10% tolerance | 0.01 |
| No color | ±20% tolerance | |

### EIA/MIL COLOR CODE

Band A — 1st significant figure
Band B — 2nd significant figure
Band C — Number of zeros or decimal multiplier
Band D — Tolerance

**Fig. 3-6** *Resistor color code.*

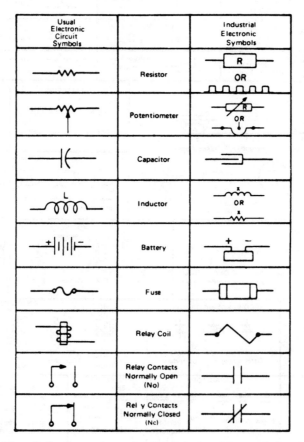

**Fig. 3-7** *Comparison of industrial electricity symbols with more common electronic circuit symbols.*

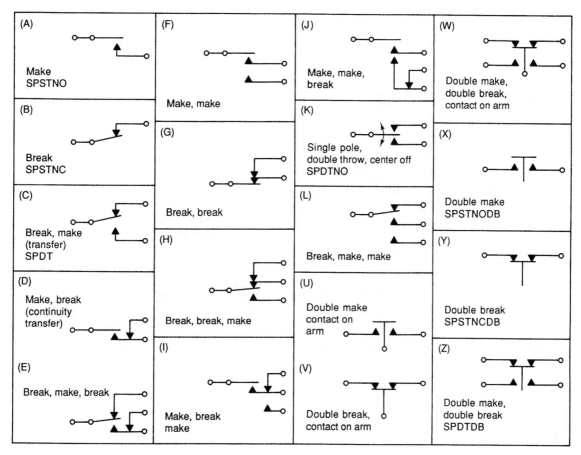

**Fig. 3-8** *Basic relay contact symbols.*

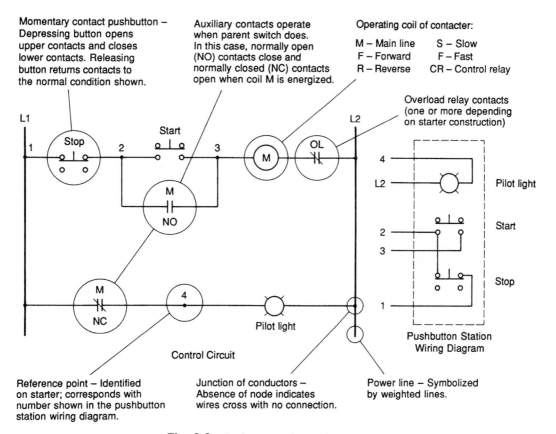

**Fig. 3-9** *Explanation of control circuit symbols.*

Review Questions 69

5. What is a ladder diagram?
6. What is the difference between a schematic and a ladder diagram?

## REVIEW PROBLEMS

The use of symbols is important in the electrical circuitry. Symbols make the task of drawing circuits and analyzing them easier. Series circuits symbols utilize symbols at the beginning level. A review of series circuits is presented here in the form of problems.

$$R_T = R_1 + R_2 + R_3 + \cdots$$

1. Three resistors of 10, 100, and 200 Ω are connected in series. What is the total resistance?
2. If the resistance in problem 1 is connected to 620 V, what is the current drawn?
3. A set of eight old-style Christmas tree lights are connected in series and each draws 0.15 A. What is the voltage drop across each lamp if the applied voltage is 120 V? What is the total current of the circuit?
4. A series resistor combination of 6, 12, and 18 Ω is connected to a power source of 72 V. What is the current flowing through an ammeter inserted between the 12- and 18-Ω resistors?
5. What is the resistance of a third resistor if the first two are 10 Ω and 20 Ω and the total resistance of the circuit measures 50 Ω.
6. The total resistance of a series circuit is 220Ω. It has two resistors of 100 Ω each. What is the resistance of the third resistor?
7. If one lamp burns out in a series string of ten 12-V lamps and you short the burned-out filament, what will be the voltage drop across each lamp?
8. What happens to the current in a series circuit if the voltage stays the same at the source but the resistance is reduced by having one resistor short?

# CHAPTER 4
# Control Circuits and Diagrams

# PERFORMANCE OBJECTIVES

After studying this chapter, you will be able to:

1. Draw a ladder diagram.
2. Explain how to use symbols in diagrams.
3. Describe undervoltage release and protection.
4. Identify a two-wire control circuit.
5. Identify a three-wire control circuit.
6. Describe the difference between a line drawing and a wiring diagram.
7. Define the role of a thermal protector.

A wiring diagram shows, as closely as possible, the actual location of all of the component parts of the device. The open terminals, which are marked by open circles, and arrows represent connections made by the user. Fig. 4-1 shows a circle with motor written in it with three arrows leaving the circle and labeled T1, T2, and T3; this represents a three-phase motor. The T stands for the terminals on the motor and are to be connected by the electrician.

**Fig. 4-1** *Wiring diagram of a three-phase motor starter.*

Since wiring connections and terminal markings are shown, this type of diagram is helpful when wiring the device or in tracing wires when troubleshooting. Note that bold lines denote the power circuit, and thin lines are used to show the control circuit. In most instances, ac magnetic equipment uses black wires for power circuits and red wiring for control circuits. It is to your advantage to learn how to read both wiring diagrams and line diagrams.

# WIRING DIAGRAMS

A wiring diagram gives the necessary information for actually wiring up a group of control devices or for physically tracing wires when troubleshooting is necessary.

Wiring diagrams or connection diagrams include all the devices in the system and show their physical relation to each other. All poles, terminals, coils, contacts, and switches are shown in their proper place on each device. These diagrams are helpful in wiring up systems, because connections can be made exactly as they are shown in the diagram. In following the electrical sequence of any circuit, however, the wiring diagram does not show the connections in a manner that can easily be followed. For this reason a rearrangement of the circuit elements to form a line diagram is desirable.

## Start Circuits

Manual starting switches are designed for starting and protecting small ac and dc motors rated at 1 hp or less where undervoltage protection is not needed. They are operated by a toggle lever mounted on the front of the switch (Fig. 4-2). Wiring diagrams do not show the

**Fig. 4-2** *Manual starting switch. On-off snap switch with thermal overload.* (Allen-Bradley)

operating mechanism since it is not electrically controlled. These motor starters consist of an on-off snap switch combined with a thermal overload device operating on the eutectic alloy ratchet principle. Terminal markings corresponding to those shown in the diagrams in Figs. 4-3 to 4-6 will be found on each switch.

**Fig. 4-3** *Single-pole switch used to start-stop a motor.* (Allen-Bradley)

**72** Control Circuits and Diagrams

**Fig. 4-4** Double-pole switch with built-in neon pilot light. (Allen-Bradley)

**Fig. 4-5** Hand-Off-Auto selector switch circuit. (Allen-Bradley)

**Fig. 4-6** Two-speed manual motor starter. (Allen-Bradley)

Other simple wiring diagrams are used to show how to connect manual starters operated by START–STOP push buttons mounted on the front of the starter. Figure 4-7 shows what the pushbutton looks like when properly packaged with the cover removed for ease in viewing the terminals. This type of starter is used where undervoltage protection is not required. Wiring diagrams do not show the operating mechanism since it is not electrically controlled. Pushing the button mechanically closes the contacts, connnecting the motor to the line.

**Fig. 4-7** Start-stop manual starter. (Allen-Bradley)

The contacts are opened by pressing the STOP button. Terminal markings corresponding to those shown on the diagrams (Figs. 4-8 to 4-11) will be found on each switch.

**Fig. 4-8** Manually operated three-phase or two-phase, three-wire starter. (Allen-Bradley)

**Fig. 4-9** Direct-current motor starter. (Allen-Bradley)

Wiring Diagrams

***Fig. 4-10*** *Single-phase, two-contact manual starter.* (Allen-Bradley)

***Fig. 4-11***. *Three-phase manual starter with pilot light.* (Allen-Bradley)

## LINE DIAGRAM OR LADDER DIAGRAM

A line diagram gives the necessary information for easily following the operation of the various devices in the circuit. It is very helpful in troubleshooting, for it shows, in a simple way, the effect that opening or closing various contacts has on other devices in the circuit.

A line diagram is sometimes referred to as an elementary diagram or a schematic diagram. No attempt is made to show the various devices in their actual positions. All control devices are shown between vertical lines which represent the source of control power, and circuits are shown connected as directly as possible from one of these lines to the other. All connections are made in such a way that the functioning of the various devices can easily be traced (Fig. 4-12).

This is a simple or elementary diagram of a switch in series with a magnetic coil (M) and overload contacts. Note how the L1 is to the left and the

***Fig. 4-12*** *Elementary two-wire control.*

number 1 is placed near the junction point of L1 with the horizontal line. Each line or point where a horizontal line is taken off the power line is numbered. This comes in handy later when the programmable controller is utilized to do the work of switching. This type of drawing is also referred to as a ladder diagram, due to its design with parallel power lines and rungs that make up the circuits.

## UNDERVOLTAGE RELEASE

Another name for undervoltage release is low-voltage release–two-wire control. These terms mean that the starter will drop out when there is a voltage failure and will pick up again as soon as power is restored (Fig. 4-12). This low-voltage release is a two-wire control using a maintained contact pilot device (limit switch, pressure, or float switch) in series with the starter coil. This arrangement is used when a starter is required to function automatically without the attention of an operator. If a power failure occurs while the contacts of the pilot device are closed, the starter will drop out. When the power is restored, the starter will pick up automatically through the closed contacts of the pilot device. The term two-wire control arises from the fact that in the basic circuit, only two wires are required to connect the pilot device to the starter.

## TWO-WIRE CONTROL CIRCUITS

There are a number of circuits that can be drawn and used to good advantage in specific situations. Figure 4-13 shows a starter operated with manual control provided by a selector switch.

A high-pressure cutout switch can be added. Both the wiring diagram and the line drawing are shown to illustrate how things are connected and arranged. The selector switch makes it possible to operate the starter manually for testing in case of failure of the automatic pilot control. When a standard across-the-line

**Fig. 4-13** *Starter operated by pressure switch or thermostat with manual control provided by a selector switch.* (Allen-Bradley)

**Fig. 4-14** *Starter controlled by a gage-type thermostat.* (Allen-Bradley)

starter without pushbuttons is used, connection Y is removed and the wiring follows the solid lines of the diagram.

Note that in the ladder diagram only L1 and L2 are used. The control circuitry uses these two for power, while the magnetic starter M has three sets of contacts that will interrupt the power to the three-phase motor in all three legs or phases.

If a high-pressure cutout switch is added, it should be inserted in the line leading from L1 to the HAND terminal of the selector switch. A low-pressure switch or thermostat can be used where A is located (Fig. 4-13). B represents the high-pressure switch or safety cutout. Note how the HAND-OFF-AUTO switch operates in both the wiring diagram and the ladder diagram.

## THERMOSTAT CONTROL

Since the contacts of a gage-type thermostat cannot handle the current to a starter coil, a thermostat relay must be used as an intermediate step between the thermostat and the starter (Fig. 4-14). The relay is energized when the close contact is made and de-energized when the open contact is made. The open contact bypasses the relay coil to de-energize it. A resistor (Res.) is built into the relay to guard against a short circuit when this is done.

The thermostat contacts must not overlap or be adjusted too close to one another, since this may burn out the resistance unit. It is also advisable to check the inrush current of the relay against the current rating of the thermostat. This arrangement can also be used with pressure controls. "A" represents the three-wire gage-type temperature control device. CR is the thermostat relay, which in this case has two sets of contacts also labeled CR.

Another method of protecting thermostat contacts from high currents can be seen in Fig. 4-15. This diagram shows the use of a thermostat whose contacts can handle even less current than the ones in the preceding example. It is advisable in such cases as this to use relays having coils that operate at a very small value of volt-amperes. This will reduce the burden on the thermostat contacts and is especially advisable where frequent operation is required. "A" represents the three-wire sensitive gage-type control. CR1 and CR2 are the low coil current relays.

When the close position is indicated it means that relay CR1 is energized. Energizing CR1 causes the contacts CR1 to close. Closing CR1 contacts in series with the control coil (M) causes it to energize, starting the motor. It also keeps itself energized by the other set of CR1 contacts. However, when open is indicated by the thermostat it completes the circuit through the

**Thermostat Control** 75

*Fig. 4-16* Low-voltage protection: three-wire control. (AllenBradley)

*Fig. 4-15* Starter controlled by a gage-type thermostat whose contacts can switch only a small amount of current. (Allen-Bradley)

closed contacts of CR1 to cause relay CR2 to energize. Once CR2 is energized it opens its contacts (CR2), which are normally closed and in series with CR1 relay coil. This causes CR1 to de-energize and open the CR1 contacts, one of which is in series with the motor control coil (M); this stops the motor. Once CR2 is energized it is quickly de-energized by its contacts opening CRTs circuit and opening the CR1 contacts, which de-energize the motor starter coil.

The difference between the line drawing and the wiring diagram is the line drawing (ladder diagram) makes it easier to see how the relays operate in sequence to control the magnetic starter coil and thereby the motor.

## UNDERVOLTAGE PROTECTION

Undervoltage protection is also referred to as low-voltage protection, three-wire control. Both terms mean the same and indicate that the starter will drop out when there is a voltage failure but will not pick up automatically when voltage returns. The control circuit is completed through the STOP button and also through a holding contact. See contact 2-3 on the starter in Fig. 4-16. When the starter drops out, these contacts open, breaking the control circuit until the START button is pressed once again.

Three wires lead from the pilot device to the starter. The START-STOP pushbutton station is usually thought of when undervoltage protection or three-wire control is mentioned. This is the most common means of providing this type of control.

The main distinction between the two-wire and three-wire controls is that with undervoltage release (two-wire) and undervoltage protection (three-wire) is the fact that one will cause the motor to start again when the power is on again and the other will not start until the START button is pressed. The designations two-wire and three-wire are used only to describe the simplest applications of the two types. Actually, in other systems, there might be more wires leading from the pilot device to the starter, but the principle of two-wire or three-wire control would still be present.

## THREE-WIRE CONTROL CIRCUITS

One of the methods used to improve your ability to read wiring diagrams and ladder diagrams is to practice. Figures 4-17 through 4-26 are three-wire control circuits that can do a variety of things. Each will be analyzed to show how it operates. Knowing how they operate will aid in troubleshooting since troubleshooting is nothing more than determining why a device is not performing the way it should normally.

One of the simplest types of three-wire control circuitry is shown in Fig. 4-17. This three-wire control has a pilot light in the circuit to indicate when the

*Fig. 4-17* Three-wire control with pilot light. *Motor running.*

76  Control Circuits and Diagrams

motor is running. The pilot light is wired in parallel with the starter coil to indicate when the starter is energized and the motor is running or at least has power applied to its terminals.

Figure 4-18 takes the three-wire control circuitry a step further. It shows a pilot light used to indicate when the motor is stopped. A pilot light may be required to indicate when the motor is stopped in some cases. This can be done by wiring a normally closed auxiliary contact located on the starter in series with the pilot light. When the starter is de-energized, the pilot light is on. When the starter picks up, the auxiliary contacts open, turning off the light.

**Fig. 4-18** *Three-wire control with pilot light.* Motor stopped.

Figure 4-19 uses a push-to-test pilot light to indicate when the motor is running. When the motor-running pilot light is not glowing, there may be doubt as to whether the circuit is open or whether the pilot light bulb is burned out. The push-to-test pilot light enables testing of the bulb simply by pushing on the switch cap. This only indicates if the bulb is working or not; it does not indicate other problems that may be in the coil or the wiring up to the motor.

**Fig. 4-19** *Three-wire control with push-to-test pilot light.* Motor running.

The illuminated pushbutton combines a start button and a pilot light (see Fig. 4-20). Pressing the pilot light lens operates the start contacts. Space is saved by requiring only a two-unit pushbutton station instead of three.

When one START-STOP station is required to control more than one starter, the arrangement shown in

**Fig. 4-20** *Three-wire control with illuminated pushbutton.* Motor running.

Fig. 4-21 can be used. A maintained overload on any one of the motors will drop out all three starters. Note how M1 contacts control M2 and M2 contacts control M3. Once the STOP button is depressed it opens the circuit to M1, and its dropout causes the other two coils to also drop out. The START switch is paralleled with contacts from M3.

**Fig. 4-21** *Three-wire control. More than one starter. One pushbutton station controls all.* (Allen-Bradley)

So far the circuits have been utilized to keep the motor running or stopped when unsafe conditions occur. Now it is time to look at what can be done to reverse the direction of motor rotation. Three-wire control of a reversing starter can be accomplished with a FORWARD-REVERSE-STOP pushbutton station, as shown in Fig. 4-22. Limit switches can be added to stop the motor at a certain point in either direction. Jumpers 6 to 3 and 7 to 5 must then be removed. The circle with F and the circle with R in the figure are the coils whose contacts actually do the switching of the leads to the motor in order to make it reverse the direction.

**Fig. 4-22** *Three-wire control. Reversing starter.* (Allen-Bradley)

In some cases more than one pushbutton station is needed. That calls for an arrangement similar to Fig. 4-23. Note how the two switches (FORWARD and REVERSE) are paralleled but arranged so that pressing FORWARD in either location will cause the F coil to energize and pushing REVERSE will cause the R coil to energize from either position. Note how the F contacts are in series with the R coil and the R contacts are placed in series with the F coil. This de-energizes the R coil when the F coil is energized.

**Fig. 4-23** *Three-wire control. Reversing starter. Multiple-pushbutton station.* (Allen-Bradley)

Sometimes it is necessary to know in which direction the motor is rotating. A pilot light can be connected to indicate the direction of rotation. Pilot lights can be connected in parallel with the forward and reverse contactor coils to indicate which contactor is energized and thus in which direction the motor is running (Fig. 4-24). Start-stop and reversing of rotation

**Fig. 4-24** *Three-wire control. Reversing starter with pilot lights to indicate that the direction motor is rotating.* (Allen-Bradley)

have been examined up to this point. Now it is time to introduce the two-speed starter. Three-wire control of a two-speed starter with a HIGH-LOW-STOP pushbutton station is shown in Fig. 4-25. The diagram allows the operator to start the motor from rest at either speed or to change from low to high speed. The STOP button must be operated before it is possible to change from high to low speed. This arrangement is intended to prevent excessive line current and shock to the motor and the driven machinery. Shock to the motor can result when motors are run at high speed and are then reconnected for a lower speed.

**Fig. 4-25** *Three-wire control, two-speed starter.* (Allen-Bradley)

Once the two-speed motor has been controlled, it is, in most instances, preferable to have some indication as to which speed is engaged. Figure 4-26 shows a three-wire two-speed starter with one pilot light to indicate that the motor is operational at both speeds. One pilot light can be used to indicate operation at both low and high speeds. One extra, normally open interlock on each contactor is required. Two pilot lights, one for each speed, could be used by connecting pilot lights in parallel with high and low coils, such as shown in Fig. 4-24.

**Fig. 4-26** *Three-wire control. Two-speed starter with one pilot light to indicate motor operation at each speed.* (Allen-Bradley)

# OVERCURRENT PROTECTION FOR CONTROL CIRCUITS

A high-quality electric motor, properly cooled and protected against overloads, can be expected to have a long life. The goal of proper motor protection is to prolong motor life and postpone the failure that ultimately takes place. Good electrical protection consists of providing both proper overload protection and current-limiting with short-circuit protection. Alternating current motors and other types of high-inrush loads require protective devices with special characteristics. Normal, full-load, running currents of motors are substantially less than the currents that result when motors start or are subjected to temporary mechanical overloads. This characteristic is shown in Fig. 4-27.

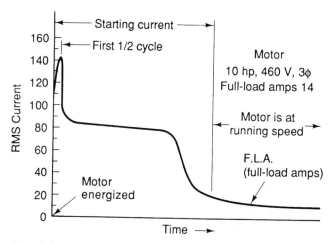

**Fig. 4-27** *Typical motor starting current characteristics.* (Bussmann Division, Cooper Industries, Inc.)

At the moment when the ac motor circuit is energized, the starting current rapidly rises to many times the normal current and the rotor begins to rotate. As the rotor accelerates and reaches running speed, the current declines to the normal running current. Thus, for this period, the overcurrent protective devices in the motor circuit must be able to tolerate the rather substantial temporary overload.

Motor starting currents can vary substantially depending on the motor type, load type, starting methods, and other factors. For the initial half-cycle, the momentary transient rms current can be 11 times higher or more. After the first half-cycle, the starting current subsides to four to eight times (typically, six times) the normal current for several seconds. This current is called the locked rotor current. When the motor reaches running speed, the current subsides to its normal running level.

The special requirements for protection of motors require that the motor overload protective device withstand the temporary overload caused by motor starting currents, and at the same time, protect the motor from continuous or damaging overloads. Overload protection is provided by three main types of devices:

- Overload relays in motor controllers. Usually, the melting alloy or bimetallic type, which are designed to simulate motor damage curve characteristics. Correctly sized, overload relays in a good-quality controller, properly maintained, provide good protection.

- Dual-element, time-delay fuses and low-peak dual-element fuses. The time-delay element in these fuses provides a minimum of 10 seconds delay at 500% load and will yield good running protection when sized correctly. In addition, short-circuit protection is afforded by the short-circuit element.

- Thermal protectors. Sensitive to heat and internally embedded in small fractional-horsepower motors and hermetic compressor motors of integral sizes; resetting can be automatic or manual.

Circuit breakers are not generally recommended for overload protection. They usually do not have sufficient time delay to permit close sizing (typically, sizing must be 200 to 250%) of motor full-load amperes (FLA) and are prone to trip out magnetically under starting conditions. The instantaneous trips of circuit breakers have to be set high enough to overcome the motor momentary transient current. Fuses other than dual-element are not recommended because they must be substantially oversized (typically, 300 to 400% of motor FLA) to permit starting.

Common control with fusing in one line only and with both lines ungrounded or, if user's conditions permit, with one line grounded are shown in Fig. 4-28. Note the fuse symbol and that it is in series with L1. L2 is grounded and not fused. Common control with fusing in both lines and with both lines ungrounded is shown in Fig. 4-29.

**Fig. 4-28** *Overcurrent protection for control circuit. One fuse.* (Allen-Bradley)

**Fig. 4-29** *Overcurrent protection for control ciruit. Two fuses.* (Allen-Bradley)

## TRANSFORMERS IN CONTROL CIRCUITS

Transformers can be used to step down or step up voltages. A step-down transformer is usually employed in control circuits where the equipment utilized to control the main operating equipment is connected to high-voltage lines. Low voltage can be used to control high voltages and high currents by energizing starter coils that have contacts that can handle the higher currents and voltages. This lower voltage allows for smaller wire to be used and for control stations to be placed in remote areas far from the actual operating device or motor.

There are a number of control circuits that use fuses for protection. Note the fusing employed in Fig. 4-30. One fuse is used to protect the transformer in this circuit, where it is placed in the secondary circuit of the transformer. However, Fig. 4-31 uses two fuses to protect the secondary transformer circuitry, while Fig. 4-32 shows how both the primary and the secondary circuits of the transformer are fused with one of the primary lines being grounded. In Fig. 4-33 the primary circuit is not grounded, so two fuses are used, one in L1 and one in L2, as well as two fuses in the secondary circuitry. Figure 4-34 shows the control circuit transformer with fusing in both primary lines, with no secondary fusing and with all lines ungrounded.

A wiring diagram and a ladder diagram are shown in Fig. 4-35, where the step-down transformer

**Fig. 4-33** *Control circuit transformer with both primary and secondary lines fused.* (Allen-Bradley)

**Fig. 4-34** *Control circuit transformer with fusing of both primary lines.* (Allen-Bradley)

**Fig. 4-30** *Control circuit transformer with one-fuse protection.* (Allen-Bradley)

**Fig. 4-31** *Control circuit transformer with two-fuse protection.*

**Fig. 4-32** *Control circuit transformer with one primary and one secondary fuse.* (Allen-Bradley)

**Fig. 4-35** *Step-down transformer in a control circuit. Wiring diagram and ladder diagram.* (Allen-Bradley)

provides low voltage for the control circuit, which is wired for three-wire control. The starter coil is operated on a voltage lower than line voltage. This is usually done for safety reasons. This also requires the use of a step-down transformer in the pilot circuit. The starter is operated from a START-STOP pushbutton station. When a control circuit step-down transformer is used with this type of starter, wiring connection X must be removed. Note that a fuse is added to the transformer secondary.

## REVIEW QUESTIONS

1. Why are wiring connections and terminal markings useful to the electrician?
2. What is the purpose of a manual start switch?
3. Describe a ladder diagram.
4. What is a maintained contact pilot device?
5. What is the difference between a two-wire and a three-wire control system?
6. What does de-energized mean with a relay?
7. What is meant by undervoltage protection?
8. What is a jumper used for?
9. How high can momentary transient rms current go?
10. What is locked rotor current?
11. What is a thermal protector?
12. Why are fuses needed in control circuits?
13. What are manual starting switches designed for?
14. What type of information does a line diagram provide?
15. What is another name for a line diagram?
16. What does L1 stand for in a diagram?
17. What does the term two-wire mean?
18. Why is a resistor built into the relay?
19. What do CR1 and CR2 mean?
20. What is the difference between a line drawing and a wiring diagram?

## REVIEW PROBLEMS

The parallel circuit is utilized in all phases of electrical work. A review of the properties of such circuits can benefit anyone working with electricity.

$$I_T = I_{R_1} + I_{R_2} + I_{R_3} + \cdots$$

$$R_T = \frac{R_1 \times R_2}{R_1 + R_2} \quad \frac{1}{R_T} = \frac{1}{R_1} + \frac{1}{R_2} + \frac{1}{R_3} + \cdots$$

1. Two resistors, one of 4 Ω and one of 8 Ω, are connected in parallel. What is the resistance of the circuit?
2. A resistance of 36 Ω is connected in parallel with a resistor of 18 Ω. What is the total resistance?
3. A 5400-, 78,000-, and 112,000-Ω resistor are connected in parallel. If 10 mA flows through the 112,000 Ω resistor, what is the current flow in the other two resistors?
4. A 10,000-, 5000-, and a 15,000-Ω resistor are connected in parallel. If 30 mA flows in the 15,000-Ω resistor, what is the current flow in the other two resistors?
5. How much is the total current in problem 3?
6. How much is the total current in problem 4?
7. How many 100,000-Ω resistors must be connected in parallel to give a combined resistance of 10,000 Ω?
8. Two resistors of 3.3 k and 4.7 kΩ are connected in parallel across a power source of 9 V. What is the current drain on the 9-V battery?

# 5
CHAPTER

# Switches

## PERFORMANCE OBJECTIVES

After studying this chapter, you will be able to:

1. Explain how a drum switch operates.
2. Describe float switch action.
3. Tell how a joystick is used.
4. Tell why interlock switches are utilized for safety reasons.
5. Describe pushbutton interlocking.
6. Identify the various types of limit switches.
7. Explain pushbutton, pressure, and selector switch operation.
8. Explain the use and action of snap switches.
9. Explain start-stop switch actions.
10. Explain how toggle switches, temperature switches, and vacuum switches operate.

## CONTROLLING ELECTRICITY

In order to make electricity useful it is necessary to control it. You will want it in the proper place at the proper time. Otherwise, it can do great damage—even kill. Electricity can be controlled by using switches, relays, or diodes. These devices are used to direct the current to the place where it will work for you. Each device is carefully chosen to do a specific job. For example, the relay is used for remote-control work, and a diode is used to control large and small amounts of current in electrical as well as electronic equipment. A diode is a device which allows current to flow in one direction only. It can be used to change ac to dc.

Industrial controls consist of nothing more than relays, diodes, and switches, which are designed for specific jobs. Switches may come in the form of thermostats or pressure-sensing devices. Relays may have one set of contacts, or any number, even 40. Diodes may be as small as a pin-head or large enough to handle high currents with the aid of heat sinks. In your car's alternator diodes are used to control the output of the generator. The alternator generates alternating current and the diodes change it to direct current before it leaves the device.

The device most often used to control electricity and all the work it accomplishes is the switch. Switches come in many sizes and shapes. They have different types of contacts for various applications. In the case of motor controls the switches have to have contacts that are able to withstand the onrush of current at the start and the inductive kickback produced by the collapsing magnetic field of the motor when it is turned off. In this chapter, we discuss those switches used to turn motors on and off and to reverse their direction of rotation. The switches have been arranged alphabetically by name. The electrician should become familiar with various manufacturers and what each has done to adapt a device to a special job.

## DRUM SWITCH

One of the most commonly used switches for motor control is the drum switch (Fig. 5-1). It has the capability to reverse a motor or turn it off. Drum switches may be used for across-the-line starting and the

***Fig. 5-1*** *Drum switch.* (Allen-Bradley)

reversing of ac poly-phase, ac single-phase, or dc motors. They are compact and inexpensive but ruggedly constructed.

## Easy Conversion

Drum switches are field convertible from MAINTAINED to MOMENTARY operation. This conversion consists of removing the handle screw and handle, turning the shaft 180°, then replacing the handle and handle screw.

A handle is used to move contacts that are mounted on an insulated rotating shaft. Moving contacts make and break contact with stationary contacts within the controller as the shaft is rotated. This drum-like rotation or moving of the contacts causes the switch to be called a *drum switch*. It is also sometimes called a *cam switch* since it uses a cam action to accomplish its switching action.

Figure 5-1 shows that the drum switch is fully enclosed and its insulated handle is used to move the contacts from point to point. Figure 5-2 shows how the six contacts of the switch are connected for off-reverse-forward positions. Note how the start and run windings connections are changed by the switching action to cause a reverse rotation.

## Three-Phase Switching

Three-phase motors can also be reversed by using the drum switch. Keep in mind that only two of the phases have to be changed to cause a three-phase motor to reverse its direction of rotation (Fig. 5-3).

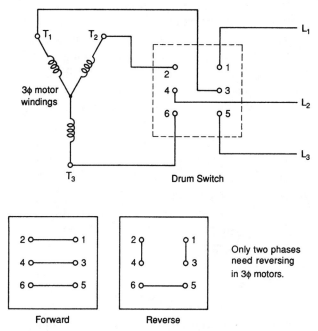

**Fig. 5-3** *Reversing a single-phase motor with a drum switch.*

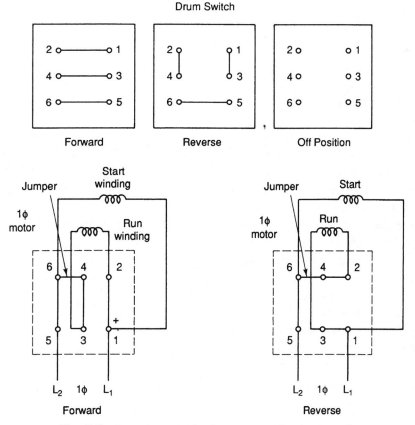

**Fig. 5-2** *Reversing a single-phase motor with a drum switch.*

Drum Switch 85

## DC Switching

Direct current motor control can also be provided by the drum switch (Fig. 5-4). The reversal is accomplished by changing the polarity. Note how the switching action causes the armature and field coil windings of the series motor to be placed in series in the forward position. Then trace it out for the reversal of polarity in the REVERSE position of the drum switch. The compound dc motor is also reversed by changing the polarity provided to the armature. The field coil maintains the same polarity in both positions of the switch. Note how the coils are connected to − and + in the same way for both positions of the switch. Then take a look at how the armature has − and + switched to make it rotate in the opposite direction. That is accomplished by 5 and 6 on the switch being the same in both the forward and reverse positions.

Some dc motors use inter-poles (commutating windings) to suppress brush arcing. Therefore, it is a good idea to keep in mind that these windings are part of the armature circuit.

## FLOAT SWITCHES

Float switches provide automatic control for motors that operate tank or sump pumps. They are built in five styles and can be supplied with accessories to provide rod or chain operation, wall or floor mounting (Fig. 5-5). Float switches are designed for automatic control of ac or dc pump motor magnetic starters, or for direct control of light motor loads.

A great variety of operating accessories are available to make float switches operate by many methods, depending on the position, depth of a tank, or sump (Fig. 5-6). Stainless steel floats, rods, chains, and stop collars are available for use in certain corrosive liquids. They are made in both normally open and normally closed configurations or both. Figure 5-7 shows how the float switch is connected in the circuit. Note the normally open and normally closed operation of the switch. This is one of the most popular means of pumping water from a basement automatically and prevents damage from seepage or flooding of any type. Once the level of the water causes the switch to close, it turns on the pump motor.

If it is a small fractional-horsepower motor, the switch can control the motor directly. If it is a larger-horsepower pump motor, the float switch is used to energize a magnetic motor starter.

## FLOW SWITCHES

A flow switch is used to detect the flow of liquids, water, oil, or other gases in a pipe or duct. It is nothing

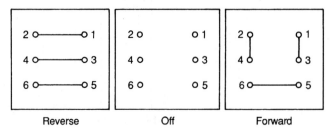

**Fig. 5-4** Direct current motor control by drum switch.

| | | WALL MOUNTING | | FLOOR MOUNTING |
|---|---|---|---|---|
| **Rod Operated**<br><br>Used with rods up to 9-feet long on the Style A and D float switches and up to 18 feet long on the Style B and C float switches. The float is fixed to one end of the rod. Adjustable stop collars at the top of the rod operate the switch. Guides may be required for vertical float movement. | | Includes two 3-foot lengths of ⅜-inch brass tubing with couplings, stop collars and copper float. | | Includes floor mounting bracket, 20-inch length of 1-inch pipe, mounting accessories, two 3-foot lengths of ⅜-inch brass tubing with couplings, stop collars and copper float. |
| **Long Rod — Free Float**<br><br>Used with rods up to 33-feet long. The double arm lever carries a counterweight to offset the weight of the rod. The float moves up and down between stops on the rod so that even though the liquid level varies greatly, the rod moves only a short distance. Top of rod is fixed to the switch lever. Guides may be required for vertical float movement. | | Includes double arm lever for float switch, counterweight, two 3-foot lengths of ⅜-inch brass tubing with couplings, stop collars and copper float. | | Includes double operating arm for float switch, floor mounting bracket, 20-inch length of 1-inch pipe, mounting accessories, two 3-foot lengths of ⅜-inch brass tubing with couplings, stop collars and copper float. |
| **Long Rod — Parallel Motion**<br><br>Used with unguided rods up to 33-feet long. The parallel lever arrangement keeps the rod vertical since the rod might otherwise tend to move sideways. The float moves up and down between stops on the rod so that even though the liquid level varies greatly, the rod moves only a short distance. Top of rod is fixed to the switch lever. | | Includes two double arms for float switch, counterweight, two 3-foot lengths of ⅜-inch brass tubing with couplings, stop collars and copper float. | | Includes two double arm levers for float switch, counterweight floor mounting bracket, 20-inch length of 1-inch pipe, mounting accessories, two 3-foot lengths of ⅜-inch brass tubing, stop collars and copper float. |

**Fig. 5-5** *Automatic float switches.* (Allen-Bradley)

| | | |
|---|---|---|
| **Chain Operated — Self-Supported Double Pulley**<br><br>The chain operated self-supported double pulley accessories have a float fixed at one end and a counterweight at the other. The adjustable stop collars move the float switch operating lever. | **WALL MOUNTING**<br>*Includes double pulley bracket, 15-foot bronze chain with stop collars, counterweight and copper float.* | **FLOOR MOUNTING**<br>*Includes double arm bracket, 15-foot bronze chain with stop collars, counterweight, copper float and 20-inch length of 1-inch pipe with mounting accessories.* |
| **Chain Operated — Self-Supported Single Sheave Wheel**<br><br>Used with the Style A float switch in the NEMA Type 1 and 4 enclosure. A single pulley is mounted on the top of the float switch. The chain has a float fixed at one end and a counterweight at the other. Adjustable stop collars on the chain move the float switch operating lever. | **WALL MOUNTING**<br>*Includes one pulley, 15-foot of bronze chain, stop collars, counterweight and copper float.* | **FLOOR MOUNTING**<br>*Includes one pulley, 15-foot of bronze chain, stop collars, counterweight, copper float and 20-inch length of 1-inch pipe with mounting accessories.* |
| **Chain Operated — Separate Pulleys**<br><br>The two pulleys included with these accessories are separate from the switch. The chain has a float fixed at one end and a counterweight at the other. Adjustable stop collars move the float switch operating lever. | **WALL MOUNTING**<br>*Included are two pulleys for separate mounting, 15-foot of bronze chain with stop collars, counterweight and copper float.* | |

**Fig. 5-5** *(Continued).*

**Fig. 5-6** *Float switch accessories.* *(Square D)*

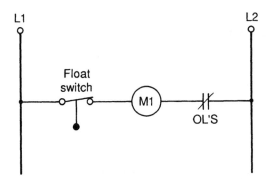

**Fig. 5-7** *Float switch connected in a circuit*

more than a normally open or normally closed switch with an operator extending from the switch enclosure (Fig. 5-8). Most are adjustable so that allowances can be made for buildup or changes in the application of the switch.

**Fig. 5-8** *Oil-tight limit switch with nylon rod.* (Allen-Bradley)

These switches are very useful in a number of locations. They may be used to detect the flow of air in a wind tunnel or in air-conditioning ductwork, or they may be used to set off alarms or start motors.

They are used to indicate the flow of water in a sprinkler system. This can be used to alert the person in charge in case a sprinkler system begins to operate without being triggered by fire. The sounding of the alarm can allow the operator to shut off the sprinkler system before too much water damage occurs (Fig. 5-9).

Clogged air filters can be detected by flow switches. Flow switches can be adjusted to indicate when a filter is clogged and is not allowing enough air through to keep a motor or other device properly cooled. Airflow detection is important in heating

**Fig. 5-9** *Flow switch sounds alarm for sprinkler malfunction.*

systems. Electrical heating systems are especially prone to damage if the proper amount of air is not allowed to circulate over the heating coils. The coils can burn out if not properly cooled by a steady flow of air. See Fig. 5-10 for the electrical circuit that puts the flow switch to work.

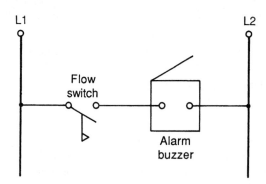

**Fig. 5-10** *Electric circuit using a flow switch for a good purpose.*

## FOOT SWITCHES

Foot switches are just what the name suggests. They are operated by pressure directed by a human foot. They can be found on stapling machines, older-model automobile headlight dimmers, and any number of machine operations that require a person to use both hands and thus operate the on-off switch by foot. This type of switch is usually a heavy-duty type and comes in a variety of sizes, voltages, and shapes. Many limit switches can be converted to foot operation.

Some floor lamps used for residential lighting are controlled by foot switches usually mounted on the base of the lamp.

## JOYSTICK

The joystick is a rather recent device and is used most commonly in conjunction with computers, although it can be used to control a number of operations (Fig. 5-11). This type of switch can be used to good advantage in the operation of a crane, where the operator has to keep his or her eyes on the load rather than on the switch.

**Fig. 5-11** *Joystick switch. (Square D)*

Two types of joysticks are available: the standard model, where the operator is free to move from position to position, and the latched lever type. The latched lever requires the operator to lift a locking ring before the lever can be moved. The switching action can be momentary or maintained contact or a combination of both.

Note the different combinations of contact positions available in Fig. 5-12. The center position on the joystick is used for the off or stop control. Figure 5-13A shows how the joystick is wired into a circuit and can operate at least four loads, depending on where the lever is located to close the switches. The switches can be adjusted to have a two-, three-, four-, or eight-position gate. The placement of the gates in the switch causes it to change its possibilities. The gates are metal plates that can be changed as needed by the person who usually makes repairs and installations.

Whenever a job requirement dictates the utilization of moving an object from left to right or up and down, the joystick can be made to control the movement needed. A good example is operating a foundry ladle or a hoist for any number of jobs. There are some advantages to joystick operation of a load. Inasmuch as no two contacts can be closed at the same time, there is a slight delay between the break of one circuit and the make of another by the switching action. This means that the load will not have to change directions quickly.

## INTERLOCK SWITCHES

Interlock switches are used for safety purposes. The idea is to make sure that the forward motion of the motor has been stopped before the reversing action is started by changing the electrical path through the motor windings. There are three ways to incorporate interlocking and the protection needed. Auxiliary contact interlocking, mechanical interlocking, and pushbutton interlocking are the three methods most commonly used.

### Mechanical Interlocking

A ladder diagram can be used to illustrate the way a mechanical interlock functions (Fig. 5-13B). The dashed lines from the forward coil to the reverse coil are used to indicate a mechanical interlock. This type of interlock switching is usually provided by the manufacturer of the starter. A piece of material is usually inserted so that when one pushbutton is depressed, the other cannot be depressed. The FORWARD pushbutton is depressed, completing the circuit to the F coil. The coil is energized, but in energizing, it closes the contacts F1. Keep in mind that there are two parts to the switch or contacts that close or touch one another. This means that the pushbutton can be released and the circuit stays energized by completing the circuit through the F1 contacts instead of the pushbutton contacts. This is often referred to as memory. The F coil stays energized until the STOP switch is depressed. Once it is depressed, the coil de-energizes and contacts (F1) open. The STOP button returns to its normally closed position and the REVERSE button can be depressed, energizing the R coil. The coil energizes and closes contacts (R) that are across the REVERSE pushbutton. The circuit stays energized until the STOP switch is depressed, causing the coil to de-energize.

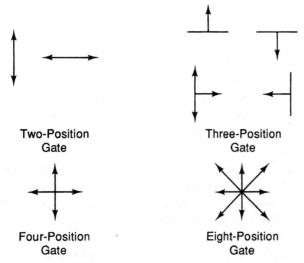

**Fig. 5-12** *Contact positions of the joystick.*

**Fig. 5-13** (A) joystick contacts. Note the black dot that indicates left, right, up, and down positions of the handle. (B) Mechanical interlocking in reverse and forward with a two-pushbutton station.

The mechanical interlock prevents both forward and reverse from being energized if the buttons were to be depressed at the same time. It also prevents the reverse power being applied while forward power is still being applied to the motor.

Keep in mind that the contacts for the mechanical interlock are not shown in a line or ladder diagram. The assembly of the switches causes one to be disabled when the other is being depressed. In this way, short circuits and burnouts are prevented. Take a look at Fig. 5-14. The three-phase motor can be reversed by changing any two of the phases. Reversing motor starters are available in vertical or horizontal construction.

## Pushbutton Interlocking

Double-circuit pushbuttons can be used for pushbutton interlocking (Fig. 5-15). This type of interlocking depends on the electrical circuit for its operation. The idea is to make sure that both coils are not energized at the same time.

Note that the STOP button is normally closed in Fig. 5-15. That means that when the FORWARD button is pressed, coil F is energized and the normally open (NO) contact $F_1$ closes to hold in the forward contactor. Because the normally closed (NC) contacts are used in the forward and reverse pushbutton units, there

**Fig. 5-14** *(A) Reversing a three-phase motor means changing two phases. (B) A one- or two-pole electrical interlock can be added to the disconnect switch or circuit breaker. (Square D)*

is no need to press the STOP button before changing direction of rotation. If the REVERSE button is pressed while the motor is running in the forward direction, the forward control circuit is de-energized and the reverse contactor is energized and held closed.

If reversing is repeated quickly, it may cause the motor to overheat and overload relays and fuses to function as designed. In most applications for motors it is best for the motor to coast to a stop before reversing. However, if plugging is called for, the starter has to be rated to handle the reversing, especially if it occurs at a rate of more than five times per minute.

### Auxiliary Contact Interlocking

Normally closed (NC) auxiliary contacts on the starter can be used for interlocking. This is done by a separate set of contacts on the reversing starter (Fig. 5-16). The three-phase motor shown has a forward (F) and reverse (R) coil that energize to close contacts in $L_1$, $L_2$, and $L_3$ to complete the power to the motor through its overload (OL) contactors, inserted inside the motor windings. The forward pushbutton is depressed. This completes the circuit to the F coil because the R normally closed contacts ($R_2$) are closed. Once coil F is energized, it holds the F contacts across the forward switch contacts closed, keeping coil F energized. When coil F is energized, it opened contact $F_2$, making it impossible for the R coil to energize. Contacts $F_2$ and $R_2$ are auxiliary contacts and are sometimes mounted on the outside of the reversing starter. In reversing the single-phase motor, keep in mind that either the run or the start winding leads are reversed, not both.

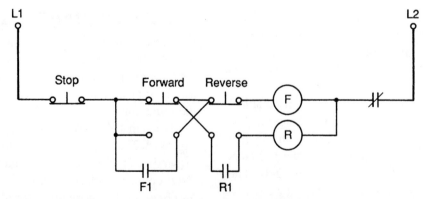

**Fig. 5-15** *Double-circuit push-button. Push-button interlocking.*

**Fig. 5-16** *Mechanical, push-button, and auxiliary contact interlocks.*

Fig. 5-17 shows how this is done with a reversing starter using single-phase.

## LIMIT SWITCHES

As the name implies, the limit switch limits the operation of a machine that is connected in line with the switching action (Fig. 5-18). Many industrial production lines use this type of switch to limit the travel of various devices on the line. A good example of the use of a limit switch is a garage door opener. When the door is lifted it must stop before it hits the motor. A limit switch turns off the motor before the door crashes into the motor that is pulling it up. Then, too, the motor must be turned off when the door is lowered, or else the motor keeps running after the door hits the floor. Limit switches do the job and make it possible for these operations to be semiautomatic, or once the action has been initiated, it is limited by limit switches to conform to the physical conditions.

Limit switches can be used to start, stop, forward, reverse, recycle, slow down, or speed up various operations. Inasmuch as they can do all these things, a variety of sizes and shapes are needed.

Limit switches are made up of two parts, the electrical contacts and the mechanical device that operates the on-off function of the contacts. The actuating mechanism may take a number of forms.

### Types of Limit Switches

There are various types of limit switches (Fig. 5-19).

*Heavy-duty precision oil-tight (type C)* provides long electrical and mechanical life together with easy installation and wiring. This type is available in a variety of head and body styles, including an explosion-proof version that is also watertight and submersible. Also available are standard, logic reed, and power reed contacts, as well as many special features.

*Miniature enclosed reed switch (type XA)* is a small, inexpensive die-cast zinc switch utilizing a hermetically sealed reed for the contact mechanism. Prewiring and potting, combined with the sealed reed make this switch a good choice where contact reliability and environmental immunity are required along with small size and low cost.

*Heavy-duty oil-tight foundry (type FT)* is used in foundries or mills where one or more of the type T features (see Fig. 5-19) are required and where hot, falling sand, or similar foreign material could cause jamming of other limit (position) switches.

**Fig. 5-17** (A) Three pole, single-phase, four-lead split-phase motor reversing starter; (B) corresponding line diagram.

94 Switches

**Fig. 5-18** Limit switches: (A) sealed contact switch; (B) oil-tight limit switch; (C) time-delay oil-tight limit switch.

| | | |
|---|---|---|
| | HEAVY DUTY PRECISION OILTIGHT **TYPE C** | The Type C provides superior electrical and mechanical life along with easy installation and wiring. Available in a variety of head and body styles, including an explosion proof version that is also watertight and submersible. Also available are standard, logic reed and power reed contacts, as well as many special features. The Type C should be the first choice for all applications. |
| | HEAVY DUTY PRECISION OILTIGHT **TYPE B** | Most Type B devices are obsolete — Select the Type C equivalent for new installations. |
| | MINIATURE ENCLOSED REED SWITCH **TYPE XA** | The Type XA is a small, inexpensive die cast zinc switch utilizing a hermetically sealed reed for the contact mechanism. Prewiring and potting combined with the sealed reed make this switch a good choice where contact reliability and environmental immunity are required along with small size and low cost. |
| | HEAVY DUTY OILTIGHT **TYPE T** | If load exceeds Type C contact ratings, if a required operating sequence is not available on the Type C or if high trip and reset forces are required, use the Type T. |
| | FOUNDRY **TYPE FT** | Use FT in foundries or mills where one or more of the above listed Type T features are required and where hot, falling sand or similar foreign material could cause jamming of other limit (position) switches. |
| | PRECISION OILTIGHT **TYPE AW** | The Type AW is suitable for applications requiring a precision, oiltight switch. It is suggested especially for applications requiring a lever type switch for use in extremely low ambients or micrometer adjustment on a plunger type switch. |
| | INTERNATIONAL POSITION SWITCHES **TYPE BD** | New Product — Meets International standards, CENELEC EN 50041. Uses same lever arms as the Type C and AW. |
| | SNAP SWITCHES | Use snap switches on applications requiring a basic contact mechanism with or without operator where an enclosure is furnished separately or not required. |

**Fig. 5-19** Limit (position) switches. *(Square D)*

*Snap switches* are used on applications requiring a basic contact mechanism with or without operator where an enclosure is furnished separately or not required.

There are limit switches designed for special applications or found to be particularly useful under certain conditions. Snap-action limit switches are designed to snap over (trip instantly) once the mechanism that operates the limit switch has traveled the required distance to trip the limit switch, regardless of the speed with which it travels. This type should be used whenever machine operation acts at a slow rate of speed or where machine motion is short. Figure 5-20 shows some of the wiring diagrams and contact configuration and terminal wire color code for limit switches.

*Gravity return limit switches* are designed for conveyor-type operations with small, lightweight

**Fig. 5-20** Wiring diagrams showing contact configurations and terminal wire color code. *(Allen-Bradley)*

**Limit Switches** 95

moving objects. They have an extremely light operating torque and use the action of gravity on the lever arm to reset the contacts (Fig. 5-21). Limit switches are also described according to how they work. They may be top push, spring return, roller type, rod type, side push, lever type, and maintained contact.

## Limit Switch Circuits

Figure 5-22 shows the limit switch in a single-station, maintained-contact configuration. The START button mechanically maintains the contacts that take the place of hold-in contacts. Pressing the START button maintains the circuit. By pressing the STOP button it breaks the circuit by opening the start contacts. If the contactor is de-energized by a power failure or overload operation, the start contacts are unaffected. The motor restarts automatically.

## PRESSURE SWITCHES

The control of pumps, air compressors, welding machines, lube systems, and machine tools requires control devices that respond to the pressure of a medium such as water, air, or oil. The control device that does this is a pressure switch. It has a set of contacts that are operated by the movement of a piston, bellows, or diaphragm against a set of springs. The spring pressure determines the pressures at which the switch closes and opens it contacts (Fig. 5-23).

Industrial pressure switches are designed for use in pneumatic and hydraulic systems in a wide variety of applications. Open types are suitable for panel mounting where permitted. They can be used on air, water, oil, or any pressure media that is compatible with the actuator material. However, piston-actuated devices used on dry gas media will have reduced seal life due to lack of seal lubrication.

Pressure controls are accurate, rugged, compact, and adaptable to a wide variety of pressure applications.

*Fig. 5-21* Gravity return limit switches: (a) slotted shaft to aid in adjustment; (b) limit switch with steel operating lever; (c) example of clockwise operation.

The start button mechanically maintains the contacts that take the place of hold-in contacts. Depressing the start button maintains the circuit; depressing the stop button breaks the circuit by opening the start contacts. If the contactor is deenergized by a power failure or overload operation, the start contacts are unaffected. The motor restarts automatically.

*Fig. 5-22* Single-station, maintained-contact buttons. (Allen-Bradley)

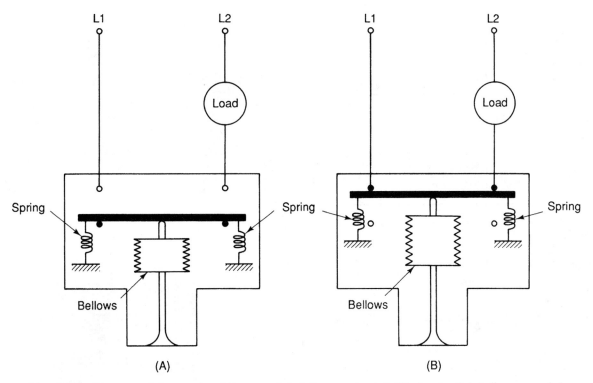

**Fig. 5-23** *Pressure switch operation: (A) open switch, bellows contracted; (B) closed switch, bellows expanded.*

The control has a snap-action precision switch equipped with silver contacts. Contact force is maintained at a high value up to the instant of snap-over. This avoids dead-center conditions. Normal industrial vibration has little effect on the positive opening and closing of the contacts. Straight in-line and relatively friction-free construction provides accurate and consistent operation regardless of the angle at which the control is mounted.

Long-life copper-alloy bellows are designed for use with air, water, oil, non-corrosive liquids, vapors, or gases in a series of pressure ranges from 30 in vacuum (Hg) to 900 psi. Stainless steel bellows are available for use with many of the more corrosive liquids or gases at pressures up to 375 psi.

Piston-type controls are designed for use on oil only and must not be used on air or gases, water, or other liquids that will rust cast iron. All pistons are provided with a drain, which should be connected to an oil return line with the reservoir vented to atmosphere. Drains should never be plugged, as back pressure on the diaphragm assembly may change the setting and could result in forcing fluid into the enclosure. Controls with seal rings generally do not require return lines, provided that a limited amount of leakage is not objectionable. Phosphate ester-based hydraulic fluids require a special diaphragm assembly. Figure 5-24 shows the symbols for the various types of pressure-operated switches.

## PUSHBUTTON SWITCHES

Pushbutton switches are used in stations to control motor operation. Figure 5-25 shows pushbutton stations fitted with two, three, and six switches. They can be used to start, stop, forward, and reverse action, as well as to reset and test or cause up-and-down movement. Pushbutton stations are used in the control circuits of magnetic starters. They are usable on the great majority of applications where compact size and dependable performance is needed. Assembled stations are available with any combination of push buttons, selector switches, pilot lights, or special-purpose devices.

Push buttons can be used in a number of ways to control motors and manufacturing processes, for instance: start-stop control, reversing control, two-speed control, jogging control, thermostat-controlled

| Symbol | | Description |
|---|---|---|
| Pressure Controls | Temperature Controls | CONTACT BLOCKS |
| **AUTOMATIC OPERATION** | | |
| ⊥ | ⊥ | Single pole double throw — automatically opens or closes on rise or fall. |
| ⊥ | ⊥ | Single pole double throw — slow acting contact with no snap action. Contacts close on rise and close on fall with an open circuit between contact closures. |
| ⊥ | ⊥ | Single pole single throw, normally open — closes on rise. |
| ⊥ | ⊥ | Single pole single throw, normally closed — opens on rise. |
| ⊥ | ⊥ | Single pole single throw, normally open — closes on rise. |
| ⊥ | ⊥ | Single pole single throw, normally closed — opens on rise. |
| ⊥ | ⊥ | Two circuit, single pole single throw, normally open — a common terminal is connected to 2 separate contacts which close on rise. |
| ⊥ | ⊥ | Two circuit, single pole single throw, normally closed — a common terminal is connected to 2 separate contacts which open on rise. |
| **MANUAL RESET** | | |
| ⊥ | ⊥ | Single pole single throw, normally open — contacts open at a predetermined setting on fall and remain open until system is restored to normal run conditions at which time contacts can be manually reset. |
| ⊥ | ⊥ | Single pole single throw, normally closed — contacts open on rise and remain open until system is restored to normal run conditions at which time contacts can be manually reset. |
| ⊥ | ⊥ | Single pole double throw, one contact normally closed — contact opens on rise and remains open until system is restored to normal run condition at which time contact can be manually reset. A second contact closes when the first contact opens. |
| ⊥ | ⊥ | Single pole single throw, normally closed — contacts close on fall and remain closed until system is restored to a higher predetermined setting. |

**Fig. 5-24** *Pressure controls symbols for automatic operation and manual reset.* (Allen-Bradley)

**Fig. 5-25** *Pushbutton stations: (A) pendant station; (B) cavity-mounted station without pilot light; (C) cavity-mounted station with pilot light; (D) six-unit custom-built station.* (Allen-Bradley)

motor, and ground detection with push-to-test pilot lights. Figure 5-26 shows a group of single stations with a MASTER STOP button. Note the difference between the ladder or line diagram and the wiring diagram of the switch. Notice how the numbers on the wiring diagram are connected in the circuit for proper operation. This circuit has a momentary contact MASTER STOP that is connected in series with a group of parallel-connected circuits. Depressing the MASTER STOP button de-energizes all the circuits.

The circuit in Fig. 5-26 is the basic start-stop circuit. Two-wire control or under voltage release circuits are not applicable here because they would be reenergized as soon as the MASTER STOP button is released.

## SAFETY SWITCHES

A safety switch can be used for motors that can be placed directly across the line for starting (Fig. 5-27). This type of switch is operated manually and usually has a handle on the outside of the enclosure to operate the switching blades. The switch enclosures are made of steel or fiberglass-reinforced polyester especially formulated to withstand attack from almost any corrosive atmosphere found in industrial applications. Electrical interlocks are available for safety purposes. A pivot arm operated from the switch mechanism breaks the control circuit before the main switch blades break. Slow-blow

**Fig. 5-26** *Group of single stations with master STOP button. (Allen-Bradley)*

**Fig. 5-27** *Safety switches. (Square D)*

fuses are usually installed inside the enclosure to handle the current surge caused by a starting motor.

Manual transfer switches are designed to transfer loads from one supply to another. They are not fusible. They are available from 30 to 600 A. Of course, the higher current switches with four-pole switching are well over $5000.

## SELECTOR SWITCHES

Selector switches can be used for determining the direction of motor rotation, for starting-stopping and jogging. They are available in two-, three-, or four-position configurations. Figure 5-28 shows the switch and the contact arrangements. Selector switches are

**Fig. 5-28** *Selector switch. (Reliance)*

available with a lever operator, coil operator, slot operator, knob operator, or key operator. They can also be obtained with a momentary contact where testing a circuit is necessary.

## SINGLE-POLE SWITCHES

Single-pole switches are simple on-off types. They are made in thousands of different shapes and styles and can be utilized as a simple on-off switch for a small motor or can be made into larger motor controllers by making the switch contacts of different materials. Many pushbutton switches are single-throw. They use two contacts, as shown in Fig. 5-29.

**Fig. 5-29** *Single-pole switch circuit.* (Allen-Bradley)

## START-STOP SWITCHES

Start-stop switches take many shapes and can be used for a number of control purposes. The main concern in a switch used to turn a motor on and off is its current rating; that is, will it be able to handle the onrush current? Figure 5-30 shows a circuit where the START button maintains the contacts that take the place of hold-in contacts. Depressing the START button maintains the circuit. Depressing the STOP button breaks the circuit by opening the start contacts. If the contactor is de-energized by a power failure or overload operation, the start contacts are unaffected. The motor restarts automatically. Note the symbol for the maintained contact switch.

## TEMPERATURE SWITCHES

Temperature-operated switches are usually referred to as thermostats. Figure 5-31 shows how the high-temperature cutout is placed in the circuit. Then note the location of the thermostat. This particular circuit has under voltage release and uses a selector switch to change from automatic operation with the thermostat in the circuit to one where the thermostat is taken out of the circuit or the circuit is de-energized by the selector switch being placed in the no-contact center position.

## TOGGLE SWITCHES

Toggle switches are available in many sizes and shapes and are made for special purposes as well as for motor control circuits (Fig. 5-32). They are used to turn various devices on and off, or to switch from one device to another. They usually have a metal handle and are mounted through a round hole. Screw terminals are usually provided for attaching wires. However, some may have wire leads furnished. A good example of a toggle switch is located in the home where it is mounted on the wall and turns on and off the overhead light.

## TRANSISTOR SWITCHING

One of the functions of the transistor is switching. It is a very efficient device when it comes to switching. Another function of the transistor is amplifying. These functions are discussed in chapter 7.

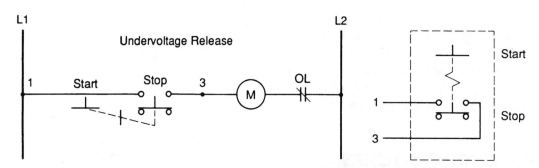

The start button mechanically maintains the contacts that take the place of hold-in contacts. Depressing the start button maintains the circuit; depressing the stop button breaks the circuit by opening the start contacts. If the contactor is de-energized by a power failure or overload operation, the start contacts are unaffected. The motor restarts automatically.

**Fig. 5-30** *Start button with maintained contacts.* (Allen-Bradley)

**Fig. 5-31** (A) Temperature switches. (Square D) (B) Thermostat-controlled motor with selector switch. (Allen-Bradley)

**Fig. 5-32** Toggle switches.

**Transistor Switching** 101

## VACUUM SWITCHES

These types of switches use a vacuum for proper operation (Fig. 5-33). They may have double-throw contacts that are normally open and normally closed so they can be used to control forward and reverse operation of a motor.

**Fig. 5-33** Vacuum switches. (Square D)

## REVIEW QUESTIONS

1. How is a drum switch converted from maintained to momentary operation?
2. What is another name for a drum switch?
3. What do some dc motors use to suppress brush arcing?
4. How are float switches used?
5. How are flow switches used in alarm systems?
6. What is a joystick? How is it used?
7. What is the purpose of an interlock switch?
8. What can the normally closed auxiliary contacts on a starter be used for?
9. How are limit switches used? List five types of limit switches.
10. What are two types of pressure switches?
11. Where are pushbutton switches useful?
12. How are safety switches incorporated into various types of equipment?
13. Where are selector switches used?
14. How are toggle switches used in motor control circuits?
15. How are vacuum switches used in motor control circuits?
16. What is one of the most commonly used switches for motor control?
17. Can the drum switch be used to reverse a three-phase motor?
18. What type of switch uses stainless steel floats?
19. What type of switch is used to indicate the flow of water in a sprinkler system?
20. How can clogged air filters be detected?

## REVIEW PROBLEMS

Switches are one of the most often used means of controlling electricity. Being able to complete the circuit or interrupt the circuit is of great importance to any electrician at one time or another. A good grasp of the drawings associated with switching improves your comprehension of circuit operation and enhances your ability to visualize the operation of switches.

1. Using Fig. P-1, draw lines to complete the circuit between the switch terminals to cause the three-phase motor to run in the *forward* direction.

**Fig. P-1**

2. Using Fig. P-2, draw lines to complete the circuit between the switch terminals to cause the three-phase motor to run in the *reverse* direction.

**Fig. P-2**

3. Using Fig. P-3, draw lines to complete the circuit between the switch terminals to cause the shunt motor to run in the forward direction.

**Fig. P-3**

4. Using Fig. P-4, draw lines to complete the circuit between the switch terminals to cause the series motor to run in the *reverse* direction.

**Fig. P-4**

5. Using Fig. P-5, draw lines to complete the circuit between the switch terminals to cause the compound motor to run in the *forward* direction.

**Fig. P-5**

6. Draw the switching circuit in Fig. P-6 so that the switches complete the circuit and cause the lamp to light.

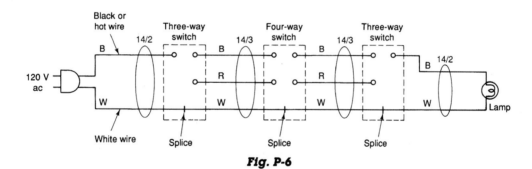

**Fig. P-6**

Review Problems  103

# CHAPTER 6

# Magnetism and Solenoids

## PERFORMANCE OBJECTIVES

After studying this chapter, you will be able to:

1. Understand how magnetism was discovered and put to work.
2. Describe a solenoid.
3. Explain the use and operation of a solenoid valve.
4. Explain the importance of the air gap and the armature in a solenoid's operation.
5. Define the role of shading coils in solenoids.
6. Identify a coil by its sealed current rating.
7. Differentiate between pickup voltage and seal-in voltage.
8. Differentiate between pickup voltage and dropout voltage.
9. Explain the role of hum in magnetic devices.
10. Identify various problems associated with magnetic coil devices and their causes.

Observation of the effects of magnetism predate the discovery of static electricity. Magnetism was discovered about 2600 B.C. Some say it was first observed by the Chinese; others believe that the Greeks were the first to observe it. There is a little doubt that primitive man also noticed its effect but lacked the knowledge to use it. Certain heavy stones or rocks had the power to attract and lift similar stones as well as pieces of iron. The material in these stones was called *magnetite*, named by the Greeks for the province of Magnesia in Asia Minor, where some of the stones were found. Its properties were later referred to as possessing *magnetism*. Magnetism is a property of stones made of magnetite (Fig. 6-1).

**Fig. 6-1** *The lodestone is a natural magnet with no definite shape but a north-south pole alignment.*

The Chinese were among the first to record observations of the stone and its properties. They hung magnetite on a string and watched it line up so that one end pointed north and the other south. This led to later development of the magnetic compass. Since this stone could lead the way for travelers, it was called the *leading stone* (later changed to *lodestone*).

Many years passed with no developments in the field of magnetism. Then William Gilbert, a physician to Queen Elizabeth I of England, began studying the mystery of the lodestone and formed the basic principles of magnetism. He also experimented with static electricity and established the use of the word *electron*. Others were also experimenting with the properties of the lodestone at the same time. Many added knowledge to the field and some have contributed their names to units of measurement in both electricity and magnetism. These pioneering workers include Alesandro *Volta*, Charles Augustin de *Coulomb*, Andre Marie *Ampere*, Luigi *Galvani*, Hans Christian *Oersted*, and Georg Simon *Ohm*.

Benjamin Franklin caused some interest in electricity in the United States by flying his kite with metal objects attached during a thunderstorm. About 80 years later Michael Faraday of England and Joseph Henry of the United States each discovered the relationship between magnetism and electron flow. This discovery became one of the most important contributions to the field of electricity and electronics.

After the establishment of this relationship, progress in the field was quite rapid. Thomas Edison did a lot of work with the electric motor and the electric light and developed a generator to produce electricity. The advent of the transformer led to developments in the modern era.

## SOLENOIDS

One device that grew out of the development of electric power and its transmission was the solenoid. The solenoid is a current-carrying coil used as a magnet. In a solenoid, there is a tendency for the core to move so that it encloses the maximum number of magnetic lines of force, each having the shortest possible path. In Fig. 6-2 the core is shown outside the coil. Because it is made of a ferromagnetic material, it presents a low-reluctance path to the magnetic lines of force at the north end of the coil. These lines of force concentrate the soft-iron core and then complete their paths back to the south pole of the electromagnet.

A movable iron core tends to be pulled to the center of a solenoid. The electromagnetic lines of force passing through the core material have thus magnetized the core. The direction of the magnetic lines has produced a south pole in the core at the north end of the electromagnet. The electromagnetic lines of force

**106** Magnetism and Solenoids

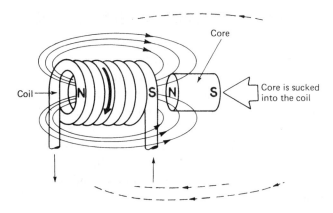

**Fig. 6-2** *A solenoid with an electric current has a tendency to pull the core into the coil.*

leave the opposite end of the iron core, and this end is the north pole of the magnetized core.

An attraction between the north pole of the coil and the south pole of the iron core tends to pull the core into the coil. Magnetic lines of force fan outward from the north pole of the magnetized core. They have a shorter magnetic path back to the south pole of the coil. As the iron core is pulled into the coil, this path becomes increasingly shorter. The magnetic lines of force travel the shortest possible distance when the core centers itself in the coil. This movement of the core into the coil can be harnessed to cause the closing of switch contacts or the opening of a valve that controls air, gas, hydraulic fluid, or any other flowing medium.

## Solenoid Valves

Solenoid valves are used on many refrigeration systems. They are electrically operated. A solenoid valve, when connected as in Fig. 6-3, remains open when current is supplied to it. It closes when the current is turned off. In general, solenoid valves are used to control the liquid refrigerant flow into the expansion valve, or the refrigerant gas flow from the evaporator when it or the fixture it is controlling reaches the desired temperature.

The most common application of the solenoid valve, in the liquid line, operates with a thermostat. With this hookup, the thermostat may be set for the desired temperature in the fixture. When the temperature is reached, the thermostat will open the electrical circuit and shut off the current to the valve. The solenoid valve then closes and shuts off the refrigerant supply to the expansion valve. The condensing unit operation should be controlled by a low-pressure switch. In other applications, where the evaporator is to be in operation for only a few hours each day, a manually operated snap switch may be used to open and close the solenoid valve.

The solenoid valve shown in Fig. 6-4 is operated with a normally closed status. A direct-acting metal ball and seat assure tight closing. The two-wire, class W coil is supplied standard for long life on low-temperature service or sweating conditions. Current failure or interruption will cause the valve to fail-safe in the closed position. The solenoid cover can be rotated 360° for easy installation. Explosion-proof models are available for use in hazardous areas.

**Fig. 6-3** *Solenoid valve connected in the suction and liquid evaporator lines of a refrigeration unit.*

**Fig. 6-4** *Solenoid valve locations and color-coded wires.*

**Uses** This solenoid valve is usable with all refrigerants except ammonia. It can also be used for air, oil, water, detergents, butane or propane gas, and other non-corrosive liquids or gases. A variety of temperature control installations can be accomplished with these valves. Such installations include bypass, defrosting, suction line, hot gas service, humidity control, alcohols, unloading, reverse cycle, chilled water, cooling tower, brine, and liquid-line stop installations and ice makers.

**Operation** The valves are normally held in the closed position by the weight of the plunger assembly and the fluid pressure on top of the valve ball. The valve opens by energizing the coil and magnetically lifting the plunger and allowing full flow by the valve ball. De-energizing the coil permits the plunger and valve ball to return to the closed position.

## Automatic Gas Furnace Solenoid

The solenoid shown in Fig. 6-5 is used to control the flow of natural gas into a furnace. When heat is desired, the thermostat in the room goes on, making contact and completing the circuit through the power supply to the valve coil (3). The energized coil creates a magnetic field that lifts the plunger (4). This allows pressurized gas to flow through the inlet at the left side of the valve body (9), beneath the plunger, and through the outlet at the right. When the room reaches the desired temperature, the thermostat breaks the circuit and the plunger drops back into the valve seat (11), shutting off the flow of gas.

**Fig. 6-5** *Cutaway of typical automatic gas furnace control.*

# CONSTRUCTION OF SOLENOIDS

The manufacture of solenoids warrants closer attention. Some parts and their interaction make for certain considerations by their manufacturer. Eddy currents can affect the operation of the solenoid. The armature gap and the coil's ability to handle current and voltage are part of the designer's concerns. They are also part of the maintenance person's concern, inasmuch as these factors contribute to the malfunctioning, in time, of the solenoids used to control motors and processes in industry and commerce.

## Eddy Currents in the Armature and Core

Eddy currents are generated any time a varying magnetic field is brought near a piece of metal (Fig. 6-6). Eddy currents can create heat and cause problems. In the ac solenoid note that the core and armature are made of laminated steel. Lamination of the steel core and armature reduce the buildup of heat produced by eddy currents. Eddy currents meet with high opposition in a lamination. Laminations are sheets of metal rather than a solid piece. Direct current solenoids are made with a solid core, inasmuch as the magnetic field varies only when the coil is energized and de-energized.

**Fig. 6-6** *Generation of eddy currents in an open core.*

## Air Gap and the Armature

The armature completes the path of the magnetic lines of force. It is important that the armature matches or fits the rest of the magnetic assembly of the solenoid. A close fit leads to less chattering. Very close tolerances are required in the fitting of the surfaces. However, some problems are associated with the armature being too closely mated to the rest of the magnetic (iron) circuit. Inasmuch as residual magnetism is generated by the electric current producing the magnetic field, it has a tendency not to let go when the current is turned off in the coil. The residual magnetism holds the armature closely even after the power has been removed. That is why a small air gap is left in the iron circuit. This allows the magnetic field to be broken and the armature to drop away after the coil is de-energized (Fig. 6-7).

**Fig. 6-7** *(A) Electromagnet and permanent magnet compared; (B) Air gap shown in center section of the magnet.*

**Fig. 6-8** *(A) Magnet assembly and armature. Note the shading coils. (B) Shading coil current versus magnet coil current.* (Square D)

## SHADING COILS

Shading coils are used on ac solenoids to reduce the chatter produced by the varying electric current through the coil and thus the varying magnetic field produced in the iron circuit. Figure 6-8 shows how the shading coil is placed on the magnetic assembly. The shading coil is a single ring of heavy-duty wire. A moving magnetic field causes current to be induced into this single turn in much the same way as in the secondary of a transformer. The induced current is slightly out of phase with the current that is flowing in the coil circuit. The result is a magnetic field, even though slightly less than the original field, that is slightly behind the rise and fall of the original field produced by the coil. By being slightly behind in its peak, the induced field created by the shading coil has a tendency to fill in when the original drops to zero. This has a tendency to hold the armature close to the magnetic circuit and reduces the chatter that would otherwise result from a varying magnetic field.

## SOLENOID COILS

The coils of solenoids have special needs. They are the source of the magnetic field when energized. Therefore, the amount of current and voltage needed for their operation is a result of their intended use. More current is needed to energize the coil than to keep it energized, especially when the coil is used on alternating current. A coil has inductive reactance that is the product of the inductance of the coil and the frequency of the current times a factor of a $2\pi$. This inductive reactance is measured in ohms ($\Omega$) and in some instances is referred to as impedance (Z). This means that the coil has a tendency to hold back the flow of current after it is energized and pulls in the core material to change its inductance. The inrushing current produces a magnetic field that extends past the first coil of wire into those on top of it, thus inducing an EMF that is in the direction opposite to that which caused it. Inasmuch as alternating current is constantly changing, in most instances 120 times per second (60 Hz), it does not always meet the physical limitations imposed by the inductance of the coil and there is an opposition generated ($X_L$) greater than the resistance of the coil's copper wire.

The inrush current is about 6 to 10 times as great as is needed to hold the armature to the magnetic assembly. Some heat is generated when the coil is energized for a period of time. The coil is rated in terms of volt-amperes (VA). Volt-amperes are the product of the volts times the amperes it takes to cause the coil to energize.

## Sealed Current Rating

If a coil has a VA rating of 300 and 30 VA sealed, the inrush current of the 120-V coils is 300 VA/120 V. This means that it takes a current of 2.5 A to produce 300 VA. If the sealed current of the same coil (that is, the current it pulls when the solenoid is completely closed) is 30/120, the sealed current will be 0.25 A. The same solenoid with a 480-V coil will draw 300/480 or 0.625 A inrush and 30/480 or 0.0625 A sealed. As you can see, the VA rating of the coil can be useful in determining the amount of current drawn at the start and at the hold-in condition of the solenoid.

## Coil Voltage

Voltage ratings of coils are determined by the manufacturer. The number of turns of wire and the size of the wire make a difference in the voltage it takes to energize the coil. Keep in mind also that the amount of current drawn by the coil is determined by the number of turns (dc resistance of the wire) and the core material. Core material and the turns as well as the length and diameter of the coil determine its inductance. The inductance is an important factor when considering ac operation of the coil.

*Pickup voltage* is the lowest voltage at which the coil armature will start to move. *Seal-in voltage* is the lowest voltage required to cause the armature to seat against the pole faces of the magnet. *Dropout voltage* is the voltage at which the armature will start to lose its attraction to the magnet and fall back to its open position (Fig. 6-9).

Pickup voltage is usually higher than seal-in voltage for most solenoids. However, there are some exceptions to this general statement, such as the bell-crank armature solenoid and magnetic assembly. Figure 6-10 shows the types of solenoid construction.

**Fig. 6-9** *Solenoid movement and current draw at various stages of plunger movement.* (Cutler-Hammer)

## Effects of Voltage Variation

Low-control voltage produces low coil currents and reduced magnetic pull. On devices with vertical action assemblies, if the voltage is greater than pickup voltage, but less than seal-in voltage, the controller may pick up but will not seal. With this condition, the coil current will not fall to the sealed value. As the coil is not designed to carry continuously a current greater than its sealed current, it will quickly get very hot and burn out. The armature will also chatter. In addition to the noise, wear on the magnet faces results (Fig. 6-11).

**Fig. 6-10** *Magnetic frame and armature assemblies.* (Square D)

## ALTERNATING CURRENT HUM

All ac devices that incorporate a magnetic effect produce a characteristic hum. This hum or noise is mainly due to the changing magnetic pull (as the flux changes) inducing mechanical vibrations. Contactors, starters, and relays could become excessively noisy as a result of some of the following operating conditions:

1. Broken shading coil
2. Operating voltage too low
3. Wrong coil
4. Misalignment between the armature and magnetic assembly—the armature is then unable to seat properly
5. Dirt, rust, filings, and so on, on the magnetic faces—the armature is unable to seal-in completely
6. Jamming or binding of moving parts; such as contacts, springs, guides, yoke bars; so that full travel of the armature is prevented
7. Incorrect mounting of the controller, such as on a thin piece of plywood fastened to a wall, so that a "sounding board" effect is produced

## REVIEW QUESTIONS

1. When was magnetism discovered?
2. What is a solenoid? How does it operate to control switching action?
3. What does a solenoid valve do?
4. How are eddy currents generated? How are eddy currents minimized?
5. Why are shading coils used on ac solenoids?
6. What is inductive reactance? In what unit is it measured?
7. What does a sealed current rating mean?
8. What effect does low control voltage have on magnetic assemblies?

**Fig. 6-11** *Magnetic starter power circuit.* (Square D)

9. What five conditions may cause excessive hum in a magnetic assembly?
10. What does the *sounding board effect* mean?

## REVIEW PROBLEMS

A solenoid has a coil. The coil has inductance and inductive reactance when used on alternating current. Impedance (Z) is also a factor in some instances when inductive reactance and resistance are both present. A quick review of these two important factors makes you aware of their presence and the need to understand their role in the proper operation of solenoids on alternating current.

$$X_L = 2\pi FL \qquad Z = \sqrt{R^2 + X_L^2}$$

1. What is the reactance of an inductor whose inductance is 6 H when the frequency of the applied voltage is 60 Hz?
2. What is the inductive reactance of a coil on 60 Hz if its inductance is 6 H?
3. What is the inductance of a coil whose resistance is 120 Ω, the reactance is 1721.7184 Ω, and the 120 V has a frequency of 60 Hz?
4. What is the maximum dc current in the inductor in problem 3?
5. What is the inductive reactance of a coil if the frequency of the applied voltage is 50 Hz and the inductance is 10 H?
6. Find the inductive reactance and the impedance of the following combinations connected in series:

|    | F (Hz) | L (H) | R (Ω) | XL | Z |
|----|--------|-------|-------|-----|---|
| a. | 100    | 10    | 5000  |     |   |
| b. | 200    | 10    | 5000  |     |   |
| c. | 300    | 10    | 5000  |     |   |
| d. | 400    | 10    | 5000  |     |   |
| e. | 500    | 10    | 5000  |     |   |

7. Find the inductive reactance and the impedance of the following combinations connected in series:

|    | F (Hz) | L (H) | R (Ω) | XL | Z |
|----|--------|-------|-------|-----|---|
| a. | 60     | 5     | 1000  |     |   |
| b. | 50     | 5     | 1000  |     |   |
| c. | 25     | 5     | 1000  |     |   |

# CHAPTER 7

# Relays

## PERFORMANCE OBJECTIVES

After studying this chapter, you will be able to:

1. Recognize a relay.
2. Identify uses for relays.
3. Describe solid-state relay actions.
4. Identify the condition of relay contacts.
5. Explain how transistors are used for switching circuits.
6. Differentiate between a triac and a diac.
7. Explain how triacs work.
8. Explain how SCRs work.
9. Understand how phase failure relays operate.
10. Describe the actions of a solid-state relay.
11. Define zero-current turn-off and zero-voltage turn-on.
12. Identify various types of solid-state relay switching.
13. Differentiate between load detector and load converter relays.
14. Explain how thermal overload relays operate.
15. Identify the various types of thermal overload relays.

## RELAYS

A relay is a device designed for remote control of another device. The word relay is defined as to pass on. That is what the electromagnetic device does. A relay uses low voltage and low current to cause switching of high voltage or high current, usually at a distant or remote location.

### Relay Solenoids

The relay puts to work the solenoid and its ability to generate an attracting magnetic force. A practical example of the solenoid principle is the relay (Fig. 7-1). One or more sets of contacts are associated with the moving armature of a relay. These serve as electrical contacts and make or break when the relay current is switched on and off.

The relay contacts can be arranged for a variety of functions, such as SPST (single-pole single-throw), SPDT (single-pole double throw), DPDT (double-pole double-throw), or other desired combinations.

The advantage of the relay is the substantial pulling power that can be developed with a small coil current. The contacts themselves can be made quite large and can handle and switch high values of electrical power. An extremely small amount of control power can be used to switch much higher voltages and currents in a safe manner.

### Uses for Relays

Relays are used to control the start and stop operations of small electric motors. Figure 7-2 shows how magnetic frames and armature assemblies are used to control the large currents used in the motor circuits. Figure 7-3 shows a DPDT relay that can be used to remotely control an electric motor or other electrical devices.

### Relay Armature

Note how the armature of the relay is connected to a set of contacts so that when the armature moves to its closed position, the contacts also closed (Fig. 7-2). The electromagnet makes the difference between a remote control possibility and a manual control. The electromagnet

**Fig. 7-2** *Magnetic frame and armature assemblies.*

**Fig. 7-1** *Solenoid used to operate a relay.*

***Fig. 7-3***  *Relay, Double-pole, double-throw.*

consists of a coil of wire placed on an iron core. When current flows through the coil, the iron bar, called the armature, is attracted by the magnetic field created by the current in the coil. To this extent, both will contact the iron bar.

The electromagnet can be compared to the permanent magnet; however, the electromagnet does have the ability to be turned on and off electrically. Interrupting the current flow to the coil or electromagnet causes it to de-energize. This means that it will cause the armature to open and return to its original position, with the contacts no longer making contact. That means that the electrical circuit connected to the contacts is then open or broken. A low voltage and small current can be used to energize the coil. This, in turn, causes the contacts to close and turn on the device connected to the relay contacts circuit. By making certain modifications, it is possible to make some rather useful motor controllers.

## Relay Contacts

If the contacts of a relay are pitted or burned, it calls for a burnishing (polishing) so that they will close tightly against one another and complete the circuit. This can be done with the power off. Place a piece of sandpaper (very fine grain) between the contacts and hold the contacts closed. Then move the sandpaper until it sands down the high points on the contacts. Use an even finer grade and polish the points further. Make sure that you do not get the contacts to the point where they no longer "mate" properly. It may be sufficient to use a tool such as that shown in Fig. 7-4. It should be noted that some manufacturers do not recommend sanding contacts, but simply replacing with a new set.

## SOLID-STATE RELAYS

Solid state refers to relays made with silicon or germanium materials that operate on the same basic principles as transistors and diodes. In most instances the relay is nothing more than a transistor, either PNP or NPN type. In other instances the solid-state relay is a silicon-controlled rectifier (SCR). Of course, the circuit arrangements are such as to allow the switching needed for the relay action. Other features, not readily available in electromechanical relays, are also available in this type of relay. As with everything, there are advantages and some limitations or disadvantages. In most instances the manufacturer will point to the advantages and you have to become aware of the limitations by closely examining the information provided by the manufacturer of the device.

One of the first differences noted between electromechanical and solid-state relays is the absence of a coil and no contacts. The solid-state relay needs very low voltage and current to to do its job of switching. The transistor or SCR do the actual switching and the change in control voltage causes the semiconductor device to conduct or not conduct according to the control voltage applied to its elements. Figure 7-5 is an example of a solid-state relay.

***Fig. 7-5***  *Solid-state relays.*

***Fig. 7-4***  *Relay contact burnishers.*

## Transistor

Inasmuch as the transistor is the device used to do the switching in a solid-state relay; it may be a good idea at this point to take a closer look at its operation

**Fig. 7-6** *Transistors in various case configurations.*

(Fig. 7-6). The transistor is the key element in this type of relay. There are three elements in a transistor: base, collector, and emitter. The *base-to-emitter* voltage of the transistor can control the current flow between the emitter and the collector. In this PNP (positive-negative-positive) type of transistor, a negative voltage on the base allows emitter-base current to flow. This is due to the properties of the silicon-doped material at the junction of the emitter and base. The emitter-base voltage can then cause the transistor to conduct current from the emitter to the collector. A positive voltage on the base and negative on the emitter prevents emitter-base current from flowing, and the transistor stops conducting. This means that it behaves as a closed contact in the first state and as an open contact in the second. This means that the current flow from emitter to collector can be controlled by a small voltage change in the base-emitter connection. There are no moving parts and no contacts to be concerned with at this time. However, there are limitations on how much current the transistor will conduct. The fact that there are no moving contacts, wear or arcing, deterioration or vibration, or dust and dirt damage makes the solid-state relay very much in demand.

## Surge Protection

Solid-state devices are sometimes used with magnetic switches. This means that a voltage transient suppressor may be needed to prevent some of the harmful electrical impulses generated by electromagnetic devices on the line. Figure 7-7 shows a transient suppressor, and Fig. 7-8 shows a transient suppressor mounted on a magnetic relay.

## Triacs

Triacs are also used in solid-state switching. Since the SCR can control current in only one direction, it is limited in its applications. The triac can conduct in both directions. The triac has the same characteristics as the SCR. This means that the triac can be thought of as two SCRs placed in parallel but connected in the opposite direction.

**Triac Construction** Examine the drawing of the triac in Fig. 7-9. The triac has three terminals: main terminal 1 ($MT_1$), main terminal 2 ($MT_2$), and gate (G). By examining the PN structure of the triac it can be seen that current can pass through a PNPN layer, or it can pass through a NPN. The device can be described as having an NPNP layer in parallel with a PNPN layer. This arrangement of four-layer material gives the triac a connection of two SCRs in parallel. This connection is shown in Fig. 7-10. The connection in Fig. 7-10 is not how the triac operates. That is because the triac gate voltage responds differently than the SCR gate voltage. Figure 7-11 shows the schematic diagram of the triac. Because the triac can conduct current in both directions, the schematic diagram contains two diodes facing in opposite directions.

**MANUAL TEST TOOL** — Provides a means of manually switching the contacts of a basic relay or timing relay and holding all contacts in their switched state until the tool is removed. This simplifies the checking of control circuits without power on the coil or contacts............................

**TRANSIENT SUPPRESSOR** — Consists of an R-C circuit designed to suppress coil generated transients to approximately 200 percent of peak voltage. It is particularly useful when switching the TYPE X relay near solid state equipment. It is designed for use on 120VAC coils only............

**Fig. 7-7** *Surge suppressors.*

**Fig. 7-8** *Surge suppressor mounted on top of a magnetic relay.*
(Square D)

**Fig. 7-9** *Layout of materials in a triac.*

**Fig. 7-10** *Two SCRs connected in parallel.*

**Fig. 7-11** *Triac symbol and various case configurations.*

**Triac Applications** Because the triac can conduct in both directions, it is best when used to control ac power. Take a look at the diagram in Fig. 7-12. In this circuit, full power is applied to the load when the gate is triggered on. When $S_1$ is open, the triac cannot conduct. This is because the voltage applied to the triac is below the break-over point. When $S_1$ is closed, the triac is triggered on, and both halves of the ac power are applied to the load. This differs from SCR operation. The SCR can apply only half of the power to the load because it conducts in only one direction. The advantage of all thyristors is that small gate currents can control large load currents.

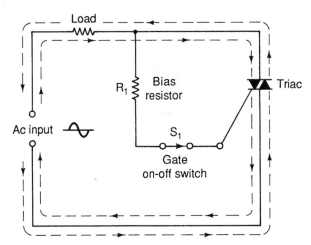

**Fig. 7-12** *A small gate current controls large current through the triac.*

The triac conducts in both directions and requires a small current to operate. It does, however, have some disadvantages compared to the SCR. The SCR has higher current ratings than the triac. The triac can handle currents up to 25 A, whereas the SCR can safely handle currents of around 800 A. That means when large currents are required the SCR is the better choice.

There are some differences in the frequency-handling abilities of the SCR and triac. The triac is usually slower in turning on when used with an inductive load, such as when it is used to control a motor. The triac is also designed to operate mainly in the low-frequency range of 30 to 400 Hz. The SCR can safely handle frequencies up to 30 kHz.

## Silicon-Controlled Rectifiers

The silicon-controlled rectifiers (SCR) is a three-junction semiconductor device that can operate as a switch (Fig. 7-13). It is basically a rectifier. It will conduct current in only one direction. The best part of SCR operation is its ability to be turned on and off. The on-off action makes the SCR very useful in controlling current.

**Solid-State Relays** 117

*Fig. 7-13* SCR symbol and various case configurations.

**Construction of the SCR** Construction details of a component provide information about how the component will operate. Solid-state devices are made by joining P and N material into junction or junctions. Bipolar transistors, diodes, and FETs (field effect transistor) are all constructed this way. The SCR is made by joining four alternating layers of P and N material. Most SCRs are made of silicon, but germanium is also used (Fig. 7-14).

*Fig. 7-14* Layer construction of the SCR.

In Fig. 7-14, note how the layers of P and N material are sandwiched together. Also note the three junctions. Leads for external connections are attached to these layers. These three connections are called the anode (A), cathode (K), and gate (G).

Now take a look at the schematic diagram of the SCR shown in Fig. 7-13. The schematic symbol is about the same as that for a rectifier diode. The main difference is the gate. In some cases the circle around the symbol is not shown. The leads may not be identified on a schematic drawing. When the leads are marked, they are identified with the letters A, K, and G.

**Operation of the SCR** Inasmuch as the SCR is a semiconductor device, it requires a biasing voltage to cause it to turn on. Figure 7-15 shows a simplified arrangement that causes the SCR to operate. A switch is used in the gate circuit to apply voltage to the gate.

*Fig. 7-15* Biasing circuit for an SCR.

Resistor R, is used to limit the current flow in the gate circuit. A second voltage source supplies the needed forward bias to the anode and cathode. A resistor is in series with the anode-cathode circuit. This resistor is also used as a current-limiting resistor. It prevents high currents from causing damage to the SCR. Without the resistor, the SCR conducts hard in forward bias and burns out after a short operating period. A specific gate current must be reached before the SCR will become a conductor. Each SCR has its own break-over voltage. This means that each SCR must have the proper forward bias applied and the proper gate current in order to operate effectively as a switch.

**Alternating Current Operation of the SCR** The SCR can be used to control dc or ac. Since it is a rectifier, it operates on only one ac alternation. The SCR conducts only when the input cycle makes the anode positive and the cathode negative (Fig. 7-16). By

*Fig. 7-16* SCR used to control voltage to RL (load resistor).

**118 Relays**

closing S, a positive voltage is developed that turns the SCR on. The series resistor is in the gate circuit for current-limiting purposes. The diode is in the circuit to protect the anode and cathode during reverse voltage operation.

If S is closed, the SCR conducts when the proper polarity appears at the anode. If the gate switch is opened, the SCR continues to conduct until the voltage between the anode and cathode falls below the break-over voltage. Once the voltage falls below this level the SCR remains off until S is closed again.

There are many advantages to the use of an SCR over an electrical switch. For instance, the SCR will not wear out. It will not develop contact arcing, nor will it stick in one position. This means that the SCR is a more reliable component than the mechanical switch, especially in high current applications. The SCR may be controlled by a switch, or it may be controlled by an electrical pulse from a computer. The most important characteristic is that a small amount of power applied to the gate controls large amounts of current to a load.

## PHASE-FAILURE RELAYS

Interchanging any two phases of a three-phase induction motor power source causes the motor to reverse its direction of rotation. This is called *phase reversal*. In elevators and other industrial operations, equipment damage and personnel injuries may result when a phase reversal occurs unexpectedly. This can happen if a fuse blows or a wire to a motor breaks while the motor is running. The motor continues to operate on single phase but that causes serious overheating. Phase failure and phase-failure reversal relays are used to protect motors against these situations.

Both voltage-sensing and current-sensing phase failure relays are available. Voltage-sensing relays may be connected at any point on the lines, but only detect abnormal conditions ahead of the point of connection. Voltage sensing offers the advantage of being able to detect abnormal conditions independent of motor running status. They are also easy to apply since motor voltage is all that is required to select the relay. Figure 7-17 shows line-side monitoring. With the relay connected before the starter, the motor cannot be started in the reverse direction. However, the motor is unprotected against phase failures between the relay and the motor.

Phase-failure relays are often used to control a shunt trip circuit breaker. When this is done, care must be taken to ensure that the shunt trip circuit always has an adequate source available. This can be accomplished by using the diagram shown in Fig. 7-18.

**Fig. 7-17** *Line side monitoring. (Square D).*

**Fig. 7-18** *Interfacing phase failure relays with a shunt trip circuit breaker. (Square D)*

When a phase failure occurs on $L_2$ or $L_3$, the shunt trip coil draws power from $L_1$ through the control relay (CR) contacts and phase-failure relay contacts (which change state on detecting a phase failure). If a phase failure occurs on $L_1$, the CR contacts change state. The shunt trip coil draws power from $L_2$ through the CR contacts and phase-failure relay contacts.

If the control relay contacts, the phase-failure relay contacts, or the shunt trip coil does not have the same voltage rating as the motor, control transformers may be interposed where needed.

Load-side monitoring is shown in Fig. 7-19. With the relay connected directly to the motor, the total feed lines are monitored. The motor may sustain a momentary bump in the reverse direction with this connection.

**Fig. 7-19** *Load-side monitoring. (Square D)*

**Fig. 7-20** *Current monitoring. (Square D)*

Current-sending relays require three externally connected current transformers. These must be sized to match the motor full-load current. Current sensing offers the advantage of being able to detect imbalances more precisely by monitoring currents. Relay selection is independent of motor voltage and requires a separate 120-V source or supply.

Three-wire control is necessary with current control to prevent the relay from "cycling" on and off when an open phase or phase reversal occurs. See Fig. 7-20. Due to the current-sensing features of this relay, the load must first draw current before a reverse-phase condition can be sensed. This means that this relay is not used for protecting motors driving equipment unable to tolerate a momentary "bump" in the reverse direction.

Figure 7-21 shows three types of phase-failure relays. These offer the reliability and accuracy of solid-state sensing circuitry with the isolation of hard output contacts. NIPS is a phase-failure and under voltage relay. DAS is a phase-failure relay. DAS and MPS are used by Square D to designate this type of relay.

**Fig. 7-21** *Phase failure relays with wiring diagrams and contact connections. (Square D)*

120   **Relays**

# SOLID-STATE MONITORING RELAYS

Solid-state monitoring relays use solid-state sensing circuitry with the isolation of hard output contacts. This type of monitoring is both accurate and reliable. Figure 7-22 shows the enclosure of several types of monitoring relays. All of these relays have an LED indication when the relay is energized.

*Fig. 7-22* Solid-state monitoring relay. (Square D)

## Voltage Relay

The relay monitors ac single-phase and dc (independent of polarity) voltages. The relay has independent adjustable controls for both pickup and dropout voltage. The relay energizes when the supply voltage is present and the monitored voltage is above the pickup setting. The relay de-energizes when the supply voltage is removed or the monitored voltage is below the dropout setting. The dropout voltage is adjustable from 50 to 95% of pickup voltage. Figure 7-23 shows how the relay is wired in the circuit and how its contacts operate.

In electromechanical type of relays the pull-in voltage means that the voltage necessary to cause the relay coil to energize is enough to pull the armature to the coil core and thereby close the contacts attached to the armature. In the semiconductor or solid-state type of relay it refers to the voltage needed to cause proper biasing of the transistor or semiconductor device to energize or conduct, thereby lowering its emitter-collector resistance or forward conduction resistance of the SCR or triac.

The term *dropout voltage* means the same in both electromechanical and semiconductor solid-state relays. The dropout voltage is lower than the voltage it takes for the relay to energize. In other words, the relay will continue to be energized between the pull-in and dropout voltages once it is energized. However, once the voltage has reached a point where there is not enough current flowing to cause a sufficient magnetic field to hold the armature to the core, it will open. In the case of the solid-state relay, it is the point at which the device stops conducting or turns off. The solid-state relay has no armature to pull in or drop out. It takes very little difference in voltage to cause the voltage monitoring relay to drop out or stop conducting. It also makes this type of relay very useful in operating with computer signals. It is very useful in digitally controlled logic circuits.

## Current Relay

This relay is almost identical to the voltage monitoring relay except that it monitors the variations in current. It is also hooked up differently (Fig. 7-24). The relay monitors ac single-phase and dc (independent of polarity) currents. The relay has independent adjustable controls for both pickup and dropout current. The relay energizes when the supply voltage is present and the

**Type DUA**

**NOTES:** $V_s$ represents supply voltage
$V_m$ represents monitored voltage
$I_m$ represents monitored current
$J$ represents externally connected thermistor
$S_k$ represents monitored contacts
Dashed lines represent optional contacts
Relays are shown in de-energized position

*Fig. 7-23* Voltage relay. (Square D)

**Type DIA**

*Fig. 7-24* Current relay. (Square D)

monitored current is above the pickup setting. The relay de-energizes when the supply voltage is removed or the monitored current is below the dropout setting. The dropout current is adjustable from 50 to 95% of the pickup current.

## Over/Under Relay

This type of relay monitors single-phase voltages and requires no additional supply voltage (Fig. 7-25). Over voltage is adjustable from 100 to 110% of nominal voltage. Under voltage is adjustable from 80 to 100% of nominal voltage. The relay energizes when voltage is between these two settings.

**Fig. 7-25** *Over/under voltage relay. (Square D)*

**Zero-Current Turn-Off** One of the unique features of a semiconductor or solid-state relay is its ability to automatically turn off when the ac load current sine wave is crossing its zero axis. This property of a solid-state device is very important when switching inductive loads, since the turn-off occurs when the current is at its minimum point on the sine wave. This reduces the inductive kickback that often sustains an arc in the electromechanical relay contacts. Inductive kickback increases voltage spikes that can damage semiconductors.

**Zero-Voltage Turn-On** Zero-voltage turn-on is not necessarily available on solid-state relays. It is one condition that can be added to a solid-state relay to provide certain features that will greatly extend the life of some types of loads.

## TYPES OF SOLID-STATE RELAY SWITCHING

The three types of relay switching that can be accomplished by the solid-state relay are the instant on (IO), the universal (US), and the zero switch (ZS). Each of these relays is designed to switch on at different locations on the sine wave and for good reasons. The type of load—resistive, inductive, lamp, or a combination of these—determines the type of relay to select for the job.

The instant-on (IO) type turns on *immediately* as soon as the control input is switched. That means that it can turn on at any point on the sine wave. However, the turn-off of the instant-on type is always at the zero point where the current crosses the zero point of the sine wave (Fig. 7-26).

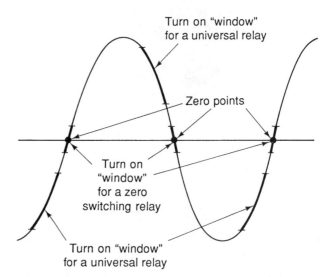

**Fig. 7-26** *Zero points and universal relay turn-on windows.*

The universal-type solid-state relay turns on within a given *window* on the ac sine wave (Fig. 7-26). The universal relay does not turn on at the zero point or at the peak of the sine wave. This type of relay can turn on only inside the given window. The universal relay turns off at zero current, just as in the other two basic types of solid-state relays.

The zero switch has an advantage in handling the turn-on surges so often associated with a circuit. This type of relay monitors the load circuit to make sure that it turns off when the sine wave passes through the zero point on the sine wave.

## Switching Relay Loads

Relays are designed to take into consideration the characteristics of inductive loads, resistive loads, and variations of these two types. One of the advantages of the solid-state relay is its ability to turn on or off at the zero point of the sine wave. Some devices benefit more than others if switched at these points. For instance, *resistive loads* such as light bulbs have a low resistance when cold and higher resistance when hot. If they are turned on cold, they can draw large amounts of current, at least more than normal operating current, which will shorten their life. When turned on at the zero point it gives the filaments time to heat up and change resistance slightly and keep the surge down to a manageable point, and in turn extends the life of the bulb or lamp. Zero-point switching also eliminates noise spikes. This means that a zero-switching relay should be used with lamps and other types of resistive loads.

Other types of load do not necessarily respond well to zero switching. *Inductive loads* are one type that need an instant-on type of relay. This is due to the nature of an inductive load presented by a motor, transformer, or solenoid coil. With an inductive load the current and voltage are nearly 90° out of phase. This means that the current is lagging the voltage by this amount and that there is no inductive reactance to increase the opposition to current flow at the instant when starting voltage is applied. This also means that there is a large current draw when the voltage is at the zero-crossing point of the sine wave. Switching at any point other than zero is preferable for an inductive load.

Loads consisting of lamps are also resistive, but the inrush current is high (up to 15 times normal operating currents) due to the low resistance of the cold filaments. This high inrush current can do great damage to relays. The main concern here is the inrush current at the start of operation. That means that a zero switching relay can be of some advantage with this type of load. However, the inrush currents should be taken into consideration when choosing the proper relay for the job.

When there is a combination of resistive and inductive loads in a circuit to be switched, it is best to use the universal relay. This type of relay does not switch at the zero point or at the peak point of the sine wave. Therefore, it is often selected for a combination load.

## Thermistor Relay

The thermistor relay operates from the signal of externally connected thermistors. See Fig. 7-27. The relay energizes when the supply voltage is present and the thermistor resistance is below 1.9 k$\Omega$ (range of 1.5 $\Omega$ to 2.3 $\Omega$). The relay de-energizes when the supply voltage is removed or the thermistor resistance increases above 3 k$\Omega$ (range of 2.5 to 3.6 k$\Omega$).

## Contact Amplifier Relay

Contact amplifier relays are used in cases where contacts do not have sufficient current and voltage ratings to switch loads such as coils, solenoids, or small motors. Typical examples are manometer contacts or supervisory relays for broken wire at wire-producing machines (Fig. 7-28).

**Fig. 7-28** *Contact amplifier relay. (Square D)*

## Load Detector and Load Converter Relays

The load on an induction motor can be monitored with the right choice of solid-state circuitry. The load detector relay and the load converter relay shown in Fig. 7-29 are used to control equipment and processes

**Fig. 7-27** *Thermistor relay. (Square D)*

**Fig. 7-29** *(A) Load detector and (B) Load converter relays. (Square D)*

by monitoring the load of an induction motor. The devices simulate the true output power of the motor by measuring the input voltage, current, power factor, and compensating for internal losses of the motor. Using these parameters, the output power is computed continuously. In an induction motor, the current varies as the input voltage varies. By computing power, the load relays are not affected by these variations and can provide much greater accuracy over a wider range of load than is possible by measuring current alone. Three relay models are available. One type provides both a maximum and a minimum trip point, a second type provides two maximum trip points, and a third type provides an analog current output.

Three types of load detectors are available. The type V load detector monitors the motor load and has two separate output relays to indicate when a trip point has been exceeded. On the type V3, the two output relays correspond to a maximum and minimum trip point. On the type V4, the two output relays correspond to two maximum trip points. Two separate pushbutton thumbwheel switches, located on the face of the device, are used to select the trip points. The load detector has two SPDT relay outputs, to indicate maximum or minimum trip points exceeded. The unit is also adjustable for startup delay, response delay, and compensation of motor losses. LEDs indicate power on, maximum trip, or minimum trip.

## Typical Applications

Either the type V3 load detector or the type G load converter is used in a crusher-conveyor situation (Fig. 7-30). The load converter can monitor the crusher load and feed the output signal to an adjustable-frequency drive controlling the feeder conveyor motor. In this way the load on the crusher is kept constant by varying the speed of the conveyor. If an adjustable drive is not used, the load detector starts and stops the conveyor motor to keep the load on the crusher within preset limits.

**Fig. 7-30** *Crusher-conveyor with load detector.* (Square D)

**Fig. 7-31** *Machine tool bit load detector.*

The type V load detector has a monitoring resolution fine enough to determine worn tools on machinery. See Fig. 7-31. When monitoring a drill motor, the drill is stopped when the drill bit is dull. This extends the life of tools and prevents breakage problems. By continuously monitoring the power consumption of a fan motor, an accurate indication of the system status can be obtained. The V load detector monitors the loading on three-phase motors and controls and protects fan systems. See Fig. 7-32.

**Fig. 7-32** *Load detector senses closed dampers.* (Square D)

The load detector can simultaneously monitor four common problems that may occur in heating, ventilating, and air-conditioning systems: broken or loose fan belts, closed dampers, blocked filters, and mechanical wear on the motor (such as bearings).

Figure 7-33 shows how the load detector and the load converter are connected in a circuit. Figure 7-34 shows the outputs of the type V3 unit that can be used to disconnect a motor when load limits are exceeded. In a three-wire control scheme, normal operating mode, with an alarm signal, the relay is wired as shown.

The typical load converter monitors the motor load and generates a 0- to 20-mA or 4- to 20-mA output signal proportional to the load on the motor. An integration adjustment on the face of the device smoothes the output signal from short-term variations in the input signals. Two separate pushbutton thumbwheel switches on the face of the device are used to select the zero point of the range and the span. The output signal can

**Fig. 7-33** *Load detector and load converter hookups.* (Square D)

**Fig. 7-34** *Three-wire control and load detector circuit.* (Square D)

be selected to increase with load or decrease as load increases. An LED indicates when power is applied to the device.

## THERMAL OVERLOAD RELAYS

Thermal overload relays sense motor current by converting this current to heat in a resistance element. The heat generated is used to open a normally closed contact in series with a starter coil causing the motor to be disconnected (Fig. 7-35).

The thermal overload relay is simple and inexpensive. It is very effective in providing motor-running over current protection. This is possible because the most vulnerable part of most motors is the winding insulation, and this insulation is very susceptible to damage by excessively high temperature.

Being a thermal model for a motor, the thermal overload relay produces a shorter trip time at higher current similar to the way in which a motor reaches its temperature limit in a shorter time at a higher current. In a high ambient temperature, a thermal overload relay trips at a lower current, or vice versa, allowing the motor to be used to its maximum capacity in its particular ambient temperature (if the motor and overload are in the same ambient).

After it is tripped, the thermal overload relay does not reset until it has cooled. This allows the motor to cool before it can be restarted.

### Types of Thermal Overload Relays

There are two types of thermal overload relays: bimetallic and melting alloy. In some types the bimetallic is available in both non-compensated and ambient-temperature-compensated versions. In both melting alloy and bimetallic, single-element and three-element overloads are available. See Fig. 7-36 for a cutaway view of a standard trip melting alloy thermal unit. With the exceptions of a few types, all thermal overloads incorporate a trip-free reset mechanism that allows the relay to trip on

**Fig. 7-35** *Thermal overload relays with replaceable contacts.* (Square D)

**Fig. 7-36** *Cutaway view of standard trip melting-alloy thermal unit.* (Square D)

an overload even though the reset lever is blocked or held in the reset position. This mechanism also prevents the control circuit contact from being reclosed until the overload relay and the motor have cooled.

## Hand-Reset Melting Alloy (NEMA Style)

Hand-reset melting alloy overload relays use a eutectic alloy solder that responds to the heat produced in a heater element by the motor current. When tripped, the overload relay is reset manually after allowing a few minutes for the motor and the relay to cool and the solder to solidify.

Repeated tripping does not affect the original calibration. Melting-alloy thermal units are available in three designs: quick trip, standard trip, and slow trip. Quick trip (class 10) units are used to protect hermetically sealed, submersible pump, and other motors that can endure locked rotor current for a very short time, or motors that have a low ratio of locked rotor to full-load current. Standard trip (class 20) units provide trip characteristics for normal motor acceleration up to approximately 7 seconds on a full-voltage start. Slow trip (class 30) units provide trip characteristics for motor acceleration up to approximately 12 seconds on a full-voltage start. The motor should be suitable for extended starting periods.

**Overload Relay Class Designation**

Class 10  Relay will trip in 10 seconds or less at a current equal to 600% of its current rating. It is used with hermetic motors, submersible pumps, or motors with short locked-rotor time capability.

Class 20  Relay will trip in 20 seconds or less at a current equal to 600 times its current rating.

Class 30  Relay will trip in 30 seconds or less at a current equal to 600 times its current rating.

General applications call for a class 20 thermal relay.

## Bimetallic Overloads (NEMA Style)

Bimetallic overload relays are used where the controller is remote or difficult to reach. Three-wire control is recommended when automatic restarting of a motor could be hazardous to personnel.

Normally, bimetallic relays are used on automatic reset. They are supplied from the factory on hand reset, but can be adjusted for either hand or automatic reset in the field. When used on hand reset, allow the motor and thermal units a few minutes to cool before resetting.

## Temperature Compensation

Ambient temperature compensation is available on some overload relays (Fig. 7-37). These relays have all the features of the non-compensated bimetallics. In addition, an extra bimetal element maintains a nearly constant trip current in relay temperatures from $-20$ to $+165°F$ for one type and $-4$ to $+131°F$ for the other type. Trip current is adjustable from 85 to 115% of the trip current ratings. A SPDT contact is standard on the 25- and 45-A sizes. The NO contact can be used in an

*Fig. 7-37* Bimetallic overloads.

alarm circuit and must be wired on the same polarity as the NC contact. Contacts are not replaceable.

**Protection Level** Protection level is the relationship between trip current rating and full-load current. Protection level is in percent and is the trip current rating divided by the motor full-load current times 100. Check with the manufacturer of the thermal relay (usually listed with instructions in tables in the back of the catalog) to make sure that the correct unit is selected for the job the motor is expected to do.

Thermal units are not included with any overload relays. They are selected and ordered and priced separately. Ideally, thermal units should be selected from the instruction sheet that is included with every starter or overload relay. If it is desirable to order thermal units along with the controller, they should be selected from a catalog with specifications listed and based on the type of controller being ordered and the nameplate full-load current of the motor. If the motor full-load current is not known at the time thermal units are ordered, an approximate selection can be made using Table 7-1 as follows:

1. Locate motor horsepower and voltage.
2. Determine approximate full load current from the table.
3. Use approximate full-load current in place of actual nameplate full-load current.

*Table 7-1* Approximate Thermal Unit Selection Based on Horsepower and Voltage[a]

| Motor Horse-power | Motor Full-Load Current | | | | | |
|---|---|---|---|---|---|---|
| | Three Phase | | | | Single-Phases | |
| | 200 V | 230 V | 460 V | 575 V | 115 V | 230 V |
| $\frac{1}{20}$ | 0.39 | 0.34 | 0.17 | 0.14 | 1.30 | 0.65 |
| $\frac{1}{12}$ | 0.55 | 0.48 | 0.24 | 0.19 | 1.90 | 0.95 |
| $\frac{1}{8}$ | 0.74 | 0.64 | 0.32 | 0.26 | 2.60 | 1.30 |
| $\frac{1}{6}$ | 0.90 | 0.78 | 0.39 | 0.31 | 3.24 | 1.62 |
| $\frac{1}{4}$ | 1.22 | 1.06 | 0.53 | 0.42 | 4.40 | 2.20 |
| $\frac{1}{3}$ | 1.52 | 1.32 | 0.66 | 0.53 | 5.47 | 2.74 |
| $\frac{1}{2}$ | 2.07 | 1.80 | 0.90 | 0.72 | 7.45 | 3.73 |
| $\frac{3}{4}$ | 2.88 | 2.50 | 1.25 | 1.00 | 10.1 | 5.07 |
| 1 | 3.68 | 3.20 | 1.60 | 1.28 | 12.6 | 6.31 |
| $1\frac{1}{2}$ | 5.18 | 4.50 | 2.25 | 1.80 | 17.2 | 8.59 |
| 2 | 6.67 | 5.80 | 2.90 | 2.32 | 21.4 | 10.7 |
| 3 | 9.66 | 8.40 | 4.20 | 3.36 | 29.1 | 14.5 |
| 5 | 15.4 | 13.4 | 6.68 | 5.35 | 42.9 | 21.4 |
| $7\frac{1}{2}$ | 22.6 | 19.6 | 9.82 | 7.86 | 58.4 | 29.2 |
| 10 | 29.7 | 25.8 | 12.9 | 10.3 | | 36.3 |
| 15 | 43.6 | 38.0 | 19.0 | 15.2 | | 49.4 |
| 20 | 57.4 | 49.9 | 24.9 | 20.0 | | |
| 25 | 70.9 | 61.7 | 30.8 | 24.7 | | |
| 30 | 84.3 | 73.3 | 36.7 | 29.3 | | |
| 40 | 111 | 96.4 | 48.2 | 38.5 | | |
| 50 | 137 | 119 | 59.6 | 47.6 | | |
| 60 | 163 | 142 | 70.8 | 56.6 | | |
| 75 | 201 | 175 | 87.6 | 70.0 | | |
| 100 | 265 | 230 | 115 | 92.0 | | |
| 125 | 327 | 284 | 142 | 114 | | |
| 150 | 389 | 338 | 169 | 135 | | |
| 200 | 511 | 445 | 222 | 178 | | |

*Source:* Courtesy of Square D

[a]Use only when motor full-load current is not known. Thermal units selected using approximate full-load currents from the table will provide a trip current between 100 and 125% of full-load current for many four-pole, single-speed, normal-torque, 60-Hz motors. Since the full-load current rating of different makes and types of motors vary so widely, these selections may not be suitable. Thermal units should be selected on the basis of motor nameplate full-load current and service factor. Thermal unit sizes originally selected on an approximate basis should always be rechecked and corrected at the time of installation if required.

*Note:* These currents should not be used for selection of fuses, circuit breakers, or wire sizes. See NEC tables 430–148 through 430–150. For motor rated 208 to 220 V, use the 230 V column. For motors rated 440 to 550 V, use 460 and 575 V columns, respectively.

**Thermal Overload Relays**

**Fig. 7-38** *Ambient temperature correction curve for thermal overload relays.*

Ambient-temperature correction curves for thermal overload relays are given in Fig. 7-38.

**Mounting Thermal Units** Always be certain the correct thermal units are installed in the starter before operating a motor. Thermal units should always be mounted so that their type designation can be read from the front of the starter (Fig. 7-39). Melting-alloy thermal units should be mounted so that the tooth of the pawl assembly can engage the teeth of the ratchet wheel when the reset button is pushed. Mounting surfaces of the starter and thermal units should be clean. Make sure that the thermal unit mounting screws are securely fastened.

# ELECTROMAGNETIC RELAYS AND MOTORS

Relays are a necessary part of many control and pilot light circuits. They are similar in design to contactors, but are generally lighter in construction, so they carry smaller currents.

Magnetic contactors are normally used for starting poly-phase motors, either squirrel cage or single phase. Contactors may be connected at any convenient point in the main circuit between the fuses and the motor. Small-diameter control wires may be run between the contactor and the point of control.

Protection of the motor against prolonged overload is accomplished by time-limit overload relays that are operative during the starting period and running period. Relay action is delayed long enough to take care of heavy starting currents and momentary overloads without tripping.

Motors for commercial condensing units on refrigeration or air-conditioning systems are normally protected by a metallic switch operated on the *thermo*, or heating principle. This is a built-in motor overload protector. It limits the motor winding temperature to a safe value. In its simplest form, the switch or motor protector consists essentially of a bimetal switch mechanism that is permanently mounted and connected in series with the motor circuit (Fig. 7-40).

When the motor becomes overloaded or stalled, excessive heat is generated in the motor winding due to

**Fig. 7-39** *Operation of melting-alloy overload relay.*

**Fig. 7-40** *Circuit for a domestic refrigerator with start relay.*

the heavy current produced by this condition. The protector located inside the motor is controlled by the motor current passing through it and the motor temperature. The bimetal element is calibrated to open the motor circuit when the temperature, as a result of excessive current, rises above a predetermined value. When the temperature decreases, the protector automatically resets and restores the motor circuit.

## Motor Winding Relays

Following the thermal overload protector in the circuit (Fig. 7-40) is the motor start relay. A motor winding relay is usually incorporated in single-phase motor compressor units. This relay is an electromagnetic device for making and breaking the electrical circuit to the start winding. A set of normally closed contacts is placed in series with the motor start winding (Fig. 7-40).

The electromagnetic coil is in series with the auxiliary winding of the motor. When the motor start and run windings are energized a fraction of a second later, the motor comes up to speed and sufficient voltage is induced in the auxiliary winding to cause current to flow through the relay coil. The magnetic force created by the current through the coil is sufficient to attract the spring-loaded armature, which mechanically opens the relay starting contacts. With the starting contacts open, the start winding is out of the circuit. The motor continues to run only on the run winding. When the control contacts open, power to the motor is interrupted. This allows the relay armature to close the starting contacts. The motor is now ready to start a new cycle when the control contacts again close.

A hermetic compressor motor relay is an automatic switching device designed to disconnect the motor start winding after the motor has attained a running speed. There are two types of motor relays used in refrigeration and air conditioning compressors: the *current-type* relay and the *potential-type* relay.

**Potential-Type Relay** The potential-type relay is generally used with large commercial air-conditioning compressors. The motors may be capacitor start-capacitor run types up to 5 hp. Relay contacts are normally closed. The relay coil is wired across the start winding. It senses voltage change. Start winding voltage increases with motor speed. As the voltage increases to the specific pickup value, the armature pulls up, opening the relay contacts and de-energizing the start winding. After switching, there is still sufficient voltage induced in the start winding to keep the relay coil energized and the relay starting contacts open. When power is turned off, the voltage drops to zero. That means that the coil is de-energized and the start contacts reset (Fig. 7-41).

**Fig. 7-41** *Potential-type relay used in refrigeration circuits.* (Tecumseh Products)

Many of these relays are extremely *position sensitive*. When changing a compressor relay, care should be taken to install the replacement in the same position as the original. Never select a replacement relay solely by horsepower or other generalized rating. Select the correct relay from the parts guidebook furnished by the manufacturer.

**Current-Type Relay** The current-type relay is generally used with small refrigeration compressors up to 4 hp. When power is applied to the compressor motor, the relay solenoid coil attracts the relay armature upward. This causes the bridging contact and stationary contact to engage (Fig. 7-42). This energizes the motor start winding. When the compressor motor attains running speed, the motor main winding current is such that the relay solenoid coil de-energizes. This allows the relay contacts to drop open, thereby disconnecting the motor start winding.

**Electromagnetic Relays and Motors** 129

**Fig. 7-42** *Current type relay used in refrigeration circuits.* (Tecumseh Products)

## ELECTROMECHANICAL RELAYS

The electromechanical relay is available in a number of different arrangements and have any number of names to indicate their purpose. The general-purpose relay can perform a number of operations in a variety of locations.

The control relay is designed specifically for use as a machine tool relay (Fig. 7-43). This relay is designed to handle the logic switching requirements of machine tools, conveyors, hoists, elevators, cranes, tire machines, and practically every type of motor-driven machinery.

The *control relay is* electromagnetically operated and held. Energizing of the magnet coil causes the normally open contacts to close and the normally closed contacts to open. De-energizing of the coil causes the contacts to switch back to their original state.

The latching relay, the control relay, and the timing relay work together to form a logic system for making automated plants possible. The control relay is designed to switch inductive and resistive loads in both ac and dc circuits. By far the greatest number of applications involve the switching of inductive loads in ac circuits. Typical loads are operating coils on such devices as other relays, timers, starters, contactors, and solenoids.

**Fig. 7-43** *Control relay.* (Square D)

The latching relay is electromagnetically operated and is held by means of a mechanical latch. Energizing the latching coil causes the normally open contacts to close and the normally closed contacts to open. The mechanical latch holds all contacts switched, even after power is removed from the latching coil. Energizing of a second coil, the unlatching coil, results in all contacts switching back to their original state. Both coils are continuously rated and require no coil clearing contacts. If the latching and unlatching coils are energized at the same time, the latching coil will override the contacts and the contacts will go to, and remain in, their switched state. Figure 7-44 shows the conventional two-coil circuit.

**Fig. 7-44** *Two-coil circuit.* (Square D)

The latch attachment has its own coil and is mounted on the control relay (Fig. 7-45). The mechanism fits on any two- to eight-pole relay and can be mounted on the relay in the field even after the relay contacts and coil are wired. The latching attachment has a self-adjusting feature that adapts the stroke of the latch to the stroke of the relay on which it is mounted to provide optimum performance.

The solid-state timer is recommended for use in control system work, with the latching relay. The solid-state timer is useful in extremely high duty cycles. This type of relay is discussed in chapter 9.

**Fig. 7-45** *Latching relay. Note the latch attachment on the front of the control relay.* (Square D)

# RELAY OPERATING CHARACTERISTICS

Ratings of relays are important. The *make* rating, the *break* rating, and the *continuous* rating have to be taken into consideration when the relay is specified.

**Resistive Ratings** This indicates the resistive load that the contacts can make, break, or carry continuously. Resistive ratings are based on 75% power factor.

**Inductive Ratings** The inductive rating refers to loads, such as coils of contactors, starters, relays, and solenoids, that the contacts can make, break, and carry continuously. Inductive rating tests are run with 35% power factor loads.

**Make Ratings** This rating applied to the current that can be handled by the contact at the time of contact closure. In inductive ac circuits, the momentary inrush current is often 10 times the sealed current, and a relay must be able to handle this inrush current as well as be able to break it in an emergency.

**Break Ratings** This rating applies to the current that can be interrupted successfully by the contact. The inductive break rating is always less than the resistive or continuous ratings. When contacts break an inductive circuit, the inductance in the load tends to maintain the current. The result is an arc across the contact that causes heating and erosion of the contacts. Because of the extra heat generated, the allowable inductive current must be less than the resistive current for equal contact life.

**Continuous Ratings** Continuous rating indicates the load that the contacts can carry continuously without making or breaking the circuit and without exceeding a certain temperature rise.

## Contact Life

The life of the control relay contact depends on the magnitude and characteristics of the electrical load, inductance, duty cycle, mechanical properties of the device in which they are used, voltage fluctuations, and environment.

When control circuit relays are operated at maximum rated load, the life of the contacts is usually less than that of the remainder of the device. If the application requires a large number of operations during the life of the contacts, the contacts must be applied at values less than their maximum make and break ratings.

NEMA (National Electrical Manufacturers Association) standards recommend that control relays for automatically operated sequencing systems be utilized with loads of less than 25% of the 60-A make and 6-A break ratings. NEMA standards do not recommend using a relay at its maximum ampere rating where the number of operations are expected to exceed substantially the 6000 operations required by the NEMA endurance test.

## Contact Construction

The relay uses a double-break contact (Fig. 7-46). This places, for practical purposes, two single breaks in series, so that two arcs occur when the contact interrupts the current flow. This division of energy in the arc materially extends the electrical life of the contact when compared to devices using single-break contacts. The stationary movable contacts are made of silver cadmium oxide material. This choice of material is important because of its resistance to welding when closing on the inrush currents normally associated with inductive loads. It also helps to reduce the contact erosion associated with repeated interruption of inductive circuits. Note how the contacts are constructed in Fig. 7-47. They provide two parallel paths per pole. The fact that both halves of the movable finger are not rigidly connected assures that all four contact points are held closed with nearly equal force. A conductive crossover saddle (Fig. 7-48) straddles the two fingers to provide a crossover path for even greater reliability.

***Fig. 7-46*** *Arcing between relay contacts. (Square D)*

***Fig. 7-47*** *Relay contacts. (Square D)*

***Fig. 7-48*** *Relay contacts with a saddle. (Square D)*

An ohmmeter test on a contact is unreliable because the most common voltage source for an ohmmeter is a 1.5-V dry cell, and to contacts being tested, this is a low-energy circuit load. Several ohmmeter readings of the same contacts may work perfectly well with a relay coil load.

## Relay Coils

Coils for relays are designed to operate satisfactorily on voltages varying as much as 15% below and 10% above the nominal values. The nominal value is the value stamped on the coil.

Most coils are stamped with a part number, operating voltage, frequency, and date code. In most relays the external part of the coil is pressure-molded epoxy (Fig. 7-49). This forms a dense protective cover that provides high strength and resistance to mechanical damage. Construction of this type also provides good heat transfer for better cooling. Moisture absorption is also reduced. Coils for control relays and latch relays are designed so that they may be operated continuously without overheating.

**Fig. 7-49** *Exploded view of control relay.* (Square D)

**Transients** The coil is relatively unaffected by switching transients. That is, because the pickup time of the relay is about 9 milliseconds, a short transient voltage is very unlikely to cause false switching of the output contacts. High-quality coil insulation plus low voltage differential between turns makes failure due to shorted turns, resulting from transients, also quite unlikely. However, when a relay coil is de-energized, a transient is generated by the relay coil that may interfere with proper operation of nearby solid-state equipment. An optional transient suppressor is available to suppress transients to approximately 200% of peak voltage (Fig. 7-7).

## Shock and Vibration

Shock and vibration show up in the form of contact bounce, false switching of contacts due to armature travel, or mechanical breakage. Most relays are designed with vibration and shock in mind, inasmuch as this is part of the operating environment of the relay. They are given consideration and compensated for by high contact pressure along with low-mass movable contacts.

## Relays and Altitude

Higher altitudes produce an atmosphere with less pressure. The low pressure in itself does not affect the relay directly. However, the higher altitudes tend to reduce the insulation value of the air as well as its cooling effect. As a result it is necessary to consider the possible need to derate electrical equipment when it is used at high altitude. The relay should be derated when used at altitudes above 6000 ft but not more than 15,000 ft. This calls for derating the relay to 75% of its normal rating.

## REVIEW QUESTIONS

1. Describe a relay.
2. What do SPST, SPDT, and DPDT mean?
3. List the parts of a relay.
4. What does *burnishing* mean?
5. What are two types of solid-state relays?
6. What are some of the advantages of solid-state relays?
7. What is a triac? Draw the symbol for a triac.
8. What is the difference between an SCR and a triac?
9. Which can handle higher currents, a triac or an SCR?
10. Draw the symbol for an SCR and label the leads.
11. Describe briefly, the operation of an SCR.
12. What is phase reversal?
13. What is dropout voltage?
14. What does an over/under relay do?
15. Why is zero-current turn-off an advantage to the solid-state relay?

16. What makes the universal-type solid-state relay different?
17. Why does an inductive load need an instant-on relay?
18. Explain how the thermistor relay operates.
19. Where are contact amplifier relays useful?
20. Where are load detectors utilized?
21. What are the two types of thermal overload?
22. What are the three designs of thermal trip overload units?
23. What does the overload relay trip class designation mean?
24. What is the difference between the operation of the current relay and the potential relay?
25. What three types of relays make it possible to have a logic system in automated plants?
26. Describe the make rating and break rating of relay contacts.
27. Why are transient suppressors needed on relay coils?
28. How does high temperature affect relay operation?
29. Does the latching relay hold all contacts closed even after power is removed?
30. What does *inductive rating* mean?
31. What does *resistive rating* mean?
32. What does *continuous rating* mean?
33. What determines relay contact life?

## REVIEW PROBLEMS

A review of series-parallel circuits can be of assistance in many electrical applications. Relays are no exception. Sometimes it is necessary to add relays in circuits with resistances in various combinations.

1. Find the current through $R_1$ in Fig. P-1.

Fig. P-1

2. Find the current through $R_4$ in Fig. P-1.
3. Find the voltage drop across $R_3$ in Fig. P-1.
4. Find the voltage drop across $R_2$ in Fig. P-1.
5. Find the current through $R_5$ in Fig. P-1.
6. Find the correct values from Fig. P-2 and place them in the following table:

Fig. P-2

| Resistor | Resistance Ω | Voltage drop V | Current A |
|---|---|---|---|
| $R_1$ | 3 | | 2.00 |
| $R_2$ | 5 | | 1.00 |
| $R_3$ | | | |
| $R_4$ | 3 | | |
| $R_5$ | | 8 | |
| $R_6$ | 8 | | 1.00 |

# CHAPTER 8

# Electric Motors

## PERFORMANCE OBJECTIVES

After studying this chapter, you will be able to:

1. Identify and classify types of motors.
2. Define counter electromotive force (CEMF).
3. Explain the advantages and disadvantages of the various types of dc motors.
4. Describe the reversing procedure for ac motors.
5. Understand reasons for compensating windings and interpoles.
6. Explain how fields rotate in ac motors.
7. Draw three-phase waveforms.
8. Explain how synchronous motors are started.
9. List the characteristics of squirrel-cage motors.
10. Define slip in a motor.
11. Understand single-phase motor operation.
12. Determine how the number of poles affects motor speed at various frequencies.
13. Calculate the horsepower of a motor given certain values.
14. Describe the classification of insulation systems used for motors.

## MOTOR CLASSIFICATIONS

There are three types of motors when classified into large groups.

- The *dc motor* runs on direct current only
- The *ac motor* runs on alternating current only
- The *universal motor* runs on ac or dc

Before you learn to control motors, it is best to know what makes them run and why they are chosen to do a particular job. The job they are designed to do and their ability to do the job are important factors in putting a motor to work at its greatest efficiency.

## DC MOTORS

DC motor is a mechanical workhorse. Many large pieces of equipment depend on a dc motor for power to move. The speed and direction of rotation of a dc motor are easily controlled. Just reverse the polarity and you reverse the direction of rotation. Change the voltage and the speed changes. This makes it especially useful for operating equipment such as winches, cranes, missile launchers, and elevators.

## Operating Principles

The operation of a dc motor is based on the principle that a current-carrying conductor placed in a magnetic field, perpendicular to the lines of flux, tends to move in a direction perpendicular to the magnetic lines of flux (Fig. 8-1). The relationship between the direction of the magnetic field, the direction of current in the conductor, and the direction in which the conductor tends to move is called the right-hand rule for motors (Fig. 8-2).

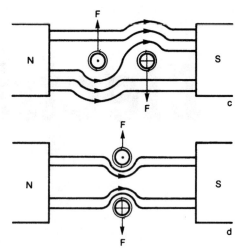

**Fig. 8-1** *Upward and downward forces created by interaction of field and armature flux.*

**Fig. 8-2** *Right-hand rule for motors.*

The right-hand rule can be used to find the direction of rotation of a motor. If the motion of a conductor is unknown, it can be found by extending the thumb, index finger, and middle finger of your right hand so that they are at right angles to each other. If the forefinger is pointed in the direction of magnetic flux (north to south) (check the polarity of the power source to determine this), and the middle finger is pointed in the direction of current flow in the conductor, the thumb will point in the direction the conductor will move.

Keep in mind that a dc motor rotates as a result of two magnetic fields interacting with one another. The armature of a dc motor acts through its coils. Since the armature is located within the magnetic field of the field poles, these two magnetic fields interact. Like magnetic poles repel each other, and unlike magnetic poles attract each other. The dc motor has field poles that are stationary and an armature that turns on bearings in the space between the field poles. The armature of a dc motor has windings that are connected to commutator segments. Figure 8-3 shows how the simple dc motor operates.

**Fig. 8-3** *DC motor armature rotation.*

Note how the brushes fitting on the commutator segments create a magnetic field in the conductor according to the polarity of the battery voltage. The changes in direction of current flow through the armature loop, caused by the switching action of the commutator segments, change the polarity of the magnetic field around the conductor. The magnetic fields repel and attract each other and the armature continues to turn. The momentum of the rotating armature carries the armature past the position where the unlike poles are exactly lined up. However, if these fields are exactly lined up when the armature current is turned on, there is no momentum to start the armature moving. In this case the motor will not rotate or start. It would be necessary to give the motor a spin to start it.

This disadvantage is eliminated when there are more turns on the armature, because then there is more than one armature field. No two armature fields can be exactly aligned with the field from the field poles at the same time.

## Counter Electromotive Force

While a dc motor is running, it acts somewhat like a dc generator. For instance, there is a magnetic field from the field poles. This means that a loop of wire is turning and cutting this magnetic field. For the moment, disregard the fact that there is current flowing through the loop of wire from the battery. As the loop sides cut the magnetic field, a voltage is induced in them just as it is in the loop sides of a dc generator. This induced voltage causes current to flow in the loop. The current induced is in the opposite direction from that which caused it (the battery). Inasmuch as this current flows in the opposite direction from that which caused it, it is called a counter electromotive force (CEMF).

In a dc motor, a counter EMF is always developed. The counter EMF cannot be equal to or greater than the applied battery voltage because then the motor would not run. The counter EMF is always a little less. Counter EMF opposes the applied voltage enough so that it is able to keep the armature current from the battery to a low value. If there were no such thing as counter EMF, much more current would flow through the armature. This is because it would have only its low dc resistance to determine the current draw. This means that the motor would run much faster. There is no way to avoid counter EMF. It makes a dc motor more economical to operate.

## Loads

DC motors are used to turn many mechanical devices: such things as water pumps, grinding wheels, fan blades, and circular saws. Keep in mind that the pump or fan blade is the load. It is the mechanical device that the motor must move. This is the motor load. This load can cause the motor to draw more current as the amount of mechanical energy demanded is increased. This, in turn, means that there is more electrical power consumed since the voltage times the current equals the power consumed. The load on a motor affects its speed, current drawn, and its efficiency.

To get the most from a motor or operate it at its most efficient point, the load and the motor characteristics and abilities must be matched. This makes for a better operating condition for the load and the motor.

## Types of DC Motors

The three categories for the dc motor are the series, shunt, and compound. Each type has distinct characteristics since each has its field coils and armature connected in different arrangements.

## Series DC Motors

In a series dc motor, the field is connected in series with the armature. The field is wound with a few turns of comparatively large diameter wire because it must carry full armature current (Fig. 8-4). There are both advantages and disadvantages in this arrangement. This type of motor develops a very large amount of turning force (torque) from standstill. Because of this characteristic, the series motor can be used to operate electrical appliances, portable electric tools, cranes, winches, hoists, and to start an automobile engine.

**Fig. 8-4** *Series-wound dc motor diagram.*

Another characteristic is that the speed varies widely between no-load and full-load. Series motors cannot be used where a relatively constant speed is needed for varying loads. A major disadvantage of the series motor is related to the speed characteristic. The speed of a series motor with no load increases to the point where the motor may become damaged. Usually, either the bearings are damaged or the windings fly out of the slots in the armature because the motor keeps increasing in speed until it self-destructs. With large motors such as cranes there is some danger to both the equipment and the personnel around it. A load must always be connected to a series motor before it is turned on. This means that there can be no belt-driven loads, since the belt may break or slip off. Small motors, such as those used in electrical hand drills, have enough internal friction (gearbox) to load themselves. Larger motors must be treated with more caution.

The series motor can be operated on ac or dc. This makes it more flexible in its use. However, it operates best on dc. The universal motor (operates on ac or dc) is discussed later.

## Shunt DC Motors

A shunt motor is connected with the field windings in parallel (shunt) with the armature windings (Fig. 8-5).

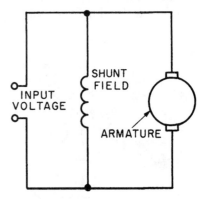

**Fig. 8-5** *Shunt-wound dc motor diagram.*

Once adjusted, the speed of a dc shunt motor remains relatively constant even under changing load conditions. One reason for this is that the field flux remains constant. A constant voltage across the field makes the field independent of variations in the armature circuit.

When the load on the motor is increased, the motor tends to slow down. This slowdown decreases the amount of counter EMF generated in the armature. This decrease in counter EMF decreases the opposition to the flow of battery current through the armature. Armature current then increases. The increased armature current causes the motor to speed up. The conditions that established the original speed are then reestablished, and the original speed is maintained.

Now, if the motor load decreases, the motor tends to increase its speed. However, the counter EMF increases. This means that the armature current decreases and the decrease in armature current causes the speed to decrease. The decrease in load and the decrease in speed cause an almost instantaneous response. This means that the speed has a tendency to appear to be constant or to have so slight a fluctuation as to be unnoticed in most cases.

## Compound DC Motors

A compound motor has two field windings (Fig. 8-6). One is a shunt field; it is connected in parallel with the armature. The other is a series field; it is connected in series with the armature. The shunt field gives the motor a constant speed advantage. The series field gives it the advantage of being able to develop a large torque when the motor is started under a heavy load. The compound motor has both shunt and series motor characteristics.

There are two types of compound motors, the long shunt and the short shunt. In the long shunt the shunt field is connected in parallel with the series field and armature (Fig. 8-6A). In the short shunt the shunt field is across the armature and the series field is in series with this parallel (shunt) arrangement (Fig. 8-6B).

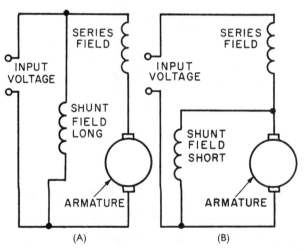

**Fig. 8-6** *Compound-wound dc motor diagram: (A) long shunt; (B) short shunt.*

## Types of Armatures

Two types of armatures are used for dc motors. The gramme-ring armature is inefficient and is not necessarily used for any purpose except to get a better understanding of the drum-wound armature. The drum-wound armature is used on ac motors.

Figure 8-7A shows an end view of the drum-wound armature as it appears cut through the middle. Current flow through the coils is indicated by a dot to show the current flowing toward you, and the + indicates that the current is flowing away from you, or that it resembles the tail feathers of an arrow as it goes away from you. Figure 8-7B is a side view of the armature and pole pieces. Notice that the length of each conductor is positioned parallel to the faces of the pole pieces. Each conductor of the armature can then cut the maximum flux of the motor field. The inefficiency of the gramme-ring type of armature is overcome by this positioning and makes it the drum-wound type.

**Fig. 8-7** *Drum-type armature: (A) end view (cross section); (B) side view.*

## Direction of Rotation

The direction of rotation of a dc motor depends on the direction of the magnetic field and the direction of current flow in the armature. If either the direction of the field or the direction of current flow through the armature is reversed, the rotation of the motor will reverse. However, if both of these factors are reversed at the same time, the motor will continue rotating in the same direction. In actual practice, the field excitation voltage is reversed to reverse the motor direction. This means that if you want a motor that can be connected to a reversing switch, the leads from the field must be brought out for easy access to the switching device.

## Motor Speed

DC motors are variable-speed motors. The speed of a dc motor is changed by changing the current in the field or by changing the current in the armature. A decrease in field current causes a decrease in the field flux. That means that the counter EMF decreases. This decrease in CEMF permits more armature current. That means that the motor speeds up. When the field current is increased, the field flux increases. More counter EMF is developed. The increase in CEMF decreases the armature current. The armature current then decreases and the motor slows down. Decreasing the applied voltage to the armature causes the armature current to decrease and the motor slows down. Increasing the armature voltage and current causes the motor to speed up.

Shunt motor speed can be controlled by a rheostat connected in series with the field windings (Fig. 8-8). Increasing the resistance of the rheostat causes the current through the field winding to decrease. This decreases the flux momentarily, which, in turn, decreases the counter EMF. The motor then speeds up. This momentary increase in speed increases the counter EMF and keeps the armature current the same. A decrease in rheostat resistance increases the current flow through the field windings and causes the motor to slow down.

*Fig. 8-8*  *Rheostat used as a speed control.*

In a series motor, the rheostat speed control may be connected either in parallel or in series with the armature windings. Moving the rheostat in a direction that allows the voltage across the armature to be lowered, allows the current through the armature to be decreased and slows the motor. Moving the rheostat in a direction that increases the voltage and current through the armature increases the motor speed. One disadvantage of putting a resistor in series with a series motor is that it destroys the torque advantage the motor characteristically possesses. A good example of this is the foot control on a sewing machine. It is a series motor that powers the sewing machine. Inserting the series resistor in the foot-speed control decreases the motor's starting torque, and to start the sewing machine the motor must be given an assist by spinning the wheel. Changing speed controls from rheostats to SCRs gives the advantage of speed control without the loss of starting torque. This type of control is discussed in chapter 7.

## Armature Reaction

Since the armature revolves in a magnetic field it is subject to the same laws of nature that control the generator. This means that it cuts the flux field produced by the field coil. Of course, it also has its own magnetic field, produced by current flowing through its windings. There is a generator reaction produced by the armature revolving in the magnetic field, as well as its own magnetic field, used to give the motor effect. This armature effect has to be compensated for in the design of the motor. There are several ways to do this, one of which is to shift the brushes after the motor has started (Fig. 8-9). Note how the armature field has distorted the flux field between the pole pieces. The effect has shifted to the neutral plane to the left. This is in opposition to the direction of rotation. As the brushes are shifted it causes the neutral plane to shift. The proper location is indicated when there is no sparking from the brushes. Another way is to place compensating windings and interpoles in the motor permanently.

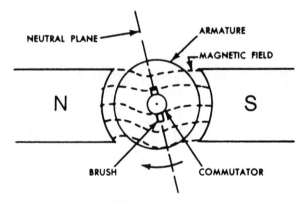

*Fig. 8-9*  *Armature reaction.*

## Compensating Windings and Interpoles

Compensating windings and interpoles cancel armature reaction in a motor. Shifting brushes reduces sparking and makes the field less effective. Canceling armature reaction eliminates the need to shift brushes.

Compensating windings and interpoles can be found in both motors and generators. Compensating windings are somewhat expensive. This means that most large dc motors depend on interpoles to correct armature reaction. Compensating windings are the same in motors as in generators. Interpoles, however, are slightly different. The difference is that a generator interpole has the same polarity as the main pole ahead of it in the direction of rotation. In a motor the interpoles have the same polarity as that of the main pole following it (Fig. 8-10).

**Fig. 8-10** *Compensating windings and interpoles in a generator.*

Interpoles are connected to carry the armature current. As the load varies, the interpole flux varies, and commutation is corrected automatically as the load changes. This means that it is not necessary to shift the brushes when the load increases or decreases. Brushes are located on the no-load neutral plane and they remain in that position for all conditions of load.

The dc motor is reversed by reversing the direction of current in the armature. This means that the current through the interpole is also reversed. The interpole, then, still has the proper polarity to provide automatic commutation.

## DC Motor Starting Resistance

The resistance of most dc motor armatures is low, somewhere between 0.05 and 0.5 Ω. Counter EMF does not exist until the armature begins to turn. This means that it is necessary to use an external starting resistance in series with the armature of a dc motor to keep the initial armature current to a safe level. As the armature begins to turn, counter EMF increases and the applied voltage is opposed by this increase in counter EMF. This means that the armature current is then reduced by its own generator effect. Then, once the motor comes up to normal speed, the external resistance in series with the armature is decreased or eliminated and full voltage can be applied across the armature.

Starting resistance can be controlled either manually, by an operator, or by any of several automatic devices. The automatic devices are usually just switches controlled by motor speed sensors. Automatic starters are covered in later chapters.

## DC Motor Characteristics and Applications

A quick reference for checking out the characteristics and applications of dc motors is provided in Table 8-1.

## Troubleshooting DC Motors

Methods used for solving problems with dc motors and a troubleshooting dc motors table are described in Chapter 22.

*Table 8-1* DC Motor Characteristics and Applications

| Speed Regulation | Speed Control | Starting Torque | Pull-out Torque | Application |
|---|---|---|---|---|
| **Series DC Motors** | | | | |
| Varies inversely as the load. Races on light loads and full voltage. | Zero to maximum, depending on control and load. | High. Varies as square of voltage. Limited by the communication, heating, and line capacity. | High. Limited by commutation, heating, and line capacity. | Where high torque is required and speed can be regulated: cranes, hoists, gates, starters. |
| **Shunt DC Motors** | | | | |
| Drops 3 to 5% from no load to full load. | Any desired range, depending on motor design and type of system. | Good with constant field, varies directly as voltage applied to the armature. | High. Limited by commutation, heating, and line capacity. | Where constant speed is needed and starting conditions are not severe: fans, pumps, blowers, conveyors. |
| **Compound DC Motors** | | | | |
| Drops 3 to 20% from no load to full load, depending on amount of compounding. | Any desired range, depending on motor design and type of control. | Higher than for shunt, depending on the amount of compounding. | High. Limited by commutation, heating, and line capacity. | Where high starting torque combined with fairly constant speed is required: Plunger pumps, punch presses, shears, geared elevators, conveyors, hoists. |

# AC MOTORS

The ac motor is less expensive than the dc motor of comparable size. The ac motor also requires less maintenance since it has, in most cases, no brushes or commutator to be maintained. Since most commercial power is generated as ac, it is only natural that this type of motor be used for doing the work that needs to be done.

An ac motor is well suited for constant-speed jobs. This is mainly because its speed is determined by the frequency of the power source. However, for some applications, the dc motor is better suited than the ac motor. The dc is more easily varied in its speed. But in some instances the speed of the ac motor can also be controlled within very narrow limits. Inverter-type drives make it easier to vary ac motor speed. It is these limits of control that we discuss here.

## Types of AC Motors

In this chapter we discuss three types of ac motors: series, synchronous, and induction. Synchronous motors may be considered as polyphase motors. They have a constant speed and their rotors are energized with dc voltage. Induction motors are commonly used as single-phase or polyphase. Their rotors are energized by induction. The series ac motor is a familiar type of motor. It is very similar to a dc motor that has already been discussed.

## Series AC Motor

The series ac motor is the same electrically as a dc series motor (Fig. 8-11). Use the left-hand rule for the polarity of coils and you can see that the instantaneous magnetic polarities of the armature and field oppose each other. This means that motor action results. By reversing the current, you reverse the polarity of the input. Note that the field magnetic polarity still opposes the armature magnetic polarity. This is because the reversal affects both the armature and the field. The ac input causes these reversals to take place continuously and the motor continues to rotate in the same direction.

Construction of the ac series motor does vary slightly from the dc series motor. Since ac is being used it calls for special metals in the pole pieces—such as silicon steel, which easily reverses its magnetic polarity without causing residual magnetism to remain after the reversal. It also means that laminations are used in the ac motor to decrease the amount of eddy currents generated in the pole pieces. DC can be used to power ac series motors efficiently, but putting ac to a series motor made for dc does not produce the same efficiencies.

Characteristically, the ac and dc series motors are similar. They both have a varying speed characteristic. Low speeds are possible for large loads and light loads produce high speeds. Speed varies directly with the size of the load. The larger the load, the slower the speed.

AC-DC types of series motors, called universal motors, are especially designed for use on both power sources. They are usually made in small horsepower sizes, usually less than 1 hp, and are used most frequently in vacuum cleaners. Universal motors cannot be operated on polyphase ac power.

*Fig. 8-11* Series ac motor.

## Magnetic Fields in AC Motors

Rotating magnetic fields are the key to the operation of ac motors. This is because the alternating current causes a continuously changing magnetic field as it rotates around a series of stator pole pairs.

The magnetic field in a stator can be made to rotate electrically. This means that it can move around and around. Then another magnetic field in the rotor can be made to chase it. This is done by having the rotor field attracted and repelled by the stator field. By allowing the rotor to turn freely, it is allowed to chase the rotating field in the stator. Rotating magnetic fields are set up in two-phase or three-phase machines. To establish a rotating magnetic field in a motor stator, the number of pole pairs must be the same as (or a multiple of) the number of phases in the applied voltage. The poles must be displaced from each other by an angle equal to the phase angle between the individual phase of the applied voltage.

# Rotating Magnetic Field: Two-Phase

A two-phase stator shows the rotating magnetic field most easily. The stator of a two-phase induction motor is made up of two windings (or a multiple of two) placed at right angles to each other around the stator (Fig. 8-12). Note that the voltages applied to phases 1-1A and 2-2A are 90° out of phase; that is, the currents that flow in the phases are displaced from each other by 90°. The magnetic fields generated in the coils are in phase with their respective current. The magnetic fields are also 90° out of phase with each other. The coil axes of these two out of phase magnetic fields are at right angles to each other. They also add together at every instant during their cycle.

*Fig. 8-13* Two-phase rotating field.

*Fig. 8-12* Two-phase motor stator.

They produce a resultant field that rotates one revolution for each cycle (hertz) of ac. Analyzing how the two-phase stator's rotating field works is one way of understanding how all ac motors work. Figure 8-13 shows the two-phase rotating field. By taking a look at the nine different locations on this figure, you will be able to see how the field moves step by step with each of the two phases.

The arrow indicates the rotor. Keeping track of the arrow shows how the rotation of the rotor is accomplished. This chart shows the voltage of each phase. The current flows in a direction that causes the magnetic polarity indicated at each pole piece. Note that from one point to the next, the polarities are rotating from one pole to the next in a clockwise manner. One complete cycle of input voltage produces a 360° rotation of the pole polarities.

## Two-Phase Waveforms

The waveforms in Fig. 8-13 are applied to the windings as shown in Fig. 8-12. Note that they are displaced by 90°. That means that when one is at its maximum, the other is at its minimum. Now take a closer look at Fig. 8-13 and position 1, indicated by the circled 1. The current flow and the magnetic field in winding 1-1A are at maximum (because the phase voltage is maximum). The current flow and magnetic field in winding 2-2A are zero (because the phase voltage is zero). The resultant magnetic field is therefore in the direction of the 1-1A axis. At the 45° point (position 2), the resultant magnetic field is midway between windings and 1-1A and 2-2A. The coil currents and magnetic fields are equal in strength. At 90°, position 3, the magnetic field in winding 1-1A is zero. The magnetic field in winding 2-2A is at maximum. Now the resultant magnetic field is along the axis of the 2-2A winding. The resultant magnetic field has rotated clockwise through 90° to get from position 1 to position 3.

Once the two-phase voltages have completed one full cycle and arrived at position 9, the resultant magnetic field has rotated through 360°. This means that by placing two windings at right angles to each other and exciting them with voltages 90° out of phase, a rotating magnetic field can be produced. Note that the arrow representing the rotor is now pointing to where it started in position 1, indicating a complete revolution of the rotor.

Two-phase current is rarely used in this country. However, it is used here to show how similar operation is used to start a single-phase motor with its start winding placed in the circuit to get the rotor turning. Single-phase and three-phase motors use the same principle of rotating magnetic fields to cause their rotors to rotate.

## Rotating Magnetic Field: Three-Phase

A three-phase induction motor also operates on the principle of a rotating magnetic field. Figure 8-14 shows how a three-phase operation is connected. This one is Y-connected. The three-phase windings can be connected to a three-phase ac input and have a resultant magnetic field that rotates (Fig. 8-15).

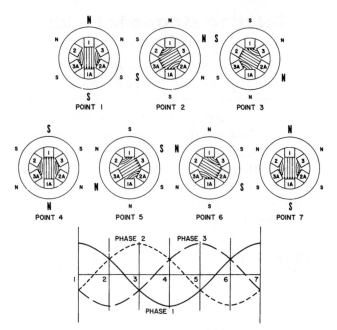

*Fig. 8-15* Three-phase rotating field polarities and input voltages.

*Fig. 8-14* Three-phase, wye connected stator.

Figure 8-14 shows how three-phase stators are connected in a wye configuration. The dot shows where connections are made to ensure a wye configuration. The pole pieces are placed 120° apart. Note that 3 times 120 equals 360. Now take a closer look at Fig. 8-15. It shows how instantaneous polarities are generated. Current flows toward the terminal number in Fig. 8-14 for positive voltages and away from the terminal number for negative voltages.

Now refer to Fig. 8-15. Note that at position 1 the magnetic field in coils 1-1A is maximum. Check the polarities. At the same time, negative voltages are present in the 2-2A and 3-3A windings. These negative voltages create weaker magnetic fields that tend to aid the 1-1A field. At position 2, maximum negative voltage is present in the 3-3A windings. This creates a strong magnetic field that is aided by the weaker fields in 1-1A and 2-2A. Now move along each point on the voltage graph. Notice that the resultant magnetic field is rotating in a clockwise direction. This means that when the three-phase voltage completes one full cycle and reaches position 7, the magnetic field has rotated through 360°.

If you place a permanent bar magnet in this rotating magnetic field with a shaft through it and allow the magnet to rotate freely in step with the rotating magnetic field, you will be able to see how the shaft rotates at the same rate as the moving magnetic field. Keep in mind that this simplified explanation of rotating fields is given here to show you how a rotating magnetic field can be utilized to cause a shaft to turn and produce usable mechanical energy from electrical energy. Motors have been designed to use a number of principles and can be utilized to do many jobs efficiently.

## Synchronous Motors

The main advantage of synchronous motors is a constant-speed characteristic. They are capable of correcting the low power factor of an inductive load when operated under certain conditions. They are often used to drive dc generators. Synchronous motors are available in sizes up to thousands of horsepower. They may be designed as either single-phase or three-phase machines (Fig. 8-16).

*Fig. 8-16* *Revolving field synchronous motor.*

Synchronous motors are often used without a load. They can be used for power factor correction. Adding the synchronous motor to the circuit can be useful in power factor correction. This means that the machine will do the same amount of work as before power correction; however, it will draw less current from the power lines. This is accomplished by the amount of excitation applied to the wound rotor in the form of low-voltage dc. Fixed condensers (capacitors) are often used in place of synchronous motors for power factor correction.

This type of motor is used whenever exact speed must be maintained or for power factor correction. Synchronous motors are more expensive than other types at the lower horsepower ratings, but may possibly be more economical for 100 hp and higher ratings (Fig. 8-16).

If three-phase ac is applied to a synchronous motor, a rotating magnetic field is set up around the rotor. The rotor is then energized with dc. That is it acts as a bar magnet since the dc produces a fixed north-south polarity for the rotor. The strong rotating magnetic field attracts the strong rotor field activated by the dc. This results in a strong turning force on the rotor shaft. The rotor is therefore able to turn a load as it rotates in step with the rotating magnetic field.

Getting the motor started is accomplished by adding a squirrel-cage winding to the rotor. It cannot start from standstill without the aid of a squirrel-cage rotor. That is because when ac is applied to the stator, a high-speed rotating magnetic effect appears immediately. This rotating field rushes past the rotor poles so quickly that the rotor does not have a chance to get started. In effect, the rotor is repelled first in one direction and then the other. In its purest form, a synchronous motor has no starting torque. It has torque only when running at synchronous speed. A synchronous motor used to correct the power factor has to operate at less than its nominal mechanical load.

## Squirrel-Cage Motor

The three-phase current with which the motor is supplied establishes a rotating magnetic field in the stator. This rotating magnetic field cuts the conductors in the rotor, inducing voltages and causing currents to flow. These currents set up an opposite-polarity field in the rotor. The attraction between these opposite stator and rotor fields produces the torque that causes the rotor to rotate. This is, in essence, how a squirrel-cage motor works.

A squirrel-cage winding can be added to the rotor of a synchronous motor to cause it to start. The squirrel cage is shown as part of the rotor in Fig. 8-17. The name comes from the shape-it looks something like a turnable squirrel cage. The windings are heavy copper bars shorted together by copper rings. A low voltage is induced in these shorted windings by the three-phase stator field. Because of the short circuit, a relatively large current flows in the squirrel cage. This causes a magnetic field that interacts with the rotating field of the stator. Because of the interaction the rotor begins to turn, following the stator field and the motor starts. The squirrel cage is also used in single-phase and other three-phase motors.

SQUIRREL-CAGE WINDING
OVER SALIENT-POLE WINDINGS

*Fig. 8-17.* *Self-starting synchronous ac motor.*

## Starting the Synchronous Motor

To start a practical synchronous motor, the stator is energized. However, the dc supply to the rotor field is not energized. The squirrel-cage windings are allowed to bring the rotor close to synchronous speed. At that point the dc field is energized. This locks the rotor in step with the rotating stator field.

Full torque is developed in the synchronous motor when it comes up to speed. Once it comes up to synchronous speed the load can be applied. A switch operated by centrifugal force is used to apply dc to the rotor as it reaches synchronous speed. The need for a dc power source for the rotor makes it an expensive type of machine to operate and maintain. The dc source may be part of the motor or it may come from an external generator.

As you can see from this discussion, synchronous motors have advantages and disadvantages, as do other types of motors. If expense is not a factor, the selection of one over another depends on the job and the power available. Table 8-2 summarizes the synchronous motor characteristics and applications.

*Table 8-2* Synchronous Motor Characteristics and Applications

| | |
|---|---|
| Speed regulation: | Constant. |
| Speed control: | None, except special motors designed for two fixed speeds. |
| Starting torque: | 40% for slow-speed to 160% for medium-speed 80%-power factor designs. Special designs develop higher torques. |
| Pull-out torque: | Unity of motors, 170%; 80%-power factor motors, 225%. Special designs up to 300%. |
| Applications: | For constant-speed service, direct connection to slow-speed machines and where power factor correction is required. |

## Induction Motors

The induction motor is the most commonly used ac motor (Fig. 8-18). It is simple in design and rugged in construction. It costs relatively little to manufacture. The induction motor rotor is not connected to an external source of voltage. The induction motor derives its name from the fact that ac voltages are induced in the rotor circuit by the rotating magnetic field in the stator. Induction in this motor is similar to the induction between the primary and secondary of a transformer.

*Fig. 8-18* Induction motor.

The rotor can be thought of as a short-circuited secondary of a transformer that is mounted on a shaft and supported by bearings that allow it to rotate freely as the rotating field in the stator moves from stator pole to stator pole (Fig. 8-19).

*Fig. 8-19* Cutaway view of three-phase motor: (A) with a half-etched squirrel-cage rotor; (B) with a cast rotor.

Induction motors are often large and permanently mounted. They drive loads at fairly constant speed. They are used in washing machines, refrigerator compressors, bench grinders, and table saws.

## Stator Construction

The stator construction of the three-phase induction motor and the three-phase synchronous motor are almost identical. However, their rotors are completely different (Fig. 8-20). An induction motor is made of a laminated cylinder with slots in its surface. The

*Fig. 8-20* Types of ac induction motor rotors.

windings in these slots are one of two types. The most common is the squirrel-cage winding. This entire winding is made up of heavy copper bars connected together at the end by a metal ring made of copper or brass. No insulation is required between the core and the bars. This is because of the very low voltages generated in the rotor bars. The other type of windings contains actual coils placed in the rotor slots. This type of rotor is then called a wound rotor (Table 8-3).

## Slip

The rotor of an induction motor cannot turn at the same speed as the rotating magnetic field. If the speeds were the same, there would be no relative motion between the stator and rotor fields. Without relative motion there would be no induced voltage in the rotor. For relative motion to exist between the two, the rotor must rotate at a slower speed than that of the rotating magnetic field. The difference between the speed of the rotating stator field and the rotor speed is called slip. The smaller the slip, the closer the rotor speed approaches the stator field speed.

Rotor speed depends on the torque requirements of the load: The greater the load, the stronger the turning force needed to rotate the rotor. Turning force increases only if the rotor-induced EMF increases. This EMF can increase only if the magnetic field cuts through the rotor at a faster rate. To increase the relative speed between the field and rotor, the rotor must slow down. This means that for heavier loads the induction motor turns more slowly than for lighter loads.

Slip is directly proportional to the load on the motor. A slight change in speed is necessary to produce

*Table 8-3* Motor Characteristics and Applications (Two- and Three-Phase)

### General-Purpose Squirrel-Cage (Class B)

| | |
|---|---|
| Speed regulation | Drops about 3% for large to 5% for small sizes. |
| Speed control: | None, except multispeed types, designed for two to four fixed speeds. |
| Starting torque: | 200% of full load for two-pole to 105% for 16-pole designs. |
| Pull-out torque: | 200% of full load. |
| Applications: | Constant-speed service where starting torque is not excessive: fans, blowers, rotary compressors, centrifugal pumps. |

### High-Torque Squirrel Cage (Class C)

| | |
|---|---|
| Speed regulation: | Drops about 3% for large to 6% for small sizes. |
| Speed control: | None, except multispeed types, designed for two to four fixed speeds. |
| Starting torque: | 250% of full load for high-speed designs to 200% for low-speed designs. |
| Pull-out torque: | 200% of full load. |
| Applications: | Constant-speed service where fairly high starting torque is required at infrequent intervals with starting current of about 400% of full load: reciprocating pumps, compressors, crushers. |

### High-Slip Squirrel-Cage (Class D)

| | |
|---|---|
| Speed regulation: | Drops about 10 to 15% from no load to full load. |
| Speed control: | None, except multispeed types, designed for two to four fixed speeds. |
| Starting torque: | 225 to 300% of full load, depending on speed with rotor resistance. |
| Pull-out torque: | 200%. Will usually not stall until loaded to maximum torque, which occurs at standstill. |
| Applications: | Constant-speed service and high starting torque, if starting is not too frequent, and for taking high-peak loads with or without flywheels: punch presses, shears, elevators. |

### Low-Torque Squirrel Cage (Class F)

| | |
|---|---|
| Speed regulation: | Drops about 3% for large to 5% for small sizes. |
| Speed control: | None, except multispeed types, designed for two to four fixed speeds. |
| Starting torque: | 50% of full load for high-speed designs to 90% for low-speed designs. |
| Pull-out torque: | 150 to 170% of full load. |
| Applications: | Constant-speed service where starting duty is light: fans, blowers, centrifugal pumps, similar loads. |

### Wound Rotor

| | |
|---|---|
| Speed regulation: | With rotor rings short-circuited, drops about 3% for large to 5% for small sizes. |
| Speed control: | Speed can be reduced to 50% by rotor resistance to obtain stable operation. |
| Starting torque: | Up to 300%, depending on external resistance to obtain stable operation. Speed varies inversely as load. |
| Pull-out torque: | 200% when rotor slip rings are short circuited. |
| Applications: | Where high starting torque with low starting current or where limited speed control is required: fans, centrifugal and plunger pumps, compressors, conveyors, hoists, cranes. |

the usual current changes required for normal changes in load. This is because the rotor windings have such a low resistance. Induction motors are referred to as constant-speed motors.

## Single-Phase Induction Motors

The single-phase induction motor is the most commonly used of all types of electric motors. It has the lowest initial cost and needs little maintenance. There are a number of descriptive titles for each type of single-phase motor: such names as split-capacitor, capacitor-start, split-phase, shaded-pole, capacitor start-capacitor run and permanent-split capacitor. Of course, other groupings include the capacitor start and resistance start as part of the overall title "split-phase."

The stator field in the single-phase motor does not rotate. It simply alternates polarity between poles as the ac voltage changes polarity (Fig. 8-21). Magnetic induction causes a voltage to be induced in the rotor, thereby producing a magnetic field in the rotor. This field is in opposition to the stator field because, as Lenz's law states: The induced voltage is in the opposite direction from that which produced it. Thus the interaction between rotor and stator fields will not produce rotation. Instead of rotation it produces a north pole directly opposite a south pole. Therefore, the rotor stands still until pushed or nudged into motion. This is shown in Fig. 8-21A.

Once the movement is started (Fig. 8-21B) the south pole on the rotor is attracted by the left-hand pole. That is, the north rotor pole is attracted to the right-hand pole. This is a result of the rotor being rotated 90° by an outside force. Once the pull exists between the two fields, it becomes a rotary force, turning the rotor toward the magnetic field of the stator. The two fields alternate continuously. They never actually line up, so the rotor will continue to rotate once started. The next problem is to design a starting method to get this single-phase motor to start without an outside force turning it each time it is started.

The basic operation of the single-phase motor is the same for all types, but the starting methods utilized create a specialized label for each. To become familiar with two of the slightly different methods used to get the single-phase motor started, we will look at the capacitor-start and the resistance-start in addition to the shaded-pole types.

## Split-Phase Induction Types

Split-phase motors are designed to use inductance, capacitance, or resistance to develop a starting torque. The capacitor-start type uses, of course, capacitance. Figure 8-22 shows how an auxiliary winding is added with a capacitor in series with it. This winding is placed in parallel with the main (run) winding and is located at right angles (90°) to it. This produces a phase difference of 90 electrical degrees between the two windings. The start or auxiliary winding is connected with a switch that is operated centrifugally. When the motor comes up to about 75% of its rated speed, opening the switch disconnects the start winding.

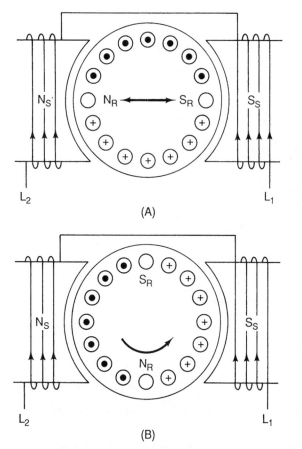

**Fig. 8-21** Rotor currents in a single-phase ac induction motor: (A) stationary; (B) rotating. $N_R$, $S_R$ rotor field; $N_S$, $R_S$ stator field.

**Fig. 8-22** Capacitor-start ac induction motor.

When the power is first applied, the starting switch is closed. This places the capacitor in series with the auxiliary winding. The capacitor is of such value that the auxiliary circuit is effectively a resistive-capacitive circuit. In this circuit the current leads the line voltage by 45°. That is because $X_C$ equals the resistance in the circuit. The main (run) winding has enough resistance and inductance to cause the current to lag the line voltage by about 45° because $X_L$ about equals the resistance. The currents in each winding are therefore 90° out of phase. The magnetic fields are thus displaced by the same amount. The effect is that the two windings act as a two-phase stator and produce the rotating field required to start the motor.

Once the motor is started, the centrifugal switch opens and takes the start winding out of the circuit; the rotor continues to rotate until power is removed from the main winding. Split-phase motors are available in small sizes, usually less than 1 hp, because they do not have sufficient starting torque to handle large loads.

Another type of split-phase induction motor is the resistance-start (Fig. 8-23). Note that there is a resistor in series with the start winding. The auxiliary circuit, consisting of the winding and the resistor, is switched in the circuit and cut out by a centrifugal switch. The electrical phase shift between the currents in the auxiliary and main windings is obtained by making the impedance of the windings unequal. Note how the main winding has a high inductance (little resistance due to windings) and the auxiliary winding has a low inductance plus a high-value series resistance. The start winding then has a lagging phase angle due to the inductance.

**Fig. 8-23** *Resistance-start ac induction motor.*

The start winding has a smaller phase angle, so that the current does not lag as much as in the main winding. There is about a 30° difference between the two windings in their phase angles. This is enough of a difference to develop enough torque to get the rotor moving if there is little or no load on the motor. The centrifugal switch opens the circuit once the rotor is up to about 75% of its normal design speed.

## Shaded-Pole Induction Motors

Shaded-pole motors are also of the induction type. They use a different type of starting arrangement. The effect of a moving magnetic field is produced by constructing the stator in a special way. These motors have projecting pole pieces similar to some dc motors. Portions of the pole-piece surfaces are surrounded by a copper strap called a shading coil.

Figure 8-24 shows a pole piece with the strap in place. The strap causes the field to move back and forth across the face of the pole piece. Note the numbered sequence and points on the magnetization curve in the figure. As the alternating stator field starts increasing from zero, step 1, the lines of force expand across the face of the pole piece and cut through the strap. A voltage is induced in the strap. The current that results generates a field that opposes the cutting action (and decreases the strength) of the main field. This produces the following:

**A.** As the field increases from zero to a maximum at 90°, the field reaches its maximum value.

**B.** A large portion of the magnetic lines of force are concentrated in the unshaded portion of the pole ❶.

**C.** At 90° the field reaches its maximum value and stops expanding. No EMF is induced in the strap. No opposition magnetic field is generated, so the main field is uniformly distributed across the poles as shown in ❷.

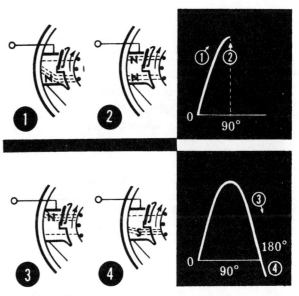

**Fig. 8-24** *Starting a shaded-pole induction motor.*

**AC Motors** 149

**D.** From 90° to 180°, the main field starts decreasing or collapsing inward. The field generated in the strap opposes the collapsing field. The effect is to concentrate the lines of force in the shaded portion of the pole face as shown in ❸.

**E.** From 180° to 360° the main field goes through the same change as it did from 0° (180°). However, note that it is in the opposite direction as shown in ❹.

The direction of the field does not affect the way the shaded pole works. The motion of the field is the same during the second half-cycle as it was during the first half of the cycle. The motion of the field back and forth between shaded and unshaded portions produces a weak torque to start the motor. Because of the weak starting torque, shaded-pole motors are built in very small sizes. They drive such things as fans, clocks, blowers, and electric razors. Timer motors on many large pieces of equipment use the shaded-pole motor for the clock mechanism.

## Speed and Slip in Squirrel-Cage Motors

The speed of a squirrel-cage motor depends on the frequency of the power source and the number of poles the motor has. The higher the frequency, the faster the motor; the more poles the motor has, the slower it runs. The smallest number of poles ever used in a squirrel-cage motor is two. A two-pole 60-Hz motor runs at approximately 3600 rpm (Table 8-4). Formula:

$$\text{Synchronous Speed} = \frac{60 \times 2f}{p}$$

where $f$ is the frequency of power supply and $p$ is the number of poles in the motor. Most standard commercial motors (143T through 445T frame sizes) are wound with a maximum of eight poles.

The actual speed of the motor is somewhat less than its synchronous speed. This difference between the synchronous and actual speeds is defined as slip. If the squirrel-cage rotor rotated as fast as the stator field, the rotor conductor bars would be standing still with respect to the rotating field. See Fig. 8-25 for a look at the rotor construction. This means that no voltage would be induced into the bars and no current would be set up to produce torque. Since no torque is produced, the rotor will slow down until sufficient current is induced to develop enough torque to keep the rotor at a constant speed. Therefore, the rotor rotates slower than the rotating magnetic field of the stator. For example, a 3600-rpm motor usually rotates at 3450 rpm and a 1800-rpm turns at 1725 rpm at rated load.

*Fig. 8-25* Rotor for a squirrel-cage motor. (The Lincoln Electric Co.)

An increased load on the motor causes the rotor to slow down; that is, the rotating field cuts the rotor bars at a higher rate than before. This has the effect of increasing the current in the bars and hence increasing the pole strength of the rotor. This increased pole strength makes it possible for the motor to carry the larger load. Slip is usually expressed in percent and can be easily computed using the formula:

$$\text{Percent Slip} = \frac{\text{Synchronous Speed} - \text{Actual Speed}}{\text{Synchronous}} \times 100$$

Squirrel-cage motors are made with the slip ranging from less than 5% to around 20%. Motors with a slip of 5% or higher are used for hard-to-start applications. A motor with a slip of 5% or less is called a normal slip motor. A normal slip motor is often referred to as a constant-speed motor because the speed changes very little with load variations.

Manufacturers usually specify the motor speed on the nameplate as that which is normal at its rated load. Actual speed is, of course, lower than the synchronous speed.

## Rotation

The direction of rotation of a polyphase squirrel-cage motor depends on the motor connection to the power

*Table 8-4* Motor Speed versus Number of Poles

| Number of Poles | 60-Hz Speed | 50-Hz Speed |
|---|---|---|
| 2 | 3600 | 3000 |
| 4 | 1800 | 1500 |
| 6 | 1200 | 1000 |
| 8 | 900 | 750 |
| 10 | 720 | 600 |
| 12 | 600 | 500 |

lines. Rotation can readily be reversed by interchanging any two input leads.

## Torque and Horsepower

Torque and horsepower are two very important motor characteristics that determine the size of the motor for a particular job. Torque is the turn effort produced by the motor. Torque is measured in pound feet (lb-ft). Horsepower is a rate of doing work. One horsepower equals 33,000 lb being lifted for a distance of 1 ft in 1 minute:

$$hp = \frac{\text{speed in (rpm)} \times 2\pi \times \text{torque}}{33,000}$$

It takes 746 watts of electrical power to produce 1 horsepower.

## Locked-Rotor Torque

An induction motor is built to supply the extra torque needed to start a load. The speed-torque curve for a typical motor is shown in Fig. 8-26. This curve shows the torque for a NEMA design B motor. The curve shows the torque to be 210% of the rated load torque.

## Breakdown Torque

Occasionally, a sudden overload will be placed on a motor. To keep the motor from stalling every time an overload occurs, these motors have what is called breakdown torque. Breakdown torque is much higher than the rated-load torque. It takes quite an overload to stall the motor. The speed-torque curve shows the breakdown torque for a typical motor to be 270% of the rated-load torque. Operating a motor overloaded for an extended period of time causes excessive heat buildup in the motor and may eventually burn up the motor windings.

Because of the variety of torque requirements, NEMA has established different designs to cover almost every application (Table 8-5). These designs take into consideration starting current and slip as well as torque. These designs should not be confused with various classes of insulation, which are also designated by letter (Table 8-6).

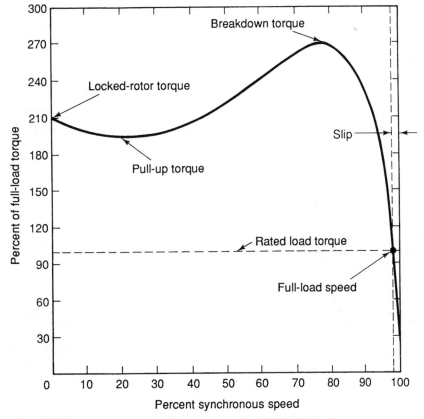

***Fig. 8-26*** *Representative speed-torque curve for NEMA design B motors.* (The Lincoln Electric Co.)

*Table 8-5* Locked-Rotor kVA/Hp

| Code Letter | kVA/hp[a] |
|---|---|
| A | 0–3.15 |
| B | 3.15–3.55 |
| C | 3.55–4.00 |
| D | 4.00–4.50 |
| E | 4.50–5.00 |
| F | 5.00–5.60 |
| G | 5.60–6.30 |
| H | 6.30–7.10 |
| J | 7.10–8.00 |
| K | 8.00–9.00 |
| L | 9.00–10.0 |
| M | 10.0–11.2 |
| N | 11.2–12.5 |
| P | 12.5–14.0 |
| R | 14.0–16.0 |
| S | 16.0–18.0 |
| T | 18.0–22.0 |
| U | 20.0–2.4 |
| V | 22.4 and up |

[a] The locked-rotor kVA/hp range includes the lower figure up to, but not including, the higher figure.

## Locked-Rotor kVA/hp

Another rating specified on motor nameplates and determined by the motor design is locked-rotor kVA per horsepower. A letter appears on the nameplate corresponding to various kVA/hp ratings (Table 8-5).

These nameplate code ratings give a good indication of the starting current the motor will draw. A code letter at the beginning of the alphabet indicates a low starting current, and a letter at the end of the alphabet indicates a high starting current for the particular horsepower rating of the motor. Starting current can be figured by using the formula:

$$\text{locked-rotor amperes} = \frac{1000 \times hp \times kVA/hp}{1.73 \times volts}$$

The starting current is important to the buyer of a motor since it indicates the amount of protection needed in the way of over-current devices. The buyer must install power lines big enough to carry the required currents and put in fuses of the proper size.

## Starting Squirrel-Cage Motors

Squirrel-cage motors are usually designed for across-the-line starting. This means connecting it directly to the power source by means of a suitable contactor. In large squirrel-cage motors and in some other types of motors the starting currents are very high. Usually, the motor is built to stand these high currents. However, since these currents are almost six times rated load current, there may be a large voltage drop in the power system. Some method of reducing the starting current has to be used to limit the voltage drop to a tolerable level. Reduced-voltage starting is described in another chapter. A summary of the mechanical and electrical formulas used with electric motors is given in Table 8-7.

## REVIEW QUESTIONS

1. What is the right-hand rule for motors?
2. Describe the armature of a dc motor.
3. What is counter EMF?
4. What affect does the load have on a motor?
5. What are the three types of dc motors?
6. List the advantages of the series motor.
7. List the advantages of the shunt motor.
8. What is the difference between short-shunt and long-shunt compound motors?
9. What controls the speed of a dc motor? How is speed control accomplished?

*Table 8-6* Classification of Insulation Systems

An insulation system is an assembly of insulating materials in association with the conductors and the supporting structural parts of a motor or generator. Insulation systems are divided into classes according to the thermal endurance of the system for temperature rating purposes. Four classes of insulation systems are used in motors and generators; classes A, B, F, and H. These classes have been established in accordance with IEEE Std I, *General Principles for Temperature Limits in the Rating of Electric Equipment.*
Insulation Systems are classified as follows:

*Class A*—A class A insulation system is one which by experience or accepted test can be shown to have suitable thermal endurance when operating at the limiting class A temperature specified in the temperature rise standard for the machine under consideration.

*Class B*—A class B insulation system is one which by experience or accepted test can be shown to have suitable thermal endurance when operating at the limiting class B temperature specified in the temperature rise standard for the machine under consideration.

*Class F*—A class F insulation system is one which by experience or accepted test can be shown to have suitable thermal endurance when operating at the limiting class F temperature specified in the temperature rise standard for the machine under consideration.

*Class H*—A class H insulation system is one which by experience or accepted test can be shown to have suitable thermal endurance when operating at the limiting class H temperature specified in the temperature rise standard for the machine under consideration.

*Table 8-7* Summary of Formulas

| Mechanical Formulas | Electrical Formulas | |
|---|---|---|
| Torque lb-ft = $\dfrac{hp \times 5252}{RPM}$  hp = $\dfrac{torque \times rpm}{5252}$ | **To Find** | **Alternating Current: Three-Phase** |
| Sync. rpm = $\dfrac{120 \times frequency}{no.\ poles}$ | Amperes when horsepower is known | $\dfrac{hp \times 746}{1.73 \times E \times Eff \times PF}$ |
| **Rules of Thumb (Approximation)*** | Amperes when kilowatts are known | $\dfrac{KW \times 1000}{1.73 \times E \times PF}$ |
| At 1800 rpm, a motor develops 3 lb-ft/hp  At 1200 rpm, a motor develops 4.5 lb-ft./hp | Amperes when kVA are known | $\dfrac{kVA \times 1000}{1.73 \times E}$ |
| At 575 V, a three-phase motor draws 1 A/hp  At 460 V, a three-phase motor draws 1.25 A/hp  At 230 V, a three-phase motor draws 2.5 A/hp  *Departs on lower hp and rpm motors. | Kilowatts | $\dfrac{1.73 \times I \times E \times PF}{1000}$ |
| **Temperature Conversion** | Horsepower (output) | $\dfrac{1.73 \times I \times E \times EFF \times PF}{746}$ |
| °C = (°F − 32) × $\dfrac{5}{9}$  F = $\left(°C \times \dfrac{9}{5}\right) + 32$ | I = current  E = voltage  Eff = efficiency | PF = power factor  kVA = kilovolt-amperes  kW = kilowatts |

10. Describe armature reaction.
11. Describe generator effect in a motor.
12. What are the three types of ac motors?
13. What type of motor can use both ac and dc?
14. What is the main advantage of a synchronous motor?
15. How did the squirrel-cage motor get its name?
16. How is a synchronous motor started?

## REVIEW PROBLEMS

Table 8-7 shows some rule-of-thumb math methods utilized to obtain various motor torques and current drain.

1. Using the rule-of-thumb approximations, what is the torque developed by a three-phase motor at 1800 rpm?
2. Using the rule-of-thumb approximations, what is the torque developed by a three-phase motor at 1200 rpm?
3. How much current does a 575-V, three-phase motor draw when producing 10 hp?
4. How much current does a 460-V, three-phase motor draw when producing 10 hp?
5. How much current does a 230-V, three-phase motor draw when producing 10 hp?
6. How many amperes does a 10-hp, three-phase motor with 230V applied draw when it has an efficiency of 80% and a power factor of 0.8?
7. How many amperes does a 100-kW motor draw if it is connected to a three-phase, 240-V line and has a power factor of 0.85?
8. If a three phase motor needs 10kVA on 240 V, what is the current draw?
9. What is the horsepower of a three-phase motor that draws 10A on a 240-V line and has an efficiency rating of 80% and a power factor of 0.75?
10. A six pole synchronous motor is connected to a 240-V, 60-Hz line. How fast does it rotate?

# CHAPTER 9
# Timers and Sensors

## PERFORMANCE OBJECTIVES

After studying this chapter, you will be able to:

1. Describe the difference between synchronous clock timers and solid-state timers.
2. Draw a ladder diagram of a control circuit with an on-delay relay.
3. List the uses of the general-purpose relay.
4. Explain how programmable timers operate.
5. Understand how thumbwheel switches are used to set time counts on electronic counters.
6. Describe a dual-in-line packaging (DIP) switch.
7. Explain the advantages of a pneumatic timing relay.
8. Explain sequence control.
9. List the uses for sensors in industry.
10. Explain how solid-state level controls work.
11. Determine dielectric constants of various materials.
12. Describe the difference between a thermistor and a thermocouple.
13. Describe strain gage operation.

## TIME AND TIMERS

Time is important in any industrial or commercial operation. Proper timing makes it possible to meet production schedules and demands. Timing the operation of a machine is an important function and needs the proper equipment to do the job. Many types of industrial timers are available to control operations of machines. In most instances it is necessary to control the motor that drives the machine. A close examination of timers shows that there are several types, viz. general-purpose, pneumatic, programmable and solid-state.

Of course, there are such individual-purpose timers as those required for welders and similar pieces of equipment. Complex operations and production schedules are timed by equipment that has been designed for doing the job and making sure it is done properly and accurately. Timers are called upon to time a fraction of a second up to hours and in some cases a year. Timers can be broken down into three categories:

- Dashpot
- Synchronous clock
- Electronic timers

Electronic timers are gaining more favor as they become less expensive and more reliable than the older mechanical types. Since there are new types being designed and released each week it is best to keep updated using the internet.

## DASHPOT OR PNEUMATIC TIME-DELAY TIMERS

The dashpot timer (Fig. 9-1) operates using a pneumatic chamber with a variable orifice. When a voltage is applied to the relay coil, the armature pushes against a diaphragm and air is forced out of the chamber. The speed at which the air exits the chamber controls the length of time delay.

*Fig. 9-1* Pneumatic time-delay relay. (Square D)

## SYNCHRONOUS CLOCK TIMERS

A synchronous motor is used to turn a clock mechanism (Fig. 9-2). The synchronous motor is very accurate since it operates on the frequency of the line voltage and is not affected by fluctuations in line voltage. This type of timer has one or more contacts that open or close depending on the position of the clock hands.

*Fig. 9-2* Synchronous motors used to power motor-driven timers.

The hands can be set to where they are to make contact. This means that it is possible to obtain from 1 minute to up to 12 hours of time delay on the clock. Some can be made to time in seconds. A good example of a clock timer is the one used in darkrooms to time the development of film.

## SOLID-STATE TIMERS

The solid-state timer is more accurate and versatile and is becoming the standard in industry. Although the price has a tendency to continue to decrease, it has become the most reliable type of timer. In most instances there are no moving parts to cause trouble or need servicing. The time delay is set by a resistance-capacitance (RC) network and the switching is usually done by an SCR. Most units are encapsulated in plastic of some type, usually an epoxy.

## TIME-DELAY RELAYS

The pneumatic-type relay has a synthetic rubber bellows for controlling the tripping time (Fig. 9-3). The timer is particularly useful for applications requiring a timer with greater accuracy and a longer timing range than is afforded by the fluid dashpot timer. These pneumatic timing relays have a range of 0.05 to 180 seconds with a repetitive accuracy of ±10%. A minimum reset time of 75 milliseconds must be provided to ensure repetitive accuracy. The times are adjustable for a period longer than 180 seconds (3 minutes) provided that 100% accuracy is not needed.

Relays can be timed so that the delay takes place with the contacts closed or open. When power is applied to the relay coil of an on-delay relay, a period of time passes before the relay contacts change state. Figure 9-3 shows the schematic symbols for time-delay relays.

A ladder diagram of a control circuit with an on-delay relay is shown in Fig. 9-4. The start button is pushed and the control relay (CR) is energized; thus all CR contacts close. The CR remains energized even after the pushbutton is released. Relay contacts CR-1 and CR-2 are closed, and a voltage is applied to the coil of the time-delay relay (TR). Even though voltage is applied to the timing relay coil, contact TR-1 does not close immediately. After a predetermined period of time, TR-1 closes and the indicator lamp turns on.

**Fig. 9-4** *Ladder diagram of control circuit with a delay relay.*

This on-delay relay gets its name from the fact that a period of time passes before the contact is closed. If the circuit also has an NC contact, the NC contact is also delayed for a period of time before it opens. When the circuit is turned off, all contacts immediately return to their normal de-energized positions.

The off-delay relay symbols are shown in Fig. 9-5. This relay has NO and NC contacts. Voltage is applied

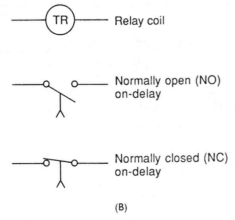

**Fig. 9-3** *(A) On-delay timer. (Allen-Bradley) (B) Schematic symbols for an on-delay and its NO and NC contacts.*

**Fig. 9-5** *Schematic symbols for an off-delay relay and its NO and NC contacts.*

to the coil of the off-delay relay. The contacts change state immediately as they do in a normal control relay. NO contacts close and remain closed as long as a voltage is applied to the coil. Even after the voltage is removed from the coil, the contacts remain in their activated states for a preset period. The contacts return to their normal states only after the relay times-out.

The off-delay relay shown in Fig. 9-6 is a ladder diagram of a control circuit. The *start* button is pushed and a voltage is applied to the relay's coil (CR). Then the START button is released, so CR-1 closes and maintains voltage to coil CR. Contact CR-2 also closes and a voltage is applied to coil TR. Contact TR 1 closes immediately and the indicator lamp turns on. The *stop* button is pushed and voltage is removed from coil CR. All CR contacts open. Voltage is also removed from coil TR. However, contact TR-1 remains closed for a preset period of time and the indicator lamp remains on. After the relay times-out, contact TR-1 opens and the light goes out.

**Fig. 9-7** *Solid-state general-purpose timing relay.* (Square D)

are counted from the keyway in a clockwise direction. However, looking at the bottom of the socket, you have to count in the opposite (counterclockwise) direction for making connections to the socket. Some timers have an 11-pin base. An adjustable knob on top of the case allows for the time delay to be adjusted from 0.1 to 10 seconds on some and other time ranges for others (Table 9-1).

**Table 9-1** *Timing Relays*

| (a) Variable Time-Delay Relays[a] ||
|---|---|
| 0.1–10 seconds | |
| 0.3–30 seconds | Integrated circuit with CMOS circuitry is |
| 0.6–60 seconds | used to provide an accurate time delay. |
| 1.2–120 seconds | |
| 1.8–180 seconds | |
| 0.1–10 minutes | |
| 0.3–30 minutes | Integrated circuit with CMOS circuitry is |
| 0.6–60 minutes | used to provide an accurate time delay. |
| 1.2–120 minutes | |

| (b) Fixed Time-Delay Relays ||
|---|---|
| Timing Mode | Timing Range (seconds) |
| On delay | 1–80 |
|  | 181–3600 |
| Off delay | 1–180 |
|  | 181–3600 |
| Interval | 1–180 |
|  | 181–3600 |
| One Shot | 1–180 |
|  | 181–3600 |
| Repeat cycle | 1–180 |
|  | 181–3600 |

[a] Fixed repeat cycle timers are supplied with the same on and off times.

**Fig. 9-6** *Ladder diagram of control circuit with off-delay relay.*

## GENERAL PURPOSE TIMING RELAYS

The general purpose timing relay has many uses. It is used for automatic control of machine tool programming, sequencing controls, heating and cooling operations, and warm-up delays. Most of this type are designed for plug-in housing (Fig. 9-7). The plug-in base resembles the eight-pin base of the older vacuum tubes. In fact, they can fit into some vacuum-tube sockets. Note how the keyway indicates the beginning or pin 1 and the location of pin 8. The pins

## CONTROL OF PROGRAMMABLE TIMERS

Programmable timers are microprocessor controlled to provide flexibility with accurate timing. The on-delay timer has five programmable timing ranges:

0.5 to 9.99 seconds

0.1 to 99.9 seconds

1.0 to 999 seconds

0.1 to 99.9 minutes

1.0 to 999 minutes

**Fig. 9-8** *Programmable timers.* (Square D)

**Fig. 9-10** *Digital solid-state timer/counter—single-stage version: (A) type P; (B) type PM.* (Square D)

A five-position rotary switch is used to select the timing ranges. Three pushbutton thumbwheels are used to select the time value (Fig. 9-8). Timing modes are shown in Fig. 9-9. These relays draw about 1.2 VA, making the solid-state relay very low in power consumption.

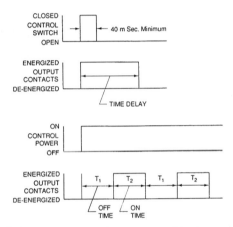

**Fig. 9-9** *Timing modes for programmable timer.* (Square D)

## DIGITAL SOLID-STATE TIMER/COUNTER

There are some extremely flexible microcomputer-based digital timer/counters. They can be field programmed to perform as a timer or counter with a variety of timing or counting modes and ranges. They are available in either a single- or dual-stage model and can be ordered with either relay or solid-state outputs and with ac or dc inputs to make them compatible with most control systems.

### Single-Stage Version

Single-stage units have a selector switch and four pushbutton thumbwheels located on the front face of the device (Fig. 9-10). The selector switch is used to select the time or count range, while the four pushbutton thumbwheels are used to set the time or count value. Operating mode selection is accomplished by programming seven internal DIP switches. The units are available with either a digital readout capable of up or down timing/counting or a LED status indicator.

### Time/Count Modes

Relay switches require a great deal of attention, which results in a lot of downtime for the machines they control. In industry, the relay has quickly been replaced by electronics, just as the telephone company uses transistor switching to replace relay switching. There are no moving parts in transistors and no contact points to be cleaned periodically. With relays, the entire circuit had to be rewired to change a program, but with a digital solid-state timer/counter, only the thumbwheels on front of the device are changed.

The four modes for the solid-state timer/counter are on-delay, off-delay, interval, and repeat cycle pulse.

*On-delay mode* Closing the initiating contact begins the delay period, and the time/count outputs are energized.

*Off-delay mode* Closing the initiating contact energizes the time/count outputs. Opening the initiating contact begins the delay period. At the end of the delay period, the time/count outputs de-energize.

*Interval* Closing the initiating contact energizes the time/count outputs and begins the delay period. At the end of the delay period, the time/count outputs de-energize.

*Repeat cycle pulse* Closing the initiating contact begins the delay period. At the end of the delay period, the time/count outputs energize for 50 milliseconds and then de-energize. The device repeats this cycle until the initiating contact is opened.

### Dual-Stage Version

Dual-stage units have eight pushbutton thumbwheels located on the face of the device (Fig. 9-11). The four left pushbutton thumbwheels are used to set the

**Fig. 9-11** *Digital solid-state timer/counterdual-stage version: (A) type P; (B) type PM.* (Square D)

time/count value for stage 1, while the four right pushbutton thumbwheels are used to set the time/count value for stage 2. A seven-position DIP switch located inside the device is used to set the time or count range and mode. The units come with two LED status indicators, one for stage 1 and one for stage 2.

*Two-stage (on-delay) mode*  Closing the initiating contact begins the stage-1 delay period. At the end of stage-1 delay period, the stage-1 time/count outputs energize and the stage-2 delay period begins. At the end of the stage-2 delay period, the stage-2 time/count outputs energize.

*Two-stage (on-delay)*  Closing the initiating contact begins the *off* stage-1 delay period. At the end of the *off* stage-1 delay period, the stage-2 time/count outputs energize and the *on* stage-2 delay period begins. At the end of the *on* stage-2 delay period, the stage-2 time/count outputs de-energize and the *off* stage-1 delay period begins. This cycle repeats until the initiating contact is opened. The stage-1 time/count outputs act as instantaneous outputs and are energized when the initiating contact is closed.

*Two-stage (off-delay)*  Closing the initiating contact energizes the stage-1 and stage-2 time/count outputs. When the initiating contact is opened, the stage-1 delay period begins. At the end of the stage-1 delay period, the stage-1 time/count outputs de-energize and the stage-2 delay period begins. At the end of the stage-2 delay period, the stage-2 time/count outputs de-energize.

*Repeat cycle (off-delay)*  Closing the initiating contact energizes the stage-1 and stage-2 time/count outputs. When the initiating contact is opened, the stage-1 time/count outputs de-energize and the *on* stage-1 delay period begins. At the end of the *on* stage-1 delay period, the stage-2 time/delay outputs de-energize and the *off* stage-2 delay period begins. At the end of the *off* stage-2 delay period, the stage-2 time/count outputs energize and the *on* stage-1 delay period begins. This cycle repeats until the initiating

contact is closed. The stage-1 outputs act as instantaneous outputs and are energized when the initiating contact is closed. These timer/counters operate on 120 V 50/60 Hz (102 to 132 V) and 240 V 50/60 Hz (204 to 264 V). Figure 9-12 shows the wiring diagrams for these timers/counters.

**Replaceable Hard Relay Outputs:**
Wiring and Contact Arrangement

| | Inductive 35% Power Factor | | | | | Resistive 75% Power Factor |
|---|---|---|---|---|---|---|
| | Make | | Break | | Continuous Amperes | Make, Break and Continuous Amperes |
| Volts | Amps | VA | Amps | VA | | |
| 120 | 30 | 3600 | 3.0 | 360 | 7 | 7 |
| 240 | 15 | 3600 | 1.5 | 360 | 7 | 7 |

**DC Contact Rating:** 7 amps max. @ 28VDC resistive or inductive

**Output Relay Operating Times:**
Pick-up 25 ms
Drop-out 25 ms

**Solid State Outputs:**
Wiring and Contact Arrangement

**Solid State Contact Ratings:**
50 mA @ 30VDC

**Fig. 9-12** *Contact points on a timer/counter.* (Square D)

## Thumbwheel Switches

Thumbwheel switches are utilized on many electronic devices to allow for the setting of time or counts. They can be ganged to make more than one. A nut driver is used to tighten the nuts on the end of the long screw that inserts into the body of the switches to hold the outside cases and any separators together. (Fig. 9-13). You can see from those shown that push button front mount switch has a pushbutton for setting each number, whereas the others have the thumbwheel that can be rotated until the correct digit comes up. Components needed to assemble a complete switch are

**PUSH-BUTTON FRONT MOUNT   INTERMEDIATE FRONT MOUNT**

**MINIATURE REAR MOUNT           MINIATURE FRONT MOUNT**

**THIN REAR MOUNT               INTERMEDIATE REAR MOUNT**

**THUMBWHEEL SWITCHES**
**HARDWARE FOR FIELD ASSEMBLY**

*Fig. 9-13* Thumbwheel switches.

- One pair of end plates
- Blank bodies to divide switches in a bank
- Divider plates to separate switch banks into framed sections
- Hardware

## DIP SWITCHES

DIP (dual-in-line packaging) is also used with integrated circuits or chips and their packaging. DIP switches are small and have contacts that resemble those of an integrated circuit (IC) or chip. They are designed to be attached to a printed circuit board and soldered in place. As you can see from Fig. 9-14, these are very small and it takes a ballpoint pen or something equally small to cause

Raised Rocker    Recessed Rocker    PIANO-DIP® Switch

**SPST SLIDE-ACTUATED SWITCHES**

*Fig. 9-14* DIP switches.

the switches to be rotated from on-to-off or off-to-on. They also come in raised rocker, recessed rocker, and piano DIP switch. A high-pressure spring-and-ball contact system causes them to make and break contact. The ball makes the complete circuit between the two contacts as shown in the cutaway views.

A high-pressure, spring-and-ball contact system is also used to make or break contact in single-pole, single-throw, slide-actuated switches, which are very small and need a very small screwdriver tip or ball point pen to close or open them. These are also mounted on printed circuit boards and are used to program a circuit or chip to function in such a manner as in the timer/counters discussed in previous paragraphs.

## PNEUMATIC TIMING RELAYS

The pneumatic timing relay has some advantages over other types. It has the advantage of not being affected by normal variations in ambient temperature and atmospheric pressure. It is adjustable over a wide range of timing periods and has good repeat accuracy. This type of relay is available with a variety of contact and timing arrangements (Fig. 9-15).

This pneumatic time-delay unit is mechanically operated by a magnet structure. The time-delay function depends on the transfer of air through a restricted orifice. This restriction is done by the use of a reinforced synthetic rubber bellow or diaphragm. The timing range is adjusted by positioning a needle valve to vary the amount of orifice or vent restriction.

**Fig. 9-15** Pneumatic timing relays: (A) type AO-10E; (B) type HO-10E. (Square D)

Energizing or de-energizing pneumatic timing relays can be controlled by devices such as pushbuttons, limit switches, and thermostatic relays. They draw very small amounts of current, so sensitive control devices are used to control the operating sequence.

This type of timing relay is used for motor acceleration and in automatic control circuits. Automatic control is needed where uses are repetitive and accuracy is required.

Two types of time delay are provided by pneumatic timers: on-delay, which means that the relay provides time delay when it is energized, and off-delay, which means that the relay is de-energized when it provides time delay. Figure 9-16 shows the on-delay arrangement in an Allen-Bradley relay. When the operating coil (O) is energized, solenoid action raises the solenoid plunger (A). The pressure on the rod (B) is released. A spring (W) is located inside the bellows (E). The spring then allows the pushrod (B) to move upward. As (B) moves upward, it causes the off-center linking mechanism (C) to move the end of the snap-action mechanism (X) upward. This action, in turn, raises (D) to operate the switch or time-controlled contacts.

The position of the needle valve at the bottom determines how fast the bellows rises. The needle valve setting determines the time interval between the solenoid closing and the rise of the bellows to operate the switch. If the needle valve is almost closed, it takes a considerable length of time for air to pass the valve and cause the bellows to rise. Gravity causes the plunger (B) to drop when the coil is de-energized. The action of the reset spring and the falling plunger resets the timer almost instantaneously.

The off-delay mode in this Allen-Bradley relay is shown in Fig. 9-16 with the solenoid rotated 180°. When the operating coil (O) is energized, plunger (A) is held down and (G) pushes against (B). That holds the

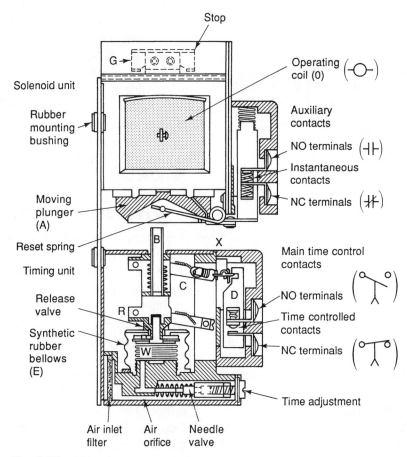

**Fig. 9-16** AC on-delay timer with coil energized and timer timed-out. (Allen-Bradley)

bellows (E) in a fully compressed position. De-energizing the coil allows the reset spring to force (A) upward. This also causes (G) to rise. That releases the downward force on (B). With the pressure taken from (B), the bellows (E) slowly expands. This forces (B) upward. As (B) moves up, it trips to toggle; this is the same as in the on-delay mode. Tripping the toggle causes the switch to be tripped. Figure 9-17 shows the symbols used in the standard elementary diagrams for timed contacts.

Contact action is retarded after the coil is:

| Energized | | De-energized | |
|---|---|---|---|
| NOTC | NCTO | NOTO | NCTC |
| On-Delay Mode | | Off-Delay Mode | |

**Fig. 9-17** *Timed contacts symbols. Contact action is retarded after the coil is: Energized De-energized.*

## MOTOR-DRIVEN TIMERS

There are many types of motor-driven timers used for various purposes. Figure 9-18 is only one of the many types utilized in industry and in commerce. This timer is operated from a sustained or momentary contact switch. In the latter case, the solenoid-actuated switch can energize a load during the time period as an interval timer. Its features are cam-actuated switches that allow programmed timing, reduce switch actuator over travel, and greatly extend switch life. A differential clutch employs hardened gears for reliable, long-life operation. Power required is 120 V and the motor consumes about 2.5 W. Scale reset is within 0.5 second. Switch ratings are 10 A resistive at 125 V ac or hp at 125 V.

**Fig. 9-18** *Reset timer.*

Many timers are easily converted from on-delay to off-delay. Figure 9-19A shows some of the types available for various jobs. Most are synchronous motor driven. These are hysteresis-synchronous motors and are self-starting but very accurate since they synchronize on frequency rather than depend on voltage for constant speed. Figure 9-19B shows a set of cams that can be programmed for their on-off periods. A number of these switches can be ganged to control a number of operations. Figure 9-20 shows how a standard start-stop station controlling an on-delay timer with a set of timed-closed contacts is arranged in an elementary diagram.

**Fig. 9-19** *Industrial timers.*

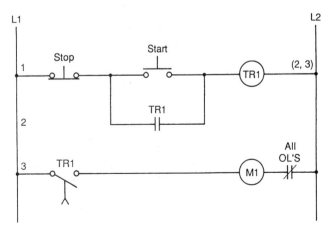

**Fig. 9-20** *Time-delay circuit.*

# SEQUENCE CONTROL

The circuit shown in Fig. 9-21 is a sequence control of two motors, one to start and run for a short time after the other stops. In this system it is desired to have a second motor started automatically when the first is stopped. The second motor is to run only for a given length of time. Such an application might be found where the second motor is needed to run a cooling fan or a pump.

**Fig. 9-21** *Sequence control.* (Allen-Bradley)

To accomplish this, an off-delay timer (TR) is used. When the *start* button is pressed, it energizes both M1 and TR. The operation of TR closes its time delay contact but the circuit to M2 is kept open by the opening of the instantaneous contact. As soon as the *stop* button is pressed, both M1 and M2 will continue to run until TR times out and the time-delay contact opens.

## PROGRAMMABLE TIMERS

With the advent of microprocessors and digitized equipment, many types of programmable timers have become available. One good example of some of the possibilities that timers offer is shown in the Watchdog Time Commander (Fig. 9-22). It is a precision, wall-mounted, 365-day timer that can be programmed to operate to an exact second. The timer consists of a clock and an enclosure with a four-circuit capacity. Another model has 16-circuit capability. Relay modules may be added to both models. Each relay module controls four electrical circuits.

**Fig. 9-22** *Watchdog Time Commander programmable timer.* (Square D)

The timer accepts 60 operational programs, each capable of controlling any or all circuits. Up to 20 holidays may be programmed, and each holiday can be a single day or an unlimited span of days. Battery backup retains programs, during power failures, for up to three years. Leap-year adjustment is automatic and the start and end of daylight savings time can be programmed. Commands may also be programmed relative to sunrise and sunset.

Some of the purposes of this type of timing are energy management, security, and a wide range of industrial and commercial applications. Lights, water heaters, fans, furnaces, and pumps may be turned on and off to best utilize energy dollars. Door locks, burglar alarms, and elevators may be cycled to ensure security during unoccupied hours. Industrial and commercial uses include sprinkler and irrigation systems, traffic control, schools, and industrial bells for life testing. One-second resolution provides precise and accurately-timed control.

Programming instructions come with the unit. The detailed instructions manual ensures that you are able to program the unit for its intended purposes.

## SENSORS

Sensors are used for many purposes in industry. They detect the presence or absence of materials or products. They can be used to set off an alarm if one or a number of conditions are not met. Sensors are made to sense temperature, pressure, and in some instances, levels of materials in tanks or storage facilities.

### Solid State Level Controls

The level control shown in Fig. 9-23 detects the level of materials by having the material's presence detected by the two probes that stick out from the body of the sensor. When the material being checked is between the two probes, it changes the characteristics of a transistorized circuit and causes it to energize a relay. A number of approaches can be taken in electronics to accomplish the sensing action. The Hall effect can be utilized (discussed in a later Chapter), the eddy current method can be used, or the capacitance between the two probes can be changed by having the dielectric altered. The sensor shown in Fig. 9-23 has the ability to respond in 0.5 second

**Fig. 9-23** *(A) Solid-state level sensor; (B) note the shape and size of the oscillating fork.* (Square D)

to level changes. It can be used to detect either rising or falling liquid levels in storage vessels, mixing tanks, or pipelines. It can be used with any liquid compatible with the stainless steel housing. It is ideally suited for applications that have material buildup, foam, gas bubbles, or suspended solids. It is unaffected by agitation, wave action, or turbulence and can be installed directly without the need for measuring chambers or bypasses. It is designed to operate between 24 and 250 V ac at 50/60 Hz, in series with the relay, solenoid, valve, or annunciator it is to control. Power should never be applied to the sensor without an external load connected in series. Figure 9-24 shows how the power input is connected and how the contacts are arranged at terminals 7 through 12. Even with the load switched off, a small current of less than 5 mA still passes through the sensor and load to maintain power to the unit's sensing electronics.

The extremely sensitive sensing head detects materials with dielectric constants as low as 1.5.

A dielectric is the material (air and vacuum or other substances) that separates the two plates of a capacitor. This material varies in its influence on the capacitance of a capacitor. Each material has a constant when compared with a vacuum, which has a constant of 1. Table 9-2 shows the constants for a variety of substances.

*Table 9-2*  Dielectric Constants

| | | | |
|---|---|---|---|
| Air | 1.00006 | Olive oil | 3.11 |
| Ammonia (liquid) | 15.5 | Paper | 2–6 |
| Asphalt | 2.68 | Paraffin | 2.0–2.5 |
| Beeswax | 2.75–3.0 | Polyethylene | 2.3 |
| Ceramics | 80–1200 | Porcelain | 6–8 |
| Glass, Pyrex | 3.8–6.0 | Rubber | 2.8 |
| Glass, Corning | 9.5 | Sulfur | 4.0 |
| Mica | 7–9 | Vacuum | 1.0 |
| Nylon | 3.5 | Water | 80 |
| Oil | 2–5 | | |

The capacitance-type level limit control shown in Fig. 9-25 has the ability to be mounted in any number of positions. The field-selectable fail-safe mode is advantageous since it allows the user to apply the unit as either normally open (material absent) or normally closed (material present). The connections shown in Fig. 9-24, which also apply here for the sensor connections, are marked next to the 115-V ac power input. The internal relay has its output connections from terminals 7 to 12.

*Fig. 9-24*  Relay operated by the level sensors. (Square D)

## Capacitance-Type Sensor

The capacitance-type level limit sensor is ideal for use in applications where control and measurement of powdered, granular, or pelletized solids is required.

## Temperature Sensing

The ability to sense the actual temperature or the ability to detect a difference between temperatures makes it possible to control the temperature of liquid baths, bearings, internal combustion engines, and large air compressors, to mention but a few devices. Figure 9-26

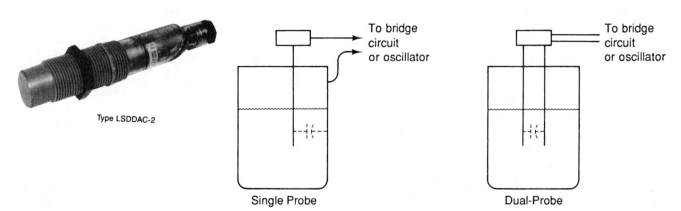

*Fig. 9-25*  Capacitance-type level sensor uses ac bridge circuits or changes frequency of an oscillator as fluid presence changes dielectric between probes or between probe and side of the tank. (Square D)

**166**  Timers and Sensors

**Fig. 9-26** *Temperature switches. (Square D)*

shows some of the temperature switches used for automatic control of temperature-maintenance equipment in industrial and general-duty applications. Temperature sensing may be done by a number of methods, including:

- thermocouple
- thermistor
- resistance temperature detector
- semiconductor temperature sensor

## Thermocouple

Thermocouple is the most widely used temperature-sensing device. It dates back to 1821, when Thomas Seebeck found that joining two wires made of different materials generate an EMF when heated at one end. The amount of heat is directly proportional to the amount of output EMF or voltage. This is a simple, rugged device that can be made inexpensively. It also has the advantage of being able to measure a 4500°F range of temperatures. It is, however, unstable and needs a reference junction to make sure that its output is useful. Figure 9-27 shows how the thermocouple works. Industry uses thermocouples for measuring the temperature of ovens and furnaces, of flowing liquids, and the core temperatures of nuclear reactors.

## Thermistor

Thermistor is a device that produces a decrease in resistance when heated. As the temperature increases, the resistance decreases. Thermistors come in many

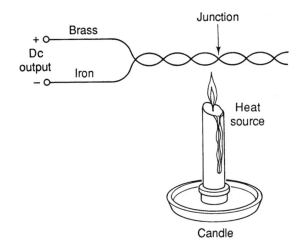

**Fig. 9-27** *Thermocouple produces the Seebeck effect.*

sizes and a multitude of shapes (Fig. 9-28). This is the most temperature sensitive device, and its resistance varies from 0.5 Ω to 80 MΩ. However, it does have disadvantages, including nonlinear response and limited temperature range. Thermistors are used where it is necessary to detect temperature changes of 1°C in chemical processes. They can also be used as a liquid- or fluid-flow control device. They detect the presence of a liquid through the temperature change in the material. Once immersed in the liquid, a thermistor can determine the temperature of the material and serves to energize the proper circuitry to cause the level to be corrected.

## Resistance Temperature Devices

The thermistor changes resistance when heated in an inverse direction, as is normal for most metals or materials. This means that gold, silver, iron, or pure elements

**Fig. 9-28** *Thermistors.*

such as these increase in resistance as they are heated. This particular characteristic can also be used in designing devices to control temperature.

Some of the best known devices are the bridge circuit (shown in Fig. 9-29), metal film resistance temperature detectors, and helical resistance temperature devices. The metal film detector is made by depositing a thin film of platinum on a flat ceramic substrate and etching it with a laser. It is then trimmed and sealed. This makes a very sensitive device for detecting temperature

**Fig. 9-29** *(A) Helical and (B) metal film resistance temperature devices with (C) a bridge circuit.*

changes. Disadvantages include the requirement of an amplifier because of the small temperature changes and the fact that response time is rather slow.

## Semiconductor Temperature Sensor

Silicon and germanium can both be used for temperature sensing. Germanium-doped crystals can be used to detect temperatures near absolute zero. The negative temperature coefficient of germanium is similar to that of the thermistor. However, silicon has a positive temperature coefficient. It is also limited in its range of sensing from $-67°F$ to $275°F$. Semiconductor temperature sensors have good linearity and are small and inexpensive.

## Pressure Sensors

There are many ways to sense pressure. The strain gage and the piezo-electrical are the two most commonly used to produce an electrical output that can be utilized to control a process. Different manufacturers use different methods. Each appears to have a specialty that has worked for years, and it is improved over the years to make use of the latest technology in the electronics field.

## Strain Gage Transducers

Figure 9-30 shows two types of pressure transducers that fulfill the requirements of many industrial users. They are reliable, accurate, and provide a continuous (analog) pressure input in process and factory automation. These strain gage-based sensors measure strain or tension. The gages are attached to one side of a pressure-sensing diaphragm, then pressure is applied to the diaphragm, producing a minute deflection that induces strain to the gages, thus changing their resistances. By applying a constant voltage across the strain gages, the change in resistance is measured. The resulting outputs are conditioned to user needs through amplifiers and other circuitry within the transducer (Fig. 9-31).

## Piezo-resistive Transducers

This type of transducer or sensor is used to sense ranges from 10 to 5000 psi. A strain sensor is incorporated in the semiconductor circuitry.

Pressure sensors are used in pneumatic systems; hydraulic systems; irrigation systems; lubrication systems; heating, air-conditioning, and ventilation systems; energy management systems; robotics; automated process equipment; plant utilities; and machine tools.

**Fig. 9-30** *Pressure transducers: (A) PTA; (B) PTB. (Square D)*

**Fig. 9-31** *Strain gage in a flow-meter.*

**Fig. 9-32** *Wiring diagrams for using a pressure switch as a sensor.*
(Square D)

Figure 9-32 shows how the transducers are wired into the circuitry.

## REVIEW QUESTIONS

1. Name the three categories of timers.
2. Explain how the dashpot timer works.
3. What is the advantage of the pneumatic-type relay over the dashpot type?
4. Explain the terms on-delay, off-delay, and times out.
5. Where is the general-purpose timing relay used?
6. How is the time delay set on electronic timers?
7. What are the four modes for solid-state timer/counters?
8. What are thumbwheel switches used for?
9. How are DIP switches set?
10. Where are pneumatic timing relays used?
11. Draw the symbols for a timer's on-delay mode and off-delay mode.
12. What is a dielectric? Where is the term useful?
13. Name four methods of temperature sensing.
14. How does the thermistor operate?
15. What are the two most commonly used methods used to detect pressure changes?
16. What does DIP mean?
17. Which timer is becoming the standard in industry?
18. What is an RC network?
19. What does CR stand for?
20. How is the time period adjusted on a general-purpose timing relay?

## REVIEW PROBLEMS

Timing has been done in electronic circuits for some time, utilizing the charge and discharge rate of a capacitor. By utilizing the time constant in a resistor-capacitor combination it is possible to time various operations. Inasmuch as many timers are now electronic based, it may be good to review the basic principles of operation of the $T = R \times C$ circuit.

1. A capacitor of 10 microfarads ($\mu F$) and a resistor of 1.5 M$\Omega$ are connected in series. What is the time constant of the circuit?
2. If a 0.5-$\mu F$ capacitor and a 2-M$\Omega$ resistor are in series and are connected across a source of 120 V dc, how long after energizing the circuit will it take for the voltage across the capacitor to reach 75.84 V?
3. If a 5.0-$\mu F$ capacitor and a 2-M$\Omega$ resistor in series are connected across a 250-V dc source, how long after energizing the circuit will it take for the voltage across the capacitor to reach 248 V?
4. What is the time constant of a 10-$\mu F$ capacitor in series with a 1-M$\Omega$ resistor?
5. What is the time constant of a 9-$\mu F$ capacitor in series with a 10-M$\Omega$ resistor?
6. What is the second time constant of a 10-$\mu F$ capacitor and a 2-Meg resistor connected in series?
7. What is the third time constant of a 10-$\mu F$ capacitor and a 2-Meg resistor connected in series?
8. What is the fourth time constant of a 10-$\mu F$ capacitor and a 2-Meg resistor connected in series?
9. What is the fifth time constant of a 10-$\mu F$ capacitor and a 2-Meg resistor connected in series?
10. How many time constants does it take for a capacitor to charge to its full value?

# 10
CHAPTER

# Sensors and Sensing

## PERFORMANCE OBJECTIVES

After studying this chapter you will be able to

1. List the classes of sensors.
2. Describe the difference between presence and non-contact sensing.
3. Explain how limit switches are used as sensors.
4. Explain how speed sensing is accomplished.
5. Describe anti-plugging.
6. Explain the operation of pressure controls.
7. Explain the operation of temperature sensors.
8. Describe the function of wells and packing glands.
9. Understand how float switches are used as sensors.
10. Understand how optical encoders operate.
11. Define binary-coded decimal (BCD) and light-emitting diode (LED).
12. Define multiplexing.
13. Explain the operation of various types of proximity switches.
14. Describe the functioning of photoelectric switches.
15. List the advantages of LEDs as a light source.
16. Describe the operation of radio-frequency (RF) identification systems.
17. Describe the operation of scanners.
18. Explain how the vision system operates.

Sensors are the foundation of the control system. They must provide usable input to the high-level controller. They must gather data and communicate the data, on a real-time basis, to the appropriate level within the control system. A wide range of sensing devices is available from many different technologies. Sensors include a broad range of technologies and physical configurations. They range from a simple electromechanical limit switch, which provides a single bit of information to the control system, to radio frequency-activated tags storing up to 2K bytes of manufacturing information.

## CLASSES OF SENSORS

Sensors may be classified as either contact or noncontact. They may be further classified as internal or external and as passive or active. Various types of sensors are added to manufacturer's catalogs every year. In order to keep up with these changes it will be necessary to use the internet to locate and read the latest information for these new devices.

## Contact Sensors

A limit switch is a contact sensor that permits a system to sense whether an object is present or missing. If the object makes contact with the limit switch, the system knows that the object is near enough to begin the next operation. If the switch is not closed, it means that the object is missing and the system must react accordingly. That usually generates what is referred to as an alarm condition.

Force, pressure, temperature, and tactile sensors all respond to contact. They all send their signals to another device or to a central location for processing.

## Noncontact Sensors

Pressure changes, temperature changes, and electromagnetic changes can all be sensed by contact methods. They usually react to a change in a magnetic field or a light pattern. If an electromagnetic field is disturbed, it is sensed and fed to the controller. The same is true of the disturbance of a light beam. Changes in the light beam—its intensity or whether or not it is present—are sent to an electronic circuit or controller for processing and then sent back to a device so that it can react according to the disturbance.

## PRESENCE SENSING

Presence sensors indicate whether a part or piece is in position. These are used to indicate an on-off condition. They are status devices that communicate a single bit of information. Limit switches, proximity switches, and photoelectric cells fall into this category. These devices are becoming smaller and faster, and have greater reliability, with increasing emphasis on non-contact sensing.

## LIMIT SWITCHES

Perhaps the most commonly used type of sensing device is the limit switch. It has been around for many years and has been used in almost all electrically operated devices. This wide usage translates into many varied types of devices. Some are designed for general purposes and others are designed for specific operations. For instance, they may be designed for corrosion resistance, with sealed contacts that are oil-tight, or they may be the plug-in style or non-plug-in type. This type of switch has been designed as a rotating-cam limit type with programming capability.

Various types of sensing devices are attached to the limit switch (Fig. 10-1). The device may be a lever type with a spring return, a lever type with a maintained

| ROLLER LEVER | | | |
|---|---|---|---|
| Type | Material | Diameter | Width |
| Non-Adjustable 3/4" Radius | Nylon | 3/4" | 9/32" |
| | Metal | 3/4" | 17/64" |
| Non-Adjustable 1 1/2" Radius | Nylon | 3/4" | 9/32" |
| | Nylon | 3/4" | 1" |
| | Steel | 3/4" | 1/4" |
| | Ball Bearing | 3/4" | 15/64" |
| | Beryllium Copper (Non-Sparking) | 3/4" | 9/32" |
| Non-Adjustable Roller on Back 1 1/2" Radius | Nylon | 3/4" | 9/32" |
| | Nylon | 3/4" | 1" |
| | Nylon | 1 1/2" | 9/32" |
| | Steel | 3/4" | 1/4" |
| | Steel | 3/4" | 3/4" |
| Adjustable 1 3/16"–3" Radius | Nylon | 3/4" | 9/32" |
| | Nylon | 3/4" | 1" |
| | Nylon | 1 1/2" | 9/32" |
| | Steel | 3/4" | 1/4" |
| | Ball Bearing | 3/4" | 15/64" |
| Fork Lever 1 1/2" Radius | Nylon L.H. Roller on Front R.H. Roller on Back | 3/4" | 9/32" |
| | Steel L.H. Roller on Front R.H. Roller on Back | 3/4" | 1/4" |
| | Nylon Both Rollers on Front | 3/4" | 9/32" |
| | Nylon Both Rollers on Front | 3/4" | 1" |
| | Steel Both Rollers on Front | 3/4" | 1/4" |
| | Nylon L.H. Roller on Back R.H. Roller on Front | 3/4" | 9/32" |
| Micrometer Adjustment 1 1/2" Radius | Nylon R.H. Adjustment | 3/4" | 9/32" |
| | Steel R.H. Adjustment | 3/4" | 1/4" |
| | Ball Bearing R.H. Adjustment | 3/4" | 15/64" |
| | Nylon L.H. Adjustment | 3/4" | 9/32" |
| | Steel L.H. Adjustment | 3/4" | 1/4" |
| | Ball Bearing L.H. Adjustment | 3/4" | 15/64" |
| | Nylon R.H. Adjustment | 3/4" | 1" |
| One-Way 1 1/2" Radius | Nylon | 3/4" | 9/32" |
| | Steel | 3/4" | 1/4" |
| | Ball Bearing | 3/4" | 15/64" |

| ROLLER LEVER | | | |
|---|---|---|---|
| Type | Material | Diameter | Width |
| Non-Adjustable Offset 1 7/16" Radius | Nylon | 3/4" | 9/32" |
| | Steel | 3/4" | 1/4" |

| ROD LEVER | | |
|---|---|---|
| Type | Material | Diameter |
| | Stainless Steel Rod 5" Long | 1/8" |
| | Stainless Steel Rod 8 1/2" Long | 1/8" |
| | Stainless Steel Rod 11 1/2" Long | 1/8" |
| | | 5/64" |
| | Nylon Rod 12" Long | 1/4" |
| | Stainless Steel Rod 5" Long | 1/16" |
| | Stainless Steel Rod 5" Long | 1/16" |
| | Nylatron Looped Rod 6" Long 2" Wide Loop | 3/16" |

**Fig. 10-1** Limit switch operating levers. *(Allen-Bradley)*

**Fig. 10-2** *Programmable limit switch.* (Allen-Bradley)

contact, a low-operating-torque type with a return spring, a push type with spring return, a wobble stick, or a cat whisker for making contact with the object being tested, counted, stopped, or started. Limit switches can also be designed for programmable operation.

Figure 10-2 is a solid-state control device used with machinery and equipment having a repetitive operation cycle where motion can be correlated to shaft rotation. Its limits are controlled by three screwdriver locations (Fig. 10-3). Chapter 5 provides many examples of limit switches.

## SPEED SWITCHES

Speed sensing can be used to sequence conveyors where it is necessary for one conveyor to be running at nearly full speed before a second conveyor is started. The switch can also be used to indicate in which direction material on a conveyor is moving from the rotation of a suitably driven shaft (Fig. 10-4). This switch is used in conjunction with automatic starters arranged for reversing or plugging duty, to provide plugging or anti-plugging of squirrel-cage motors. The pilot device can be used as a speed-sensing switch or indicate direction of rotation from the driven shaft.

**Fig. 10-3** *Rotating-cam limit switch adjustment screws.* (Allen Bradley)

**Fig. 10-4** *Speed switches: (A) speed switch with lockout solenoid; (B) speed-switch-less lockout solenoid; (C) speed switches with various mountings.* (Allen-Bradley)

Figure 10-4 shows a speed switch used in a plugging operation. A switch with normally open (NO) contacts is used. It is designed to interrupt reverse braking power automatically as the motor approaches zero speed. The speed at which the contacts operate can be adjusted so as to avoid coasting or reverse rotation of the motor. A speed switch can be wired for plugging in either or both directions.

Anti-plugging can also be accomplished by using this switch. Then a switch with normally-closed (NC) contacts is used. It is designed to keep the reverse contacts open until the machine being driven has slowed to a predetermined safe speed. At this speed the contacts are designed to close, permitting reversing or breaking by a designated method. The switch can be used for anti-plugging in either or both directions.

## Operation of the Speed Switch

When the shaft of the switch is rotated, a magnet induction linkage operates a contact. One contact is provided for forward operation and one for reverse. In plugging circuits, the forward or reverse contact is closed (depending on the direction of rotation) at any speed above the point at which the contact is set to operate. However, as the shaft speed is reduced, the electromagnetic torque holding the contact closed is also reduced, until a point is reached where the contact returns to the normally-open position.

The contacts can be adjusted to determine the speed at which they will operate. This is done by altering two external screws, one for each set of contacts. After the operating temperature has been reached, the screw is turned to adjust the point at which the contacts are to operate. Changes in inertia of moving equipment may require readjustment of the set points.

In some cases an accidental turn of the shaft may close the switch contacts and start the motor. To prevent this, the switch can be equipped with a lockout solenoid. The solenoid mechanically keeps the contacts from operating unless the lockout coil is energized. This is a factory-mounted feature or can be field mounted from a kit. If a lockout solenoid is used, a slight change in the wiring is necessitated. Figure 10-5 shows which terminals are used for the lock out (LO) solenoid.

**Fig. 10-5** *Lockout solenoid connections.*

In Fig. 10-6 the control circuit for forward-direction plugging (with lockout protection) is optional. Operation is as follows—pushing the *start* button closes the forward contactor and the motor runs *forward*. The normally-closed contact (F) opens the circuit to the reverse contactor. The *forward* contact on the speed switch closes.

**Fig. 10-6** *Typical plugging circuit.* (Allen-Bradley)

Pushing the *stop* button drops out the forward contactor. The reverse contactor is energized and the motor is plugged. The motor speed decreases to the preset speed setting of the switch, at this point the speed switch contact opens and drops out the *reverse* contactor.

Speed Switches 175

# PRESSURE CONTROLS

Pressure controls operate on the principle of responding to changes in pneumatic (air or gas) or hydraulic (water or oil) pressure applied to a bellows or piston. This force is opposed by a main spring. Varying the force on the main spring (by turning the range adjustment screw) allows setting the contacts to trip at the upper pressure setting. Turning the differential adjusting screw (when provided) varies the force on a secondary spring and allows setting the lower pressure setting, where the contacts reset to their static state. Many ranges and differentials can be achieved by using bellows or pistons of different sizes to meet various requirements (Fig.10-7).

Refrigeration-type controls may have low pressures ranging from 20 inHg vacuum to 120 psi. High pressure cutout controls have ranges from 100 to 500 psi. They are available with or without maximum-limit range stops to meet high-pressure safety adjustment requirements. Figure 10-9 shows the trip and reset pressure settings with minimum adjustment and maximum adjustment as well as differentials. Figure 10-10 shows the selection range for general applications. Bellows are used in low-pressure ranges up to 650 psi. Piston assemblies are used in high-pressure ranges up to 5000 psi. A diaphragm is used on the vacuum controls.

**Fig. 10-7** *Pressure controls: (A) style A, NEMA type-1 enclosure; (B) style C, NEMA type-1 enclosure; (C) style C, NEMA types-7 and -8 enclosure.* (Allen-Bradley)

Copper alloy bellows are used with air, water, oil, noncorrosive liquids, vapors, or gases in a series of pressure ranges from 30 in vacuum (Hg) to 900 psi. Stainless steel bellows are used with many of the more corrosive liquids or gases at pressures up to 375 psi.

Contact blocks are single-pole, double-throw and can be wired to open or close on increasing or decreasing pressures. Contact blocks can be obtained as a replacement kit. See Fig. 10-8.

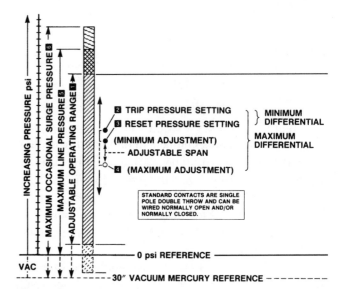

**Fig. 10-9** *Tripping points for pressure controls: 1. The total span within which the contacts can be adjusted to TRIP (2) and RESET (3) when adjusted to minimum difference. 2. TRIP—the higher pressure (temperature) setting, 3. the contacts change state. RESET—the lower pressure (temperature) setting 4. the contacts return to their normal state. [Note: Differential is the difference between TRIP (2) and RESET (3) (minimum adjusted setting) to (4) (maximum adjusted setting).] 5. The maximum sustained pressure that can be applied to the bellows without permanent damage. The control should not be cycled at this pressure. [Note: Does not apply to piston-type controls.] 6. A transient(s) [pulse(s)] that can occur in a system prior to reaching a steady-state condition, expressed in milliseconds. Complex electronic instrumentation is required to measure the varying amplitude, frequency, and duration of this waveform. Frequent occurrence of extreme surge pressures could reduce bellows life. Surge pressure within published value generated during startup or shutdown of a machine or system, not exceeding eight times per day, are negligible. 7. Vac.—vacuum (negative pressure) inches of mercury. 8. Copper alloy bellows may be used on water, air, and other liquids or gases not corrosive to this alloy. Type 316 stainless steel bellows are available for the more corrosive liquids or gases.* (Allen-Bradley)

**Fig. 10-8** *Contact block replacement kit for pressure controls.* (Allen-Bradley)

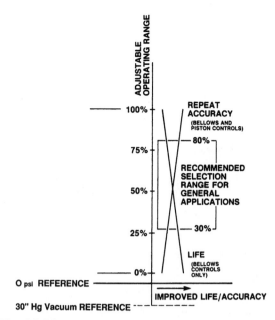

**Fig. 10-10** *Pressure controls operating points.* (Allen-Bradley)

## TEMPERATURE CONTROLS

Temperature controls may have direct vertical immersion bulbs for sensing changes in temperature or they may have remote capillary and bulb (Fig. 10-11). Temperature controls are similar to pressure controls. They differ in that a closed chemically filled bellows system is used. The pressure in the system changes in proportion to the temperature of the bulb.

The temperature response medium in the system is a liquid whose vapor pressure increases as the temperature of the bulb rises. As the temperature of the bulbs falls, the vapor pressure decreases. The pressure change is transmitted to the bellows through a capillary tube operating the control at a predetermined setting. Temperature controls using this accurate and long-life vapor-pressure method of sensing temperatures are available in a series of ranges from $-150$ to $+570°F$.

Armored capillary is available for added protection of all bulb and capillary-type temperature controls. Copper bulb and capillary, bronze armor, and controls with stainless steel bulb and capillary are available for various environments.

## Wells and Packing Glands

Wells are used to mount and protect as well as permit removal of the bulb when necessary without discharging the system. Glands form seals at any desired position along the standard capillary. Armored capillary wells and glands include a setscrew to hold the armor so as not to expose the capillary. Wells are also available with a retaining nut to secure the bulb (Fig. 10-12). Figure 10-13 shows some contact blocks for these controls.

**Fig. 10-11** *Temperature controls.* (Allen-Bradley)

**Fig. 10-12** *(A) Thermostat immersion wells; packing gland assembly; (B) Immersion well* (Allen-Bradley)

**Temperature Controls** 177

## CONTACT BLOCKS

| Symbol | | Description | Rating |
|---|---|---|---|
| Pressure Controls | Temperature Controls | | |
| **AUTOMATIC OPERATION** | | | |
| | | Single pole double throw — automatically opens or closes on rise or fall. | Non-inductive 5A, 240V 3A, 600V Control Circuit Rating AC-125VA, 24 to 600V DC-57.5VA, 115 to 230V |
| | | Single pole double throw — slow acting contact with no snap action. Contacts close on rise and close on fall with an open circuit between contact closures. | Control Circuit Rating AC-125VA, 24 to 250V |
| | | Single pole single throw, normally open — closes on rise. | 1 H.P., 230V AC .5 H.P., 115V AC Control Circuit Rating AC-125VA, 24-110V AC-345VA, 110-600V DC-57.5VA, 110-250V |
| | | Single pole single throw, normally closed — opens on rise. | |
| | | Single pole single throw, normally open — closes on rise. | 1 H.P., 115V AC 1.5 H.P., 230V AC Control Circuit Rating AC-600VA, 110-600V DC-57.5VA, 110-250V |
| | | Single pole single throw, normally closed — opens on rise. | |
| | | Two circuit, single pole single throw, normally open — a common terminal is connected to 2 separate contacts which close on rise. | Non-inductive 5A, 240V 3A, 600V Control Circuit Rating AC-125VA, 24 to 600V DC-57.5VA, 115 to 230V |
| | | Two circuit, single pole single throw, normally closed — a common terminal is connected to 2 separate contacts which open on rise. | |
| **MANUAL RESET** | | | |
| | | Single pole single throw, normally open — contacts open at a predetermined setting on fall and remain open until system is restored to normal run conditions at which time contacts can be manually reset. | Non-inductive 5A, 240V 3A, 600V Control Circuit Rating AC-125VA, 24 to 600V DC-57.5VA, 115 to 230V |
| | | Single pole single throw, normally closed — contacts open on rise and remain open until system is restored to normal run conditions at which time contacts can be manually reset. | |
| | | Single pole double throw, one contact normally closed — contact opens on rise and remains open until system is restored to normal run condition at which time contact can be manually reset. A second contact closes when the first contact opens. | |

**Fig. 10-13** *Pressure and temperature controls.* (Allen-Bradley)

## FLOAT SWITCHES

Float switches are also sensing devices. They provide automatic control for motors that pump liquids from a sump or tank. They have both motor and pilot duty ratings (Fig. 10-14).

**Fig. 10-14** *Float switches: (A) style A, wall or floor mount for tank or sump, NEMA type 4; (B) style C and (C) style B for one- or three-phase or dc; (D) style D and DS.* (Allen-Bradley)

Float switches have a snap-action mechanism for quick-make and quick-break contact operation. This feature provides high snap-through forces once the mechanism has traveled the required distance. This type of switch is available in copper, brass, and stainless steel to allow for wall or floor mounting to accommodate different tank or sump depths. Figure 10-15 shows various types of float switch arrangements.

## ENCODERS

Industrial optical encoders are designed to convert mechanical shaft rotation to an accurate electrical output in binary-coded decimal (BCD) or optional Gray code format. Gray code is defined as sequential numbers by binary values in which only one value changes at a time. This type of encoder (Fig. 10-16) is designed for use in industrial environments. It is a noncontacting optical design that uses high-speed, low-torque operation. Long life and good reliability are a result of using electronic components. They include a light-emitting diode as the source of light. A metal code disk is used with the single-LED light source, utilizing a fiber-optic light guide. Operating speed for BCD is 800 rpm and for Gray code 2000 rpm.

Many digital circuits, including some combinational logic circuits, are designed to handle BCD data. Binary-coded decimal is a code that uses a 4-bit binary number to represent each digit in the decimal system. This is a conventional code for taking decimal information from a device such as a calculator keyboard and converting it to binary information for processing by digital circuits. It is equally useful for converting the binary output from digital circuits to decimal information for displaying on an output device such as a seven-segment light-emitting diode (LED) or liquid crystal display (LCD) (Table 10-1). This information generated by the encoder can be fed to circuits that are designed to utilize the information and present it as a speed in rpm, and it can also be utilized to control the speed of rotation by adjusting the motor controller output.

### Optical Programmable Controller Encoders

Encoders are also designed for interface with programmable controllers. The encoder contains all the electronics necessary to provide a latched output on command from a programmable controller. A data-ready output signals the programmable controller when the encoder data are latched. The encoders are capable of being multiplexed, allowing one programmable controller with one set of input cards/modules to accept data from many encoders. The multiplex and latch circuits operate independently.

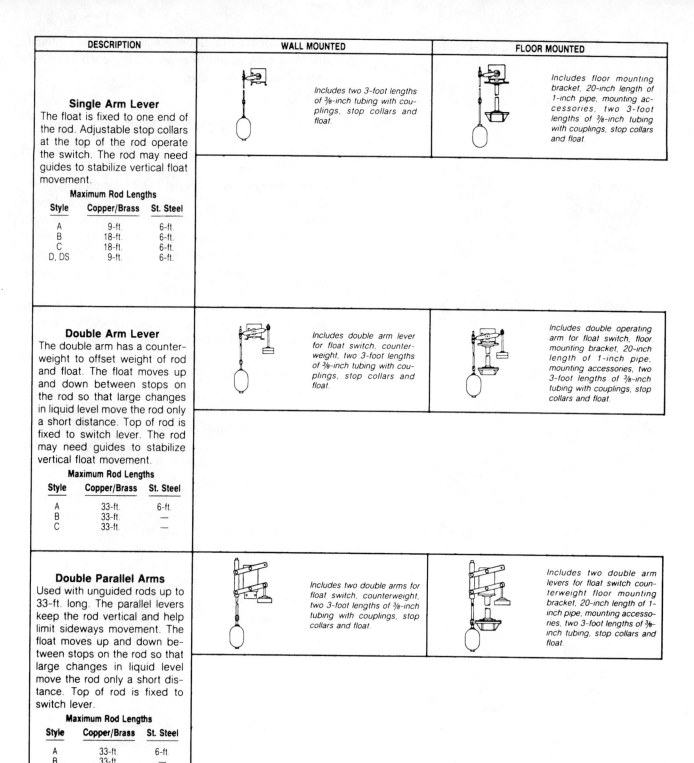

**Fig. 10-15** *Automatic float switches.* (Allen-Bradley)

| DESCRIPTION | WALL MOUNTED | FLOOR MOUNTED |
|---|---|---|
| **Double Arm Lever, Double Pulley** The double pulley is self-supported. The chain has a float fixed at one end and a counterweight at the other. The adjustable stop collars on the chain/cable move the switch operating lever. | *Includes double pulley bracket, 15-foot chain with stop collars, counterweight and float.* | *Includes double arm bracket, 15-foot chain with stop collars, counterweight, float and 20-inch length of 1-inch pipe with mounting accessories.* |
| **Double Arm Lever, Single Sheave Wheel** Used with Style A switch. A single pulley is mounted on the top of the float switch. The chain/cable has a float fixed at one end and a counterweight at the other end. The adjustable stop collars on the chain/cable move the switch operating lever. | *Includes one pulley, 15-foot chain, stop collars, counterweight and float.* | *Includes one pulley, 15-foot chain, stop collars, counterweight, float and 20-inch length of 1-inch pipe with mounting accessories.* |
| **Single Arm Lever, Separate Pulleys** The two pulleys are separate from the switch. The chain/cable has a float fixed at one end and a counterweight at the other end. The adjustable stop collars on the chain/cable move the switch operating lever. | *Included are two pulleys for separate mounting, 15-foot chain with stop collars, counterweight and float.* | |

 **FLOAT SIZE TABLE**

| Float | Sphere Diameter | Elongated Sphere Diameter x Length |
|---|---|---|
| A | 6" | — |
| B | 7" | — |
| C | 8" | — |
| D | 9" | — |
| E | 10" | — |
| F | — | 7" x 12.5" |

**Note:** Float size dimensions do not imply that switch operator assemblies can be interchanged.

**Fig. 10-15** *(Continued).*

Encoders 181

**Fig. 10-16** *Optical single-turn absolute encoder.* (Allen-Bradley)

**Fig. 10-17** *Self-contained proximity switches: (A) top sensing; (B) front sensing.* (Allen-Bradley)

*Table 10–1* BCD Code to Decimal Digit

| BCD Code | Decimal Digit |
|---|---|
| 0000 | 0 |
| 0001 | 1 |
| 0010 | 2 |
| 0011 | 3 |
| 0100 | 4 |
| 0101 | 5 |
| 0110 | 6 |
| 0111 | 7 |
| 1000 | 8 |
| 1001 | 9 |

## Multiplexing

Placing a logic zero (0) on the multiplex control line will force all outputs to high-impedance state, allowing the programmable controller to scan multiple encoders using one set of input cards/modules. A multiplexer is a type of combinational digital circuit that is readily available in an IC package. Multiplexing allows a single conductor (bus line) to carry signals alternately from a variety of signal sources. Multiplexers are also called data selectors. They operate like an electrically controlled rotary switch. The output line is like a pole of a rotary switch, and the input lines are like the positions of a rotary switch. Data selectors are available with as many as 16 input lines.

## PROXIMITY SWITCHES

Proximity switches are designed for industrial environments in places where it is required to sense the presence of metal objects (ferrous and nonferrous) without touching them. These are self-contained, two-wire devices used in 120-V ac control circuits (Fig.10-17).

## Operation

Proximity switches are designed for general-purpose use, with a high output rating of 1 A maximum continuous and 10 A maximum inrush. They are capable of energizing such external loads as relays, contactors, motor starters, and solenoids. Switches with programmable normally-open/normally-closed output have two LED indicators: a power LED indicator that is *on* when power is applied to the device, and an output LED indicator that is *on* when the switch output is conducting current.

## Solid-State Switches

Switches for solid-state applications interface directly with programmable controllers, hard-wired, solid-state logic, and similar high-impedance loads. In addition, they have circuit protection against overload and short-circuit conditions. In switches with a normally-open, fixed output, the LED indicator is *on* when a metal object is within the sensing field and the load is energized.

Self-contained proximity switches include both front-sensing and top-sensing types. The front-sensing version can be arranged to sense in any of the four directions: front, rear, or either side. The switches come with a conduit coupler, threaded conduit opening, pre-wired receptacle, or pre-wired cable base.

## Inductive Cylindrical Switch

Inductive cylindrical switches are self-contained and solid state (Fig. 10-18). They are also designed for industrial environments where it is required to sense the presence of metal objects, both ferrous and nonferrous, without touching them. The switches are equipped with a red LED to indicate the presence (NO output

**Fig. 10-18** Inductive cylindrical proximity switches. (Allen-Bradley)

**Fig. 10-20** Inductive proximity switch. (Allen-Bradley)

configuration) or absence (NC output configuration) of a target in the sensing field.

This type of switch interfaces with programmable controller input modules without the use of an external loading resistor. Switches can also be used to energize relays, contactors, and motor starters. The switch housing is made of nickel-plated brass. The electronic circuitry is potted for protection against shock, vibration, and contaminants.

## Extended-Sensing-Range Inductive Proximity Switch

The extended-sensing-range switch is also self-contained and solid state (Fig. 10-19). It is designed for industrial use such as materials handling (transfer lines, roller-belt conveyors, and the like). The position of objects traveling these lines cannot always be accurately controlled, therefore the need to sense an extended distance. This switch interfaces with programmable controllers without the use of loading resistors.

**Fig. 10-19** Extended range inductive proximity switch. (Allen-Bradley)

## Compact Inductive Proximity Switch

The compact inductive proximity switch is square, is PC compatible, and has a 15-mm sensing range (Fig. 10-20). The two-wire device energizes and de-energizes an external load when metal targets are sensed.

## Self-Contained Low Profile Proximity Switch

The self-contained low profile proximity switch will operate with PCs and sense the presence of metal objects (Fig. 10-21). There are two types: weld field and general purpose. The weld-field type has a built-in tolerance to electromagnetic fields generated by resistance welding equipment and can be mounted within 1 in. of a 20,000-A current-carrying bus.

**Fig. 10-21** Self-contained proximity switches. (Allen-Bradley)

## PHOTOELECTRIC SWITCHES

The photoelectric industrial switch is self-contained and has solid-state electronics as part of the package (Fig. 10-22). The industrial-type photoelectric switch is capable of sensing the presence of a wide variety of objects without making contact.

**Fig. 10-22** *Self-contained photoelectric switches: (A) type RA uses a modulated infrared LED light source and detector, (B) type SAC source (left) and type DA detector (right); (C) type RL operates loads between 10 and 30 V dc.* (Allen-Bradley)

The switch shown in Fig. 10-22 uses a modulated infrared LED light-source detection system. It is modulated or pulsed in the near-infrared portion of the electromagnetic spectrum. This switch is highly tolerant of ambient light, including sunlight.

## Operation

This switch is a retro-reflective device that operates loads between 10 and 30 V dc. It can be wired in current sink or current source mode, for energizing a variety of loads. The operating range is 18 ft for a 3-in reflector. Other types use the through-beam photoelectric system. The through-beam pair of 120-V ac devices is especially well suited for demanding applications that require a maximum range of 25 ft. These switches can be used to operate relays, contactors, and motor starters and interface directly with most programmable controllers. The output condition can be programmed between light and dark operation by means of a recessed rocker switch. A rotatable sensing head can be arranged to sense in any of four directions.

## Features

Reliable performance is the needed feature on any sensing operation or system. One means of ensuring proper operation of a photoelectric switch is to use a stability indicator. A green stability LED is provided on some switches to indicate reliable system performance. An unstable indication from the LED means that the light intensity level at the receiver is staying within +20% of the switch trip point, resulting in marginal switch operation (Fig. 10-23). Common situations causing such an indication include:

1. Sensitivity adjustment set too low
2. Switch misalignment, particularly when the emitter halo rather than the emitter beam is detected (Fig.10-23)
3. Target or reflector set too far from the switch
4. Dust or other residue buildup on the lens or reflector
5. Target not large enough to block the light beam completely
6. Insufficient contrast between the target and background (in diffusing applications)

**Fig. 10-23** *Emitter (LED) light pattern.* (Square D)

This setup and use of a photoelectric switch system without a stability indicator can result in premature system failure. Failure modes can include erratic tripping or no switching response.

Figure 10-24 shows selectable light-dark operation of the photoelectric switch. Polarized lenses are found only on retro-reflective devices; a photoelectric switch with polarized lenses detects only light signals bounced off a retro-reflector. The switch is immune to any undesirable light reflection from targets such as metals, foils, mirrors, tiles, shrink wrap, and shiny plastics.

**Fig. 10-24** *Selectable light/dark operation.* (Square D)

## Photoelectric Light Sources

Generally, switch emitters are either an incandescent light or an LED light source. LED light sources have many advantages compared to incandescent. LEDs have:

- Longer life
- Virtually no heat dissipation
- Resistance to shock and vibration
- Longer sensing ranges
- Cost-effectiveness
- The ability to be pulsed at high frequencies

Pulsing an LED beam creates a stronger light beam capable of longer sensing ranges. By modulating the pulse frequency, emitters can be keyed to the light receiver. This helps eliminate ambient-light interference problems.

LEDs are available in four basic color bands: infrared, red, green, and yellow. All but yellow are used as photoelectric switch emitters. Of the three color bands used, infrared LED emitters are found in a large majority of photoelectric switches. Infrared light does the best job of penetrating dust, fog, and other airborne particles that may interfere with a photoelectric beam.

Red LED emitters can be found in fiber-optic units using plastic fiber cable and in retro-reflective units with polarized lenses. Mark sensors use either a red or green LED emitter.

## Mark Sensors

Selecting a mark sensor with the proper LED emitter color depends on both the mark color and its background color. Table 10-2 is a useful guide for determining if a mark sensor will

- Require a green or red LED emitter
- Function stably in the light-operate or dark-operate mode

*Table 10-2* Mark Sensor Guide[a]

| Background Color | Emitter LED COLOR | Color of Mark | | | | | | |
|---|---|---|---|---|---|---|---|---|
| | | White | Black | Transparent | Red | Yellow | Green | Blue |
| White | Green | – | • | • | • | – | ▲ | • |
| | Red | – | ù | • | – | - | • | • |
| Black | Green | ■ | – | - | – | ■ | ■ | – |
| | Red | ■ | – | - | ■ | ■ | – | - |
| Transparent | Green | ■ | – | - | – | ■ | ■ | - |
| | Red | ■ | – | - | ■ | – | – | - |
| Red | Green | ■ | – | - | – | ■ | ■ | • |
| | Red | ■ | ù | • | - | - | • | • |
| Yellow | Green | ■ | ù | • | ▲ | – | - | • |
| | Red | – | ù | • | - | - | ■ | • |
| Green | Green | ♦ | • | • | • | – | - | ▲ |
| | Red | ■ | – | - | ■ | ♦ | – | - |
| Blue | Green | ■ | – | - | ■ | ■ | ♦ | – |
| | Red | ■ | – | - | ■ | ■ | – | - |

[a] In the dark-operate mode;
• means detects the mark stably
▲ means may detect the mark.
In the light-operate mode;
■ means detects the mark stably and
♦ means may detect the mark.

**Fig. 10-25** *Miniature specialty photoelectric switches.* (Square D)

Mark sensors detect (diffuse setting) color marks on most materials, both opaque and transparent (including film). They have either green or red emitter LEDs and can detect a wide variety of color marks on different backgrounds. Special optics and adjustable lensing allow precise setting of the switch focal point. This feature is useful in pinpoint detection (narrow visibility types) or for setting a precise limit to a detecting distance (wide-visibility types). Special applications for this device include detecting objects as shown in Fig. 10-25. A miniature specialty focusable diffuse switch is shown in Fig. 10-26.

**Fig. 10-26** *Focusable diffuse switch.* (Square D)

## Clear Material Detection

The retro-reflective type photoelectric switch is ideal for transparent bottle detection. A special optical system permits stable detection of clear materials not previously possible with conventional retro-reflective photoelectric switches. The 1-m sensing range of this switch makes the system flexible compared to standard diffuse photoelectric switches normally used in these applications.

## SWITCH APPLICATIONS

Figure 10-27 shows the typical photoelectric switch applications. Note the diffuse mode, specular mode, through-beam mode, retro-reflective mode, and diffuse mode. Sensing ranges are based on clean lenses and a clean air environment. Air contaminants and dirty lenses affect switch performance.

## AUTOMATIC IDENTIFICATION

Automatic identification keeps a control system or operation under scrutiny. It supplies the control system and data bases with information they need. Automatic identification locates everything and tells you where it is going. There are systems already tested that can keep track of everything in production. They consist of a radio-frequency (RF) identification system, bar code readers and decoders, and multiplexing.

**Fig. 10-27** *Typical photoelectric switch uses.* (Square D)

All of these involve hardware, software, and networking communications. They require sensors made for special systems applications. Identification devices can provide a few characters of data in the case of a bar code system, or up to 2K bytes of data when a radio-frequency system is used.

## RADIO-FREQUENCY IDENTIFICATION SYSTEMS

Figure 10-28 shows the Allen-Bradley radio-frequency (RF) system as it is set up for operation. The system keeps track of parts and follows them through the manufacturing process, without the need for a paper chase or factoring in an operator error. Data can be transmitted or received by an RF tag whenever they pass an antenna. Keep in mind that the actual units may have changed in shape as newer configurations are available.

The system is based on radio waves that can penetrate plastic, wood, and various types of factory contaminants. It can be installed and operated in a wide range of industrial environments. It is dependable in transferring its information to a control system (Fig.10-29). Note how the system consists of three main pieces of hardware: the antenna, RF tags, and the universal identification interface (UII).

**Fig. 10-28** *Radio-frequency (RF) identification systems.* (Allen-Bradley)

Using this method it is possible to use a single method of product identification in all phases of production and processing. Data can be stored on a read/write RF tag and can be added to, altered, or deleted whenever the tag passes an antenna.

**Radio-Frequency Identification Systems** 187

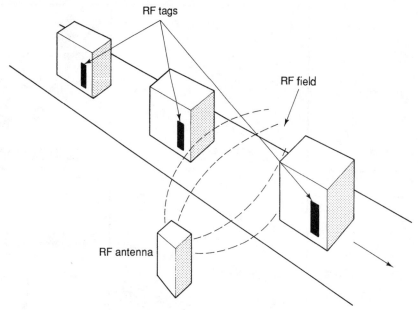

**Fig. 10-29** Radio frequency identification. (Allen-Bradley)

## The Antenna

The antenna contains both the transmit and receive antennas (Fig. 10-30A). It decodes the data stored in the tags and transmits this information to the UII or the host, depending on the control architecture utilized. Housed in a die-cast aluminum enclosure, the unit can be installed on the shop floor or in a wide range of industrial environments.

## The RF Tag

There are two RF tags available: the read/write and the programmable (Fig. 10-30B). The read/write tag is a small battery-powered tag that can store up to 2K bytes of data and is reusable.

The tag is activated when it enters the antenna's RF field. At that time the information can be read from or written into the tag's memory (Fig. 10-30).

The programmable tag is small and rugged. It is designed for industrial applications and is also reusable. The tag can be exposed to temperatures as high as 180°C for up to 1 hour, without damaging data stored in the tag. Up to 40 bytes of data can be programmed into the tag whenever the tag enters an antenna's RF field.

## BAR CODE READERS AND DECODERS

Bar coder readers offer fast entry of data, with speed and accuracy not humanly possible through keyboard entry (Fig. 10-31 shows a typical bar code). Scanning devices can be stationary or portable. They can be automatic or manually operated. Allen-Bradley has a variety of scanners to meet industrial needs.

**Fig. 10-30** (A) RF antenna (B) RF tags (C) UII multiplexer. (Allen-Bradley)

*Fig. 10-31* Bar code. *(Allen-Bradley)*

## Hand-Held Scanners

This reader (Fig. 10-32A) accepts input from a variety of bar code scanners, offers high-performance decoding, and full auto-discrimination of most popular symbologies. Its RS-232 port makes it compatible with most computers. The reader works with either a wand or a moving-beam laser scanner. The laser scanner (Fig. 10-32B) is a non-contact device that can operate in harsh environments. It can be dropped on concrete and still survive to work again. The scanning rate is 36 scans per second. Labels can be decoded within 1 to 23 in. from the scanner, depending on the density of the bar code symbol.

## Moving-Beam Scanners

This type of scanner (Fig. 10-32C) operates at 175 scans per second and can decode symbols up to 20 in. away. The scanner head is separate from the decoder console, so it can be installed in areas where space is a problem. The decoder automatically decodes Code 39, interleaved 2-of-5, UPC/EAN, and Codabar symbols.

## MULTIPLEXING

The universal identification interface (UII) is a new idea in automatic identification control. With one UII you can multiplex up to eight sensors (Fig. 10-33). This type of sensor utilization makes it possible for programmable controllers and robots to function as intended.

## VISION SYSTEM

Sensing involves the five senses: touch, feel, smell, taste, and seeing. The vision system makes use of the size and shape of objects for its operation. Of course, the ideal inspection rate is 100% at production-line speeds. With a vision system it is possible to inspect,

*Fig. 10-32* (A) Hand-held reader; (B) Hand-held laser scanner; (C) decoder console and scanner head. *(Allen-Bradley)*

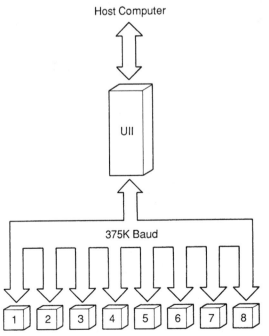

*Fig. 10-33* Multiplexing with UII. *(Allen-Bradley)*

**Fig. 10-34** *Programmable vision system.* (Allen-Bradley)

visually, all products coming off a line. In fact, it is also possible to inspect each product as it passes through an operation heading toward completion (Fig. 10-34).

Identification, location, and sorting of objects by pattern recognition is possible with the utilization of video imagery and computer technology. The system offers a means of reducing product waste, inspection errors, and human fatigue. This inspection system can operate 24 hours a day and 7 days a week without endangering product integrity

## REVIEW QUESTIONS

1. How are sensors classified?
2. What is another name for a limit switch?
3. What type of switches respond to contact?
4. What types of switches are used for presence sensing?
5. Other than speed, what can a speed switch sense?
6. How do you adjust pressure controls for a given point?
7. Why are wells and packing glands needed?
8. What is an optical encoder?
9. What does BCD stand for?
10. What is multiplexing?
11. Where are proximity switches used?
12. What is the most needed feature of a sensing system?
13. In what colors are LEDs available?
14. What are the three pieces of hardware that make up the RF identification system?
15. What does UII stand for?
16. How are scanners used to sense and input information into a system?
17. What type of switch is used for anti-plugging?
18. What does LO stand for?
19. What is the Gray code?
20. What does placing a logic "0" on the multiplex control line do?

# 11
CHAPTER

# Solenoids and Valves

## PERFORMANCE OBJECTIVES

After studying this chapter, you will be able to:

1. Explain how a solenoid provides mechanical motion.
2. Identify various types of industrial solenoids.
3. List applications for solenoid devices.
4. Select a solenoid according to its class.
5. Describe how to service solenoid coils.
6. Explain how solenoid valves work in circuits.

## SOLENOIDS

Solenoids are devices that turn electricity, gas, oil, or water on and off. Solenoids can be used, for example, to turn the hot water on and the cold water off, to get a proper mix of warm water in a washing machine. To control the hot water solenoid, a thermostat is inserted in the circuit.

A solenoid where the length is greater than the diameter is one of the most common types of coil construction used in electricity and electronics. The field intensity is the highest at the center in an iron-core solenoid. At the ends of the air-core coil, the field strength falls to a lower value.

A solenoid that is long compared to the diameter has approximately half the field intensity at the ends than at the center. If the solenoid has a ferromagnetic core, the magnetic lines pass uniformly through the core.

Mechanical motion can be produced by the action of a solenoid or a solenoid can generate a voltage that is a result of some mechanical movement. The term solenoid commonly means a coil of wire with a moving iron core that can center itself lengthwise within the coil when current is applied to the coil. Then if a ferromagnetic core is properly suspended and under suitable tension, it can be moved in and out of a solenoid coil form with the application of current to the coil. This is the operating basis of some relays and a number of other electromechanical devices. When an outside force is used to move the ferromagnetic core physically, it is possible to induce a voltage in the solenoid coil.

There is a tendency in a solenoid for the core to move so that it encloses a maximum number of magnetic lines of force. Each line of force has the shortest possible length (Fig. 11-1). In the illustration the core is outside the coil. Because it is a ferromagnetic material, the coil presents a low-reluctance path to the magnetic lines of force at the north end of the coil. These lines of force concentrate the soft-iron core and then complete their paths back to the south pole of the electromagnet.

**Fig. 11-1** *Solenoid pulls core into the coil. Sucking effect of a coil.*

Electromagnetic lines of force that pass through the core magnetize it. That means that the induced magnetic field in the core has a south pole near the coil's north pole. Inasmuch as unlike poles attract, the core is attracted toward the hole in the solenoid coil. This attraction tends to pull the core into the coil. As the iron core is pulled into the coil, the magnetic field becomes increasingly shorter and the magnetic lines of force travel the shortest possible distance when the core centers itself in the coil.

By attaching a spring to the core, it is possible to have the core return to its outside position once the power is interrupted to the coil. When the power is then turned on again, it pulls the core back into the coil. It is this type of movement that is utilized in the construction of industrial solenoids that operate switch contacts in relays and motor starters and valves in gas, air, and liquid lines of various types.

## INDUSTRIAL SOLENOIDS
### Tubular Solenoids

There are various uses for solenoids. Figure 11-2 shows tubular solenoids. Notice the type, voltage rating, coil resistance, and the minimum and maximum lifts and strokes. Some are pull types and others are push types. They are also specified as intermittent and continuous duty.

### Frame Solenoids

The frame-mounted types of solenoids (Fig. 11-3) are available in intermittent and continuous duty as well as usable on either ac or dc. Types 11 and 28 can operate on ac or dc. The other types are identified as to whether they operate best on ac or dc.

**T DENOTES PULL TYPE; TP DENOTES PUSH TYPE; LT DENOTES LONG LIFE PULL TYPE.**

## T, TP AND LT SERIES SOLENOIDS

**T** and **TP series** are UL recognized.
Duty: I=intermittent. C=Continuous.

| Type | Duty-Volt. | Coil Resis. Ohms | Lifts and Strokes | | | |
|---|---|---|---|---|---|---|
| | | | Min. | | Max. | |
| | | | Oz. | @in | Oz | @in |
| T3.5X9 | C-12D | 60.2 | 4½ | 1/32 | .3 | 9/16 |
| | I-12D | 31.1 | 7 | 1/32 | .7 | ½ |
| | I-24D | 122 | 7 | 1/32 | .7 | ½ |
| | C-24D | 254 | 4½ | 1/32 | .3 | 9/16 |
| LT3.5X9 | C-12D | 52.4 | 2.5 | 1/32 | .2 | 3/8 |
| | C-24D | 221 | 2.5 | 1/32 | .2 | 3/8 |
| TP3.5X9 | I-24D | 122 | 6 | 1/32 | .5 | ½ |
| | C-24D | 254 | 4 | 1/32 | .3 | 9/16 |
| T4X7 | I-24D | 131 | 7 | 1/32 | 1 | ¼ |
| | C-24D | 270 | 5 | 1/32 | .5 | ¼ |
| LT4X7 | C-12D | 63.3 | 2.5 | 1/32 | 1 | .15 |
| | C-24D | 264 | 2.5 | 1/32 | 1 | .15 |
| TP4X7 | I-24D | 131 | 6.5 | 1/32 | .75 | .15 |
| | C-24D | 270 | 3.5 | 1/32 | .50 | 1.5 |
| T4X16 | C-12D | 45.1 | 4.5 | 1/16 | 1.2 | ½ |
| | I-12D | 17.7 | 7 | 1/16 | 2.5 | 1 |
| | C-24D | 173 | 4.5 | 1/16 | 1.2 | ½ |
| | I-24D | 72.7 | 7 | 1/16 | 2.5 | 1 |
| LT4X16 | C-12D | 42.5 | 4 | 1/16 | 2 | ½ |
| | I-12D | 14 | 8 | 1/16 | 2.5 | ½ |
| | C-24D | 168 | 4 | 1/16 | 2 | ½ |
| | I-24D | 69.1 | 8 | 1/16 | 2.5 | ½ |
| TP4X16 | C-24D | 173 | 3.8 | 1/16 | 1 | ½ |
| | I-24D | 72.7 | 5.5 | 1/16 | 2 | ½ |
| T4X12 | C-24D | 195 | 6 | 1/32 | 1.5 | .15 |
| | I-24D | 96.7 | 9 | 1/32 | 2 | .15 |
| LT4X12 | C-12D | 49.3 | 7.5 | 1/32 | .5 | ½ |
| | C-24D | 19.2 | 7.5 | 1/32 | .5 | ½ |
| TP4X12 | I-24D | 96.7 | 7 | 1/32 | 1.5 | .15 |
| | C-24D | 195 | 5 | 1/32 | 1.0 | .15 |
| T6X12 | C-12D | 31.7 | 15 | 1/16 | 1 | ¾ |
| | I-12D | 12.1 | 23 | 1/16 | 3 | ¾ |
| | C-24D | 121 | 15 | 1/16 | 1 | ¾ |
| | I-24D | 60.6 | 23 | 1/16 | 3 | ¾ |
| LT6X12 | C-12D | 35 | 18 | 1/16 | 1 | ½ |
| | I-12D | 13.8 | 35 | 1/16 | 2.5 | ½ |
| | C-24D | 138 | 18 | 1/16 | 1 | ½ |
| TP6X12 | C-24D | 121 | 12 | 1/16 | .8 | ¾ |
| | I-24D | 60.6 | 18 | 1/16 | 2.5 | ¾ |
| T8X9 | C-24D | 135 | 18 | 1/16 | 2.5 | 7/16 |
| | I-24D | 44 | 45 | 1/16 | 5 | 7/16 |
| LT8X9 | C-24D | 109 | 18 | 1/16 | 1 | ½ |
| | I-24D | 44.6 | 35 | 1/16 | 3 | ½ |
| TP8X9 | I-24D | 44 | 36 | 1/16 | 4 | 7/16 |
| | C-24D | 135 | 14 | 1/16 | 2 | 7/16 |
| T8X16 | I-12D | 9.3 | 58 | 1/16 | 5 | .60 |
| | C-12D | 28.3 | 30 | 1/16 | 2.5 | .60 |
| | I-24D | 36.1 | 58 | 1/16 | 5 | .60 |
| | C-24D | 110 | 30 | 1/16 | 2.5 | .60 |
| LT8X16 | I-12D | 6.2 | 67 | 1/16 | 12 | ½ |
| | C-12D | 19.3 | 33 | 1/16 | 2.5 | ½ |
| | I-24D | 29.7 | 62 | 1/16 | 10 | ½ |
| | C-24D | 77.2 | 33 | 1/16 | 2.5 | ½ |
| T12X13 | I-24D | 28.4 | 100 | 1/8 | 15 | .60 |
| | C-24D | 90.4 | 40 | 1/8 | 5 | .60 |
| LT12X13 | I-12D | 5.9 | 110 | 1/8 | 10 | .5 |
| | C-12D | 18.3 | 35 | 1/8 | 1 | .5 |
| | I-24D | 22 | 110 | 1/8 | 10 | .5 |
| | C-24D | 71.2 | 35 | 1/8 | 1 | .5 |
| TP12X13 | I-24D | 28.4 | 80 | 1/8 | 12 | .60 |
| | C-24D | 90.4 | 32 | 1/8 | 4 | .60 |
| T12X19 | I-24D | 22.1 | 130 | 1/8 | 40 | ¾ |
| | C-24D | 68 | 70 | 1/8 | 10 | ¾ |
| LT12X19 | I-12D | 4.66 | 130 | 1/8 | 15 | ¾ |
| | C-12D | 14.8 | 70 | 1/8 | 5 | ¾ |
| | I-24D | 18.6 | 130 | 1/8 | 15 | ¾ |
| | C-24D | 71.8 | 70 | 1/8 | 3 | ¾ |
| TP12X19 | I-24D | 22.1 | 100 | 1/8 | 30 | ¾ |
| | C-24D | 68 | 55 | 1/8 | 8 | ¾ |

**Fig. 11-2** *Tubular solenoids.*

Type 2  Type 2HD  Type 4  Type 4HD

Type 11

Type 11HD

Type 12

Type 14

### DC VOLTAGE MODEL SOLENOIDS

| Type | Duty | Volts | Ohms | Oz. @ Inch of Stroke Minimum | Oz. @ Inch of Stroke Maximum |
|---|---|---|---|---|---|
| 2HD | Intermittent | 24 | 22.6 | 96 @ 1/8" | 15 @ 1/2" |
| 2HD | Continuous | 24 | 71 | 48 @ 1/8" | 5 @ 1/2" |
| 4 | Intermittent | 24 | 15.8 | 100 @ 1/8" | 20 @ 1" |
| 4 | Continuous | 24 | 61.3 | 60 @ 1/8" | 7 @ 4/5" |
| 4 | Intermittent | 110 | 296 | 100 @ 1/8" | 20 @ 1" |
| 4 | Continuous | 110 | 1215 | 60 @ 1/8" | 7 @ 4/5" |
| 4HD | Intermittent | 24 | 18.9 | 130 @ 1/8" | 25 @ 3/4" |
| 4HD | Continuous | 24 | 57.5 | 80 @ 1/8" | 5 @ 3/4" |
| 4HD | Intermittent | 110 | 354 | 130 @ 1/8" | 25 @ 3/4" |
| 4HD | Continuous | 110 | 1140 | 80 @ 1/8" | 5 @ 3/4" |
| 11 | Intermittent | 6 | 1.88 | 45 @ 1/8" | 10 @ 1/2" |
| 11 | Continuous | 6 | 4.69 | 30 @ 1/8" | 4 @ 1/2" |
| 11 | Intermittent | 24 | 29.1 | 45 @ 1/8" | 10 @ 1/2" |
| 11 | Continuous | 24 | 93.1 | 30 @ 1/8" | 4 @ 1/2" |
| 11HD | Intermittent | 24 | 29.3 | 70 @ 1/8" | 5 @ 3/4" |
| 11HD | Continuous | 24 | 76.3 | 30 @ 1/8" | 2 @ 3/4" |
| 11P | Continuous | 24 | 93.1 | 24 @ 1/8" | 3.2 @ 1/2" |
| 22 | Intermittent | 6 | 5.8 | 17 @ 1/16" | 2 @ 0.3" |
| 22 | Continuous | 6 | 11.5 | 11 @ 1/16" | 1 @ 0.3" |
| 22 | Intermittent | 24 | 93.2 | 17 @ 1/16" | 2 @ 0.3" |
| 22 | Continuous | 24 | 182 | 11 @ 1/16" | 1 @ 0.3" |
| 28 | Intermittent | 6 | 3.03 | 40 @ 1/16" | 3 @ 1/2" |
| 28 | Continuous | 6 | 7.5 | 23 @ 1/16" | 2 @ 1/2" |
| 28 | Intermittent | 12 | 11.9 | 40 @ 1/16" | 3 @ 1/2" |
| 28 | Continuous | 12 | 29.8 | 23 @ 1/16" | 2 @ 1/2" |
| 28 | Intermittent | 24 | 47.4 | 40 @ 1/16" | 3 @ 1/2" |
| 28 | Continuous | 24 | 116 | 23 @ 1/16" | 2 @ 1/2" |

Type 16

Type 18

Type 22

### AC VOLTAGE MODEL SOLENOIDS

| Type | Duty | Volts | Ohms | Minimum | Maximum |
|---|---|---|---|---|---|
| 2 | Intermittent | 120 | 60 | 45 @ 1/8" | 11 @ 7/8" |
| 2 | Continuous | 120 | 166 | 14 @ 1/8" | 3 @ 7/8" |
| 2HD | Intermittent | 120 | 36 | 70 @ 1/8" | 16 @ 3/4" |
| 2HD | Continuous | 120 | 113 | 25 @ 1/8" | 6 @ 3/4" |
| 4 | Intermittent | 120 | 37 | 36 @ 1/8" | 26 @ 1" |
| 4 | Continuous | 120 | 133 | 8 @ 1/8" | 7 @ 1" |
| 11 | Intermittent | 120 | 85 | 21 @ 1/8" | 11 @ 3/4" |
| 11 | Continuous | 120 | 200 | 12 @ 1/8" | 6 @ 3/4" |
| 11HD | Continuous | 120 | 165 | 12 @ 1/8" | 3 1/2 @ 1" |
| 11P | Continuous | 120 | 200 | 9.6 @ 1/8" | 4.8 @ 3/4" |
| 12 | Intermittent | 120 | 100 | 48 @ 1/8" | 9 @ 7/8" |
| 12 | Continuous | 120 | 150 | 28 @ 1/8" | 6 @ 7/8" |
| 14 | Intermittent | 120 | 11 | 108 @ 1/8" | 56 @ 1 1/2" |
| 14 | Continuous | 120 | 18 | 75 @ 1/8" | 40 @ 1 1/2" |
| 16 | Intermittent | 120 | 41 | 110 @ 1/8" | 28 @ 3/4" |
| 16 | Continuous | 120 | 85 | 63 @ 1/8" | 15 @ 3/4" |
| 16 | Continuous | 240 | 350 | 63 @ 1/8" | 15 @ 3/4" |
| 16P | Intermittent | 120 | 41 | 88 @ 1/8" | 22.5 @ 3/4" |
| 16P | Continuous | 120 | 85 | 50.5 @ 1/8" | 12 @ 3/4" |
| 18 | Intermittent | 120 | 8.8 | 350 @ 1/8" | 208 @ 7/8" |
| 18 | Continuous | 120 | 19.7 | 152 @ 1/8" | 100 @ 7/8" |
| 18 | Intermittent | 240 | 45 | 350 @ 1/8" | 208 @ 7/8" |
| 18 | Continuous | 240 | 78 | 152 @ 1/8" | 100 @ 7/8" |
| 18P | Intermittent | 120 | 8.8 | 315 @ 1/8" | 187 @ 7/8" |
| 18P | Continuous | 120 | 19.7 | 137 @ 1/8" | 90 @ 7/8" |
| 24 | Continuous | 120 | 500 | 10 @ 1/8" | 2 @ 5/8" |
| 28 | Continuous | 24 | 17.4 | 24 @ 1/16" | 5 @ 5/8" |
| 28 | Continuous | 120 | 400 | 24 @ 1/16" | 5 @ 5/8" |

Type 24  Type 28

• **All Models are UL Recognized**

These intermittent and continuous duty solenoids are available in AC and DC versions, and in three constructions: box frame, U-frame and laminated. **Types 2, 2HD, 4, 4HD, 11, 11HD, 11P, 22** and **28** are box frame. **Types 12, 14, 16, 16P, 18** and **18P** are laminated. **Type 24** is U-frame. Suffix **P** indicates a push type model. Suffix **HD** indicates a heavy duty model. All box frame models have quick connect terminals.

**Fig. 11-3** *Frame solenoids.*

## APPLICATIONS

Solenoids are devices that turn electricity, gas, oil, or water on and off. Solenoids can be used, for example, to turn the cold water on, and the hot water off to get the proper mix of warm water in a washing machine. To control the hot water solenoid, a thermostat is inserted in the circuit.

Figure 11-4 shows a solenoid for controlling natural gas flow in a hot-air furnace. Note how the coil is wound around the plunger. The plunger is the core of the solenoid. It has a tendency to be sucked into the coil whenever the coil is energized by current flowing through it. The electromagnetic effect causes the plunger to be attracted upward into the coil area. When the plunger is moved upward by the pull of the electromagnet, the soft disc (10) is also pulled upward, allowing gas to flow through the valve. This basic technique is used to control water, oil, gasoline, or any other liquid or gas.

**Fig. 11-4** *Solenoid for controlling natural gas flow to a hot air furnace.* (Honeywell)

The starter solenoid on an automobile uses a similar procedure except that the plunger has electrical contacts on the end that complete the circuit from the battery to the starter. The solenoid uses low voltage (12 V) and low current to energize the coil. The coil in turn sucks the plunger upward. The plunger, with a heavy-duty copper washer attached, then touches heavyduty contacts that are designed to handle the 300 A needed to start a cold engine. In this way, low voltage and low current are used, from a remote location, to control low voltage and high current.

## Solenoids as Electromagnets

An electromagnet is composed of a coil of wire wound around a core of soft iron. A solenoid is an electromagnet. When current flows through the coil, the core becomes magnetized.

The magnetized core can be used to attract an armature and act as a magnetic circuit breaker (Fig. 11-5). Note how the magnetic circuit breaker is connected in series with both the load circuit to be protected and with the switch contact points. When excessive current flows in the circuit, a strong magnetic field in the electromagnet causes the armature to be attracted to the core. A spring attached to the armature causes the switch contacts to open and break the circuit. The circuit breaker must be reset by hand to allow the circuit to operate properly again. If the overload is still present, the circuit breaker will "trip" again. It will continue to do so until the cause of the short circuit or overload is found and corrected.

**Fig. 11-5** *Magnetic circuit breaker.*

## Solenoid Coils

The coil is the most important part of the solenoid, inasmuch as the valve or switch contacts that it operates cannot work unless the coil is capable of being energized. There are at least three types of coils you should be aware of in solenoids used in air conditioning, refrigeration, and heating circuits. For various applications they are divided into classes, as outlined in Table 11-1 (see Fig. 11-6).

**Table 11-1** *Classes of Solenoid Coils*

| Class | Application |
|---|---|
| A | Moisture-resistant coil for normal use of gas or fluid up to 175°F. |
| B | Ambient and fluid temperature up to 200°F |
| H | Temperatures up to 365°F, high steam pressure, rapid valve cycling, high voltage, fungus-proof. |
| BW | Same as coil B, and waterproof, fungus-proof, plastic-encapsulated for temperatures up to 200°F |
| W | Same as coil A, and waterproof, fungus-proof, plastic-encapsulated for temperatures up to 175°F. |

**Fig. 11-6** *Solenoids for valves.*

## Servicing Coils

Coils can be replaced when they malfunction. Excessive heat causes coil malfunction. Make sure that the valve is not heated to a temperature above the coil rating. When replacing a coil, reassemble the solenoid correctly. A missing part or improper reassembly causes excessive coil heat. See the exploded view in Fig. 11-7.

Applied voltage must be at the coil-rated frequency and voltage. A damaged plunger tube or tube sleeve causes heat and can prevent the solenoid from operating. For applications requiring greater resistance or different electrical requirements, use the proper coil in the solenoids. Do not change from ac to dc or dc to ac without changing the entire solenoid assembly (coil, plunger, plunger tube, and base fitting).

When replacing a coil, first be sure to turn off the electric power to the solenoid. It will not be necessary in most instances to remove the valve from the pipeline. Disconnect the coil leads. Disassemble the solenoid carefully and reassemble in reverse order. Failure to reassemble the solenoid properly can cause coil burn out.

Surge suppressors are available to protect the coil from unusual line surges. Figure 11-8 shows how the coil leads can be connected to allow for 120- or 210-V operation. These are referred to as dual-voltage coils.

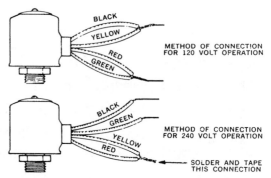

**Fig. 11-8** *Dual-voltage coil wiring diagrams.*

**Fig. 11-7** *Exploded view of balanced diaphragm valve.*

The valve shown in Fig. 11-7 is a series-balanced diaphragm solenoid valve that provides on-off control for domestic and industrial furnaces, boilers, conversion burners, and similar units using thermostats, limit controls, or similar control devices. The valve uses a balanced diaphragm for high operating pressure with low electrical power consumption. It is suitable for use with all gases and comes in a variety of sizes, capacities, and pressures.

Presence of a low, barely audible hum is normal when the coil is energized. If the valve develops a buzzing or chattering noise, check for proper voltage. Thoroughly clean the plunger and the interior of the plunger tube. Make sure that the plunger tube and solenoid assembly are tight. See Fig. 11-9.

## SOLENOID VALVES IN CIRCUITS

Solenoid valves are used on multiple installations in refrigeration systems. They are electrically operated as shown in Fig. 11-10. When connected as shown in the

**Fig. 11-9** *Solenoid coil with cover removed.*

**Fig. 11-10** *Solenoid valve connected in the suction and liquid evaporator lines of a refrigeration system.*

illustration, the valve remains open when current is supplied to it. It closes when the current is turned off. In general, solenoid valves are used to control the liquid refrigerant flow into the expansion valve or the refrigerant gas flow from the evaporator when it or the fixture it is controlling reaches the desired temperature. The most common application of the solenoid valve is in the liquid line and operates with a thermostat. With this hookup, the thermostat is set for the desired temperature in the fixture. When this temperature is reached, the thermostat opens the electrical circuit and shuts off the current to the valve. The solenoid valve, closes and shuts off the refrigerant supply to the expansion valve. The condensing unit operation is controlled by the low-pressure switch. In other applications, where the evaporator is in operation for only a few hours each day, a manually operated snap switch is used to open and close the solenoid valve.

## Refrigeration Valve

The solenoid valve shown in Fig. 11-11 is operated with a normally closed status. A direct-acting metal ball and seat assure tight closing. The two-wire, class W coil is supplied standard for long life on low-temperature service or sweating conditions. Current failure or interruption causes the valve to fail-safe in the closed position. Explosion-proof models are available for use in hazardous areas.

This solenoid valve is usable with all refrigerants except ammonia. It can also be used for air, oil, water, detergents, butane or propane gas, and other non-corrosive liquids or gases.

A variety of temperature control installations can be accomplished with these valves. Such installations include bypass, defrost, suction line, hot gas service, humidity control, alcohols, unloading, reverse cycle, chilled water, cooling tower, brine, and liquid-line stop installations and ice makers.

The valves are held in the normally closed position by the weight of the plunger assembly and fluid pressure on top of the valve ball. The valve is opened by energizing the solenoid or coil. This magnetically lifts the plunger and allows full flow by the valve ball. De-energizing the coil permits the plunger and valve ball to return to the closed position.

**Fig. 11-11** *Schematic of a refrigeration installation.*

**Solenoid Valves in Circuits**

# REVIEW QUESTIONS

1. Define solenoid.
2. What is meant by the term sucking of a solenoid?
3. What does the armature of an electromagnet do?
4. What is the most important part of the solenoid?
5. List the five classes of solenoids.
6. What are dual-voltage coils? How are they wired?
7. Where are series-balanced diaphragm solenoid valves used?
8. What does a barely audible hum indicate when a coil is energized?
9. Where are solenoid valves used?
10. How are valves made fail-safe?

# 12 CHAPTER

# Motor Starting Methods

# PERFORMANCE OBJECTIVES

After studying this chapter, you will be able to:

1. Describe how a split-phase motor can be reversed.
2. List uses for the split-phase motor.
3. Identify the repulsion-induction motor.
4. Identify the capacitor-start motor.
5. Explain how to reverse the direction of rotation of a capacitor-start motor.
6. Describe the best use for the permanent split-capacitor motor.
7. Explain how a shaded-pole motor operates.
8. List various motor starter methods.
9. Explain how primary resistor starting operates.
10. Explain how part-winding starting is accomplished.
11. List the advantages and disadvantages of part-winding starters.
12. Describe how wye-delta and star-delta starters work.
13. List the advantages and disadvantages of wye-delta and star-delta starters.
14. Describe the operation of consequence-pole motor controllers.
15. Identify the least expensive type of starter.
16. List basic characteristics of five types of starting methods.
17. Select a starter for a desired characteristic.

# ELECTRIC MOTORS

Electric motors are designed to deliver their best overall performance when operated at the design voltage shown on the nameplate. However, this voltage is often not maintained. Instead, it varies between minimum and maximum limits over what is termed voltage spread. The voltage spread is usually due to the wiring and transformers of the electrical distribution system and varies in proportion to motor or load currents.

In most modern plants using load-center power distribution systems, variations in voltage normally will be within recommended limits of 110 to 120, 220 to 240, 440 to 480, and 550 to 600 for single-phase and three-phase squirrel-cage and synchronous motors. However, there are older plants throughout the country with large low-voltage systems. Long low-voltage feeders often cause voltage drops that result in below-standard voltages at the motor terminals, especially during motor starting, when currents may be up to six times normal full-load. Table 12-1 shows the effect of voltage variations on the performance of polyphase induction motors.

*Table 12-1* Voltage Variations

| Rated Voltage | Lower Limit | Upper Limit |
|---|---|---|
| 220 | 210 | 240 |
| 440 | 420 | 480 |
| 550 | 525 | 600 |
| 2300 | 2250 | 2480 |
| 4000 | 3920 | 4320 |
| 4600 | 4500 | 5000 |
| 6600 | 6470 | 7130 |

Single-phase and polyphase motors call for different approaches or methods to start them under various conditions of operation. Most single-phase motors are started by the turning on of an on-off switch or a magnetic starter.

# STARTING THE MOTOR

One of the most important parts of the electric motor is the start mechanism. A special type is needed for use with single-phase motors. A centrifugal switch is used to take a start winding out of the circuit once the motor has come up to within 75% of its run speed. The split-phase, capacitor-start, and other variations of these types all need the start mechanism to get them running.

The stator of a split-phase motor has two types of coils, one called the run winding and other the start winding. The run winding is made by winding the enamel-coated copper wire through the slots in the stator punchings. The start winding is made in the same way except that the wire is smaller. Coils that form the start windings are positioned in pairs in the stator directly opposite each other and between the run windings. When you look at the end of the stator, you see alternating run windings and start windings (Fig. 12-1).

*Fig. 12-1* Split-phase motor windings. (Bodine Electric Co.)

The run windings are all connected together, so the electrical current must pass through one coil completely before it enters the next coil, and so on through all the run windings in the stator. The start windings are connected together in the same way and the current must pass through each in turn (Fig. 12-2).

**Fig. 12-2** *Single phase induction motor.*

The two wires from the run windings in the stator are connected to terminals on an insulated terminal block in one end bell where the power cord is attached to the same terminals. One wire from the start winding is also tied to one of these terminals. However, the other wire from the start winding is connected to the stationary switch mounted in the end bell. Another wire then connects this switch to the opposite terminal on the insulated block. The stationary switch does not revolve but is placed so that the weights in the rotating portion of the switch, located on the rotor shaft, will move outward when the motor is up to speed and open the switch to stop electrical current from passing through the start winding.

The motor then runs only on the main winding until such times as it is shut off. Then, as the rotor decreases in speed, the weights on the rotating switch again move inward to close the stationary switch and engage the start winding for the next time it is started.

## Reversibility

The direction of rotation of the split-phase motor can be changed by reversing the start winding leads.

## Uses

This type of motor is used for fans, furnace blowers, oil burners, office appliances, and unit heaters.

## REPULSION-INDUCTION MOTOR

The repulsion-induction motor starts on one principle of operation and, when almost up to speed, changes over to another type of operation. Very high twisting forces are produced during starting by the repulsion between the magnetic pole in the armature and the same kind of pole in the adjacent stator field winding. The repulsing force is controlled and changed so that the armature rotational speed increases rapidly, and if not stopped, would continue to increase beyond a practical operating speed. It is prevented by a speed-actuated mechanical switch that causes the armature to act as a rotor that is electrically the same as the rotor in single-phase induction motors. That is why the motor is called a repulsion-induction motor.

The stator of this motor is constructed very much like that of a split-phase or capacitor-start motor, but there are only run or field windings mounted inside. End bells keep the armature and shaft in position and hold the shaft bearings.

The armature consists of many separate coils of wire connected to segments of the commutator. Mounted on the other end of the armature are governor weights, which move pushrods that pass through the armature core. These rods push against a short-circuiting ring mounted on the shaft on the commutator end of the armature. Brush holders and brushes are mounted in the commutator end bell, and the brushes, connected by a heavy wire, press against segments on opposite sides of the commutator (Fig. 12-3).

**Fig. 12-3** *Brush-lifting, repulsion-start, induction-run, single-phase motor.*

When the motor is stopped, the action of the governor weights keeps the short-circuiting ring from touching the commutator. When the power is turned on and current flows through the stator field windings, a current is induced in the armature coils. The two brushes connected together form an electromagnetic coil that produces a north and south pole in the armature, positioned so that the north pole in the armature is next to a north pole in the stator field windings. Since like poles try to move apart, the repulsion produced in this case can be satisfied in only one way, by the armature turning and moving the armature coil away from the field windings.

The armature turns faster and faster, accelerating until it reaches what is approximately 80% of the run speed. At this speed the governor weights fly outward and allow the pushrods to move. The pushrods, which are parallel to the armature shaft, have been holding the short-circuiting ring away from the commutator. Now that the governor has reached its designed speed, the rods can move together electrically in the same manner that the cast aluminum disks did in the cage of the induction motor rotor. This means that the motor runs as an induction motor.

*Fig. 12-4* Single-phase diagram for the AH air conditioner and heat pump compressor.

## Uses

The repulsion-induction motor can start very heavy, hard-to-turn loads without drawing too much current. They are made from 1/2 to 20 hp. This type of motor is used for applications such as large air compressors, refrigeration equipment, large hoists, and are particularly useful in locations where low-line voltage is a problem. This type of motor is no longer used in the refrigeration industry. Some older operating units may be found with this type of motor still in use.

## CAPACITOR-START MOTOR

The capacitor motor is slightly different from a split-phase motor. A capacitor is placed in the path of the electrical current in the start winding (Fig. 12-4). Except for the capacitor, which is an electrical component that slows any rapid change in current, the two motors are the same electrically. A capacitor motor can usually be recognized by the capacitor can or housing that is mounted on the stator (Fig. 12-5).

Adding the capacitor to the start winding increases the effect of the two-phase field described in connection with the split-phase motor. The capacitor means that the motor can produce a much greater twisting force when it is started. It also reduces the amount of electrical current required during starting to about 1.5 times the current required after the motor is up to speed. Split-phase motors require three or four times the current in starting than they do in running.

## REVERSIBILITY

An induction motor will not always reverse while running. It may continue to run in the same direction but at a reduced efficiency. An inertia type load is difficult to reverse. Most motors that are classified as reversible while running will reverse with a noninertial-type load. They may not reverse if they are under no-load conditions or have a light load or an inertial load.

RIGID BASE

RESILIENT BASE

*Fig. 12-5* Capacitor-start motor.

One of the problems related to the reversing of a motor while it is still running is the damage done to the transmission system connected to the load. In some cases it is possible to damage a load. One of the ways to avoid this is to make sure that the right motor is connected to a load.

Reversing (while standing still) the capacitor-start motor can be done by reversing its start winding connections. This is usually the only time that it will work on a motor. The available replacement motor may not be rotating in the direction desired, so the electrician will have to locate the start winding terminals and reverse them in order to have the motor start in the desired direction.

Figure 12-6A shows a capacitor-start, induction-run motor used in a compressor. This type uses a relay to place the capacitor in and out of the circuit. More details regarding this type of relay will be given later. Figure 12-6B shows how the capacitor is located outside the compressor.

## Uses

Capacitor motors are available in sizes from 1/6 to 20 hp. They are used for fairly hard-starting loads that can be brought up to run speed in under 3 seconds. They may be used in industrial machine tools, pumps, air conditioners, air compressors, conveyors, and hoists.

| Start capacitor* sizes | |
|---|---|
| Compressor 1/8 hp | Capacitor size is 95 to 200 μF |
| Compressor 1/6 hp | Capacitor size is 95 to 200 μF |
| Compressor 1/4 hp | Capacitor size is 200 to 300 μF |
| Compressor 1/3 hp | Capacitor size is 250 to 350 μF |
| Compressor 1/2 hp | Capacitor size is 300 to 400 μF |
| Compressor 3/4 hp | Capacitor size is 300 to 400 μF |

Black case (Bakelite)

**Fig. 12-6** *(A) Capacitor start motor used for a compressor. (B) location of start capacitor in a compressor circuit.*

## PERMANENT SPLIT-CAPACITOR MOTOR

The permanent split-capacitor (PSC) motor is used in compressors for air-conditioning and refrigeration units. It has an advantage over the capacitor-start motor inasmuch as it does not need the centrifugal switch and its associated problems.

The PSC motor has a run capacitor in series with the start winding. Both run capacitor and start winding remain in the circuit during start and after the motor is up to speed. Motor torque is sufficient for capillary and other self-equalizing systems. No start capacitor or relay is necessary. The PSC motor is basically an air-conditioner compressor motor. It is very common upto 3-hp, but is also available in the 4- and 5-hp sizes (Fig. 12-7).

**Fig. 12-8** *Shading of the poles of a shaded-pole motor.*

**Fig. 12-7** *Permanent split-capacitor motor schematic.*

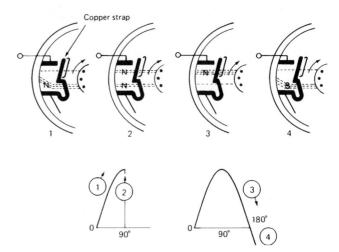

**Fig. 12-9** *Shaded poles as used in shaded pole ac motors.*

## SHADED-POLE MOTOR

The shaded-pole induction motor is a single-phase motor. It uses a unique method to start the rotor turning. The effect of a moving magnetic field is produced by constructing the stator in a special way (Fig. 12-8).

Portions of the pole-piece surfaces are surrounded by a copper strap called a shading coil. The strap causes the field to move back and forth across the face of the pole piece.

In Fig. 12-9, the numbered sequence and points on the magnetization curve are shown. As the alternating stator field starts increasing from zero (1), the lines of force expand across the face of the pole piece and cut through the strap. A voltage is induced in the strap. The current that results generates a field that opposes the cutting action (and decreases the strength) of the main field. This action causes certain actions: As the field increases from zero to a maximum of 90°, a large portion of the magnetic lines of force are concentrated in the unshaded portion of the pole (1). At 90° the field reaches its maximum value. Since the lines of force have stopped expanding, no EMF is induced in the strap, and no opposite magnetic field is generated. As a result, the main field is uniformly distributed across the poles as shown in (2).

From 90° to 180° the main field starts decreasing or collapsing inward. The field generated in the strap opposes the collapsing field. The effect is to concentrate the lines of force in the shaded portion of the poles as shown in (3).

Note that from 0° to 180°, the main field has shifted across the pole face from the unshaded to the shaded portion. From 180° to 360°, the main field goes through the same change as it did from 0° to 180°. However, it is now in the opposite direction (4). The direction of the field does not affect the way the shaded pole works. The motion of the field is the same during the second half-hertz as it was during the first half-hertz.

The motion of the field back and forth between shaded and unshaded portions produces a weak torque. This torque is used to start the motor. Because of the weak starting torque, shaded-pole motors are built in only small sizes. They drive such devices as fans, clocks, and blowers.

## Reversibility

Shaded-pole motors can be reversed mechanically. Turn the stator housing and shaded poles end for end. These motors are available from 1/250 to 1/2 hp.

## Uses

As mentioned previously, this type of motor is used as a fan motor in refrigerators and freezers. They can also be used as fan motors in some types of airconditioning equipment where the demand is not too great. They can also be used as part of the timing devices used for defrost timers and other sequenced operations.

The fan and motor assembly are located behind the provisions compartment in a refrigerator, directly above the evaporator in the freezer compartment. The suction fan pulls air through the evaporator and blows it through the provisions compartment air duct and freezer compartment fan grille (Fig. 12-10). This is a shaded-pole motor with a molded plastic fan blade. For maximum air circulation the location of the fan on the motor shaft is most important. Mounting the fan blade too far back or too far forward on the motor shaft, in relation to the evaporator cover, will result in improper air circulation. The freezer compartment fan must be positioned with the lead edge of the fan 1/4 in in front of the evaporator cover.

The fan assembly shown in Fig. 12-11 is used on the top-freezer, no-frost, fiberglass-insulated model refrigerators. The freezer fan and motor assembly is located in the divider partition directly under the freezer air duct.

**Fig. 12-11** *Fan and fan motor bracket assembly.*

**Fig. 12-10** *Fan, motor, and bracket assembly.*

## SPLIT-PHASE MOTOR

Instead of rotating, the field of a single-phase motor merely pulsates. No rotation of the rotor takes place. A single-phase pulsating field may be visualized as two rotating fields revolving at the same speed but in opposite directions. The rotor will revolve in either direction at nearly synchronous speed if it is given an initial impetus in either one direction or the other. The exact value of this initial rotational velocity varies widely with different machines. A velocity higher than 15% of the synchronous speed is usually sufficient to cause the rotor to accelerate to the rated or running speed. A single-phase motor can be made self-starting if means can be provided to give the effect of a rotating field.

To get the split-phase motor running, a run winding and a start winding are incorporated into the stator of the motor. Figure 12-12 shows the split-phase motor with the end cap removed so that you can see the starting switch and governor mechanism.

**Fig. 12-12** *Single-phase starting switch and governor mechanism.*

**Fig. 12-14** *Noncombination magnetic motor starter.* (Westinghouse)

This type of motor is difficult to use with airconditioning and refrigeration equipment inasmuch as it has very little starting torque and will not be able to start a compressor since it presents a load to the motor immediately upon starting. This type of motor, however, is very useful in heating equipment. (Fig. 12-13).

**Fig. 12-13** *Single-phase, split-phase motor.*

## POLYPHASE MOTOR STARTERS

The simple manual starter works for single-phase motors and also, in some instances, for polyphase motors. Most of the polyphase manual starters consisting of an on-off switching arrangement are designed for motors of 1 hp or less. Figure 12-14 shows the magnetic motor starter designed for across-the-line control of squirrel-cage motors or as primary control for wound-rotor motors. They are available for nonreversing, reversing, and two-speed applications. The drawing in Fig. 12-15A shows the difference between the single- and three-phase nonreversing type of starter. Figure 12-15B shows the reversing drawing; Fig. 12-15C is the two-speed, one-winding starter, and Fig. 12-15D is the two-speed, two-winding starter for motors up to 100 hp.

During across-the-line starting, motor input current is five to eight times the normal full-load current. This can cause an excessive temporary voltage drop on power lines that causes lights to flicker or may even interrupt the service.

To control these temporary voltage drops, power companies have restrictions such as

- A specific maximum starting current (or kVA)
- A specific limit on kVA/hp
- A maximum horsepower motor size which can be started across-the-line
- A specific maximum line current that can be drawn in steps (increment starting)

The specified restrictions vary considerably between power companies, even within one company's service area. It is wise to check local power company restrictions before making a larger motor installation.

## REDUCED-VOLTAGE STARTING METHODS

Reduced-voltage starters operate such that input current and consequently torque are reduced during starting. Table 12-2 briefly describes the various methods of starting and gives features and limitations of each.

When motors are started at reduced voltage, the current at the motor terminals is reduced in direct proportion to the voltage reduction, while the torque is reduced as the square of the voltage reduction. For example, if the "typical" motor were started at 65% of line voltage, the starting current would be 42% and the torque would be 42% of full-voltage values. Thus reduced-voltage starting provides an effective means of reducing both current and torque (Fig. 12-19).

(A)

(B)

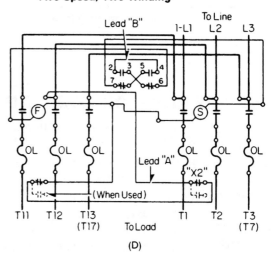

(C)                                    (D)

**Fig. 12-15** *(A) Nonreversing single-phase and nonreversing three-phase diagrams; (B) reversing three phase wiring diagram. (C) two-speed, one winding, three-phase delta; (D) two-speed, two-winding, three-phase starter diagram.* (Westinghouse)

**Reduced-Voltage Starting Methods** 207

*Table 12-2  Starting Method Characteristics*[a]

| Starting Method | Operation | | Starting Current (% of Locked Rotor Current) | Starting Torque (% of Locked Rotor Torque) | Open or Closed Transition | Basic Characteristics | |
|---|---|---|---|---|---|---|---|
| | | | | | | Advantages | Limitations |
| Across-the-line | Initially connects motor directly to power lines. | | 100% | 100% | None | 1. Lowest cost. 2. Higher starting torque. 3. Used with any standard motor 4. Least maintenance | 1. High starting current. 2. High starting torque 3. May shock driven machine |
| Primary resistance reduced voltage | Inserts resistance units in series with motor during first step(s) | | 50–80% | 25–64% | Closed | 1. Smoothest starting. 2. Least shock to driven machine. 3. Most flexible in application 4. Used with any standard motor. | 1. High power loss because of heating resistors 2. Heat must be dissipated. 3. Low torque per ampere input. 4. Higher cost. |
| Autotransformer reduced voltage | Uses autotransformer to reduced voltage applied to motor. | Tap 50% 65% 80% | 25% 42% 64% | 25% 42% 64% | Closed | 1. Best for hard to start loads. 2. Adjustable starting torque 3. Used with any standard motor. 4. Less strain on motor | 1. May shock driven machine 2. High cost |
| Wye-Delta | Starts motor with windings wye connected, then reconnects them in delta connection for running. | | 33% | 33% | Open or closed | 1. Medium cost. 2. Low starting current. 3. Low starting torque. 4. Less strain on motor. | 1. Low starting torque. 2. Requires delta-wound motor. |
| Part Winding | Starts motor with only part of windings connected, then adds remainder for running. | | 70–80% | 50–60% Minimum pull pull-up torque 35% of full-load torque | Closed | 1. Low cost. 2. Popular method for medium staring torque applications. 3. Low maintenance. | 1. Not good for frequent starts 2. May require special wound motor. 3. Low pull-up torque. 4. May not come up to speed on first-step when started with load applied. |

NOTE: The reduced starting torque (LRT) indicated in this table for the various reduced starting methods can prevent starting high inertia loads and must be considered when sizing motors and choosing starters.

## PRIMARY RESISTOR STARTING

In primary resistor starting, a resistor is connected in each motor line (in one line only in single-phase starters) to produce a voltage drop due to the motor starting current. A timing relay shorts out the resistors after the motor has accelerated. Thus the motor is started at reduced voltage but operates at line voltage.

Figure 12-16 shows two types of motor starter resistors. The resistance element will retain its mechanical and electrical properties both during and after repeated heating and cooling. All metal parts are either plated with or fabricated of corrosion-resistant material for overall corrosion protection. Under certain conditions operating temperatures may reach 600°C and not

*Fig. 12-16*  Wire-wound resistors used in primary resistor starter circuits. (Westinghouse)

change the resistance value. The motor starter resistors are 11, 14, 17, and 20 in long and come in wattage ratings of 450 to 1320. Table 12-3 shows the resistance

*Table 12-3* Resistor Ranges and Properties

| | Low R-High Current | | | High R-Low Current | | |
|---|---|---|---|---|---|---|
| Unit Length (in.) | Resistance Range (Ω) | Current Range (A) | Heat Dissipation (Watts per Unit[a]) | Resistance Range (Ω) | Current Range (A) | Heat Dissipation (Watts per Unit[a]) |
| 11 | 0.051–433 | 11–104 | 450–630 | 4.0–2000 | 0.46–10.3 | 426 |
| 14 | 0.069–5.7 | 11–104 | 620–820 | 5.0–2500 | 0.48–10.8 | 575 |
| 17 | 0.085–7.1 | 11–104 | 770–1080 | 5.0–2500 | 0.53–12.0 | 700 |
| 20 | 0.10–8.6 | 11–104 | 900–1320 | 6.4–4000 | 0.47–11.8 | 900 |

[a]Approximate only

ranges and other factors. Note the current-handling ability of the resistors.

Primary resistor starters are sometimes known as cushion starters. The main reason for the name is their ability to produce a smooth, cushioned acceleration with closed transition. However, this method is not as efficient as other methods of reduced-voltage starting, but it is ideally suited for applications such as conveyors, textile machines, or other delicate machinery where reduction of starting torque is of prime consideration.

## Operation

Figure 12-17 is the reduced-voltage magnetic starter that uses resistors to operate a three-phase motor properly at

*Fig. 12-17* Primary resistor type of magnetic starter. (Westinghouse)

*Fig. 12-18* Wiring diagram for a primary resistor type of starter. (Westinghouse)

start. Closing the *start* button or other pilot device energizes the start contactor (S) shown in Fig. 12-18. This connects the motor in series with the starting resistors for a reduced-voltage start. The contactor (S) is now sealed in through its interlock (Sa). Timing relay (TR) is energized, and after a preset time interval its contacts $TR_{TC}$ close. This energizes the run contactor, *run*, which seals through its interlock $run_a$. The contacts (*run*) close, bypassing the starting resistors, and the motor will now be running at full voltage. The contactor (S) and timing relay (TR) are de-energized when the interlock $run_b$ opens.

An overload, which opens the *stop* button or other pilot device, de-energizes the (*run*) contactor. This removes the motor from the line.

Primary resistor starters provide extremely smooth starting due to the increasing voltage across the motor terminals as the motor accelerates. Since motor current decreases with increasing speed, the voltage drop across the resistor decreases as the motor accelerates and the motor terminal voltage increases. Thus if a resistor is shorted out as the motor reaches maximum speed, there is little or no increase in current or torque.

## AUTOTRANSFORMER STARTING

Autotransformer starters provide reduced-voltage starting at the motor terminals through the use of a tapped, three-phase autotransformer. Upon initiation of the controller pilot device, a two-and a three-pole contactor close to connect the motor to the preselected autotransformer taps. A timing relay causes the transfer of the motor from the reduced-voltage start to line-voltage operation without disconnecting the motor from the power source. This is known as closed *transition starting*.

Taps on the autotransformer provide selection of 50%, 65%, or 80% of line voltage as a starting voltage. Starting torque will be 25%, 42%, or 64%, respectively, of line-voltage values. However, because of transformer action, the controller line current will be less than motor current, being 25%, 42%, or 64% of full-voltage values. This autotransformer starting may be used to provide maximum torque available with minimum line current, together with taps to permit both of these factors to be varied. Figure 12-19 shows torque and voltage tap points.

Manual autotransformer starters are used to start squirrel-cage polyphase motors when the characteristics of the driven load or power company limitations require starting at reduced voltage (Fig. 12-20).

NEMA (National Electrical Manufacturers Association) permits one start every 4 minutes, for a total of four starts followed by a rest period (2 hours). Each starting period is not to exceed 15 seconds. Figure 12-21 shows a autotransformer type of starter. Note the location of the taps on the starting transformer.

The autotransformer provides the highest starting torque per ampere of line current. Thus it is an effective means of motor starting for applications where the inrush current must be reduced with a minimum sacrifice of starting torque. This type of starter arrangement features closed-circuit transition, an arrangement that maintains a continuous power connection to the motor during the transition from reduced to full voltage. This avoids the high transient switching currents characteristic of starters using open-circuit transition. It provides smoother acceleration as well.

## Operation

Operating an external *start* button or pilot device closes the neutral and start contactors, applying reduced voltage to the motor through the autotransformer. After a

**Fig. 12-20** (A) Autotransformer type of magnetic starter; (B) corresponding wiring diagram. (Allen-Bradley)

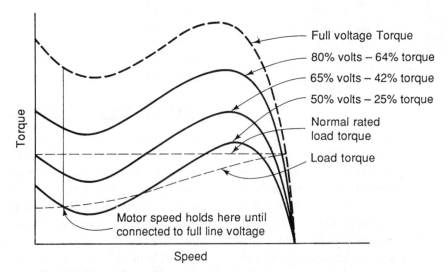

**Fig. 12-19** Autotransformer starting-speed versus torque. (The Lincoln Electric Co.)

**Fig. 12-21** *Typical wiring diagram for an autotransformer type of reduced-voltage starter.* (Allen-Bradley)

preset interval, the timer contacts drop out the neutral contactor, breaking the autotransformer connection but leaving part of the windings connected to the motor as a series reactor. The *run* contactor then closes to short out this reactance and apply full voltage to the motor. Transition from reduced to full voltage is accomplished without opening the motor circuit.

For starters rated up to 200 hp you should allow a 15-second operation out of every 4 minutes. This procedure is followed for 1 hour with a rest period of 2 hours. For starters rated above 200 hp, you should allow three 30-second operations separated by 30-second intervals followed by a rest period of 1 hour. The major disadvantages of this type of starter are its expense for lower horsepower ratings and its low power factor.

## PART-WINDING STARTING

Part-winding motors have two sets of identical windings—intended to be operated in parallel—which can be energized in sequence to provide reduced starting current and reduced starting torque. Most (but not all) dual-voltage 230/460-V motors are suitable for part-winding starting at 230 V.

When one winding of a part-winding motor is energized, the torque produced is about 50% of "both winding" torque, and line current is 60 to 70% (depending on motor design) of comparable line voltage values. Thus, although part-winding starting is not truly a reduced voltage means, it is usually also classified as such because of its reduced current and torque.

When a dual-voltage delta-connected motor is operated at 230 V from a part-winding starter having a three-pole start and a three-pole run contactor, an unequal current division occurs during normal operation resulting in overloading of the starting contactor. To overcome this defect, some part-winding starters use a four-pole starting contactor and a two-pole run contactor. This arrangement eliminates the unequal current division obtained with a delta-wound motor, and it enables wye-connected part-winding motors to be given either a one-half or two-thirds part-winding start.

The class 8640 starters have a start contactor, a timing relay, a run contactor, and necessary overload relays. Closing the pilot device contact causes the start contactor to close to connect the start winding and to initiate the time cycle. After expiration of the preset timing, the run contactor closes to connect the balance of the motor windings. A time setting of 1 second is recommended. Most motor manufacturers do not permit energization of the start winding alone for longer than 3 seconds. Part-winding starters provide closed transition starting.

## Operation

The part-winding type of starter is shown in Fig. 12-22. The parts are located for ease in understanding the operation. By taking a look at the schematic in Fig. 12-23 you can see how the starter operates. Closing the *start* button or other pilot device energizes the start contactor (1M) that seals in through its interlock (1M$_a$) and energizes the timer (TR). The (1M) contacts connect the first half-winding of the motor across the line. After a preset time interval, timer TRTC contacts close the energizing contactor (2M). The (2M) contacts connect the second half-winding of the motor across-the-line.

**Fig. 12-23** *Typical wiring diagram for part-winding type of starter.* (Westinghouse)

An overload, which opens the *stop* button or other pilot device, de-energizes contacts 1M, 2M, and timer TR, removing the motor from the line. The three-pole contactor (1M) connects only the first halfwinding of the motor for reduced inrush current on starting. A three-pole contactor (2M) connects the second half-winding of the motor for running.

## Advantages and Disadvantages

Part-winding starters are the least expensive reduced-voltage controller. They use closed transition starting and are small in size. The disadvantages are that they are unsuitable for long acceleration or frequent starting, require special motor design, and that there is no flexibility in selecting starting characteristics.

### WYE-DELTA OR STAR-DELTA STARTERS

Wye-delta or star-delta starters are used with delta-wound squirrel-cage motors that have all leads brought out to facilitate a wye connection for reduced-voltage starting. This starting method is particularly suitable for applications involving long accelerating times or frequent starts. Wye-delta starters are typically used for high-inertia loads such as centrifugal air-conditioning units, although they are applicable in cases where low starting torque is necessary or where low starting current is necessary and low starting torque is permissible.

When 6- or 12-lead delta-connected motors are started star-connected, approximately 58% of fullline voltage is applied to each winding and the motor develops 33% of full-voltage starting torque and draws 33% of normal locked-rotor current from the line. When the motor has accelerated, it is reconnected for normal delta operation.

**Fig. 12-22** *Part-winding type of magnetic starter.* (Westinghouse)

## Operation

Operating an external *start* button energizes the motor in the wye connection (Fig. 12-24). This applies approximately 58% of full-line voltage to the windings. At this reduced voltage, the motor will develop about 33% of its full-voltage starting torque and will draw about 33% of its normal locked rotor current. After an adjustable time interval,

**Fig. 12-25** *Typical wiring diagram for wye-delta starter, closed-circuit transistion.* (Allen-Bradley)

**Fig. 12-24** *Typical wiring diagram for wye-delta starter, open-circuit transistion.* (Allen-Bradley)

the motor is automatically connected in delta, applying full line voltage to the windings. In starters with open-circuit transition the motor is momentarily disconnected from the line during the transition from the wye to delta. With closed transition (Fig. 12-25) the motor remains connected to the line through the resistors. This avoids the current surges associated with open-circuit transition.

## Advantages and Disadvantages

The advantages are moderate cost and its suitability for high-inertial, long-acceleration loads. It does have torque efficiency. However, the disadvantages are that it requires special motor design, starting torque is low, and it is inherently open transition—closed transition is available at added cost. There is no flexibility in selecting starting characteristics.

## Star-Delta (Wye-Delta) Connections

There is a 12-lead motor wound for wye-delta (Y-Δ). starting operation on either low voltage or high voltage (Fig. 12-26). A six-lead single-voltage motor is also

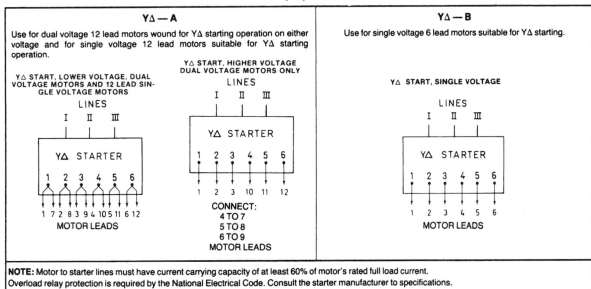

**Fig. 12-26** *Star-delta connections.* (The Lincoln Electric Co.)

suitable for wye-delta starting. Figure 12-26B shows the connection to the lines for the six-lead motor. Keep in mind that overload relay protection is required by the National Electrical Code. The size of the protection is determined by the manufacturer of the motor (Table 12-4).

## MULTISPEED STARTERS

Multispeed starters are designed for the automatic control of two-speed squirrel-cage motors of either the consequent pole or separate winding types. These starters are available for constant-horsepower, constant-torque, or variable-torque three-phase motors. Multispeed motor starters are commonly used on machine tools, fans, blowers, refrigeration compressors, and many other types of equipment.

### Low-Speed Compelling Relay

When added to a standard starter, the low-speed compelling relay compels the operator always to start the motor in low speed before switching to a higher speed. This is a safety feature where damage to equipment may result when the motor is started at high speed (Fig. 12-27).

### Automatic Sequence Accelerating Relay

The automatic sequence accelerating relay will control the sequence of acceleration from low speed to high speed.

### Automatic Sequence Decelerating Relay

The automatic sequence decelerating relay is used with large-inertia loads. The braking effect caused by a sudden change from high to low speed may cause damage to the motor or to the driven machine. To avoid this danger, the operation should give the motor sufficient time to slow down by pushing the *stop* button and then waiting a short interval before pushing the button for a lower speed.

To help provide correct operation, multispeed starters can be equipped with an automatic sequence decelerating relay for each lower-speed step. This relay automatically interposes a time delay between the speed steps and makes it unnecessary to press the *stop* button when switching to a lower speed.

*Table 12-4* Selection of a Controller Best Suited for a Particular Characteristic

| Characteristic Wanted | Type of Starter to Use (Listed in Order of Desirability) | Comments |
|---|---|---|
| Smooth acceleration | 1. Solid state (class 8660)<br>2. Primary resistor (class 8647)<br>3. Wye-delta (class 8630)<br>4. Autotransformer (class 8606)<br>5. Part-winding (class 8640) | Little choice between 3 and 4. |
| Minimum line current | 1. Autotransformer (class 8660)<br>2. Solid state (class 8660)<br>3. Wye-delta (class 8630)<br>4. Part winding (class 8640)<br>5. Primary resistor (class 8647) | |
| High starting torque | 1. Autotransformer (class 8606)<br>2. Solid state (class 8660)<br>3. Primary resistor (class 8647)<br>4. Part winding (class 8647)<br>5. Wye-delta (class 8630) | |
| High torque efficiency (torque vs. line current) | 1. Autotransformer (class 8606)<br>2. Wye-delta (class 8630)<br>3. Part winding (class 8640)<br>4. Solid state (class 8660)<br>5. Primary resistor (class 8647) | Little choice between 3, 4, and 5. |
| Suitability for long acceleration | 1. Wye delta (class 8630)<br>2. Autotransformer (class 8606)<br>3. Solid state (class 8660)<br>4. Primary resistor (class 8647) | For acceleration time greater than 5 seconds, primary resistor requires non-standard resistors. Part-winding controllers are unsuitable for acceleration time greater than 2 seconds. |
| Suitability for frequent starting | 1. Wye-delta (class 8630)<br>2. Solid state (class 8660)<br>3. Primary resistor (class 8647)<br>4. Autotransformer (class 8606) | Part winding is unsuitable for frequent starts. |
| Flexibility in selecting starting characteristics | 1. Solid state (class 8660)<br>2. Autotransformer (class 8606)<br>3. Primary resistor (class 8647) | For primary resistor, resistor change required to change starting characteristics. Starting characteristics cannot be changed for wye-delta or part-winding controllers. |

Source: Courtesy of Square D.

**Fig. 12-27** (A) Multispeed starter and twospeed consequent pole starter without enclosure; (B) typical wiring diagram for two-speed separate winding motor starter; (C) general-purpose enclosure with cover removed. (Allen-Bradley)

## CONSEQUENT-POLE MOTOR CONTROLLER

By increasing the number of poles a motor has it is possible to change its speed. By increasing the number of poles, the speed of the motor is decreased. Inasmuch as a motor is wound and mounted rather permanently on a frame, it is not easily possible to take out or put in poles or the associated windings. Therefore, an electrical means must be found if the speed of the motor is to be changed by using the number of poles method to do so. One method of doing this is the consequent-pole arrangement. This method can be used for two-speed, one-winding motors or four-speed, two-winding motors.

The reversal of some of the currents in the windings has the same effect as physically increasing or decreasing the number of poles. Three-phase motors are wound, in some cases, with six leads brought out for connection purposes. It is possible to connect the windings, using combinations of the terminals for connection purposes, either in series delta or in parallel wye (Fig. 12-28). By tapping the windings it is possible to send current in two different directions, effectively creating more poles and decreasing the speed of the motor. The number of poles is doubled by reversing through half a phase. Two speeds are obtained by producing twice as many consequent poles for lowspeed operation as for high speed.

Figure 12-29 shows how the controller is wired to produce consequent poles for constant torque or variable torque. The wiring diagram and the line drawing (Fig. 12-30) illustrate connections for the following

**Fig. 12-28** Connections made by the consequent-pole starter for constant torque or variable torque. (Allen-Bradley)

**Fig. 12-29** Wiring diagram for a two-speed, consequent-pole, constant-horsepower motor. NEMA size 0-4. (Square D)

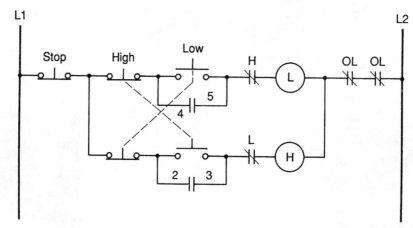

**Fig. 12-30** Line diagram for a two-speed motor. (Allen-Bradley)

method of operation: The motor can be started in either *high* or *low* speed. The change from *low* to *high* or from *high* to *low* can be made without first pressing the *stop* button.

Figure 12-31 shows pilot devices with connections that can be made to obtain different sequences and methods of operation. The series delta arrangement produces high speed. It also produces the same horsepower rating at high and low speeds.

The torque rating is the same for both speeds if the winding is such that the series delta connection gives the low speed and the parallel wye connection gives the high speed. Consequent-pole motors that have a single winding for two speeds have the extra tap at the midpoint of the winding. This permits the various connection possibilities. However, the speed range is limited to a 1:2 ratio of or 600/1200 or 900/1800 rpm.

Figure 12-32 shows the motor terminal markings and connections for a constant-horsepower delta. The wiring diagram (Fig. 12-33) and the line drawing

| CONNECTIONS MADE BY STARTER | | | | |
|---|---|---|---|---|
| Speed | Supply Lines L1 L2 L3 | | Open | Together |
| Low | T1 T2 T3 | | None | T4, 5, 6 |
| High | T6 T4 T5 | | T1, 2, 3 | None |

**Fig. 12-32** Connections made by the starter for constant horsepower. (Allen-Bradley)

(Fig. 12-34) illustrate connections for the following method of operation: Motor can be started in either *high* or *low* speed. The change from *low* to *high* can be made without first pressing the *stop* button. When changing from *high* to *low*, the *stop* button must be

**Fig. 12-31** Pilot device diagrams show connections that can be made to obtain different sequences and methods of operation.

**216** Motor Starting Methods

**Fig. 12-33** Wiring diagram for two-speed, consequent-pole, constant-or variable-torque motor starter. NEMA size 0-4. (Square D.)

pressed between speeds. The pilot devices shown in Fig. 12-35 show the other connections that can be made to obtain sequences and methods of operation.

Four-speed, two-winding consequent-pole motor controllers can be used on squirrel-cage motors that have two reconnectable windings and two speeds for each winding. This type of motor does need a special type of starting sequence. This means that it must use the properties of the compelling relay, accelerating relay, and decelerating relay to operate correctly.

Figure 12-36 shows the two-speed consequent-pole starter with variable-torque and constant-torque connections. Figure 12-37 shows how the four-speed, two-winding controller is connected for the possible arrangements using this type of motor.

## FULL-VOLTAGE CONTROLLERS

The least expensive of the starters is the full-voltage type. There is no limit to the horsepower, size, voltage rating, or type of motor that can be started on full voltage when the power is available.

Full-voltage starters are always the first choice when the power system can supply initial inrush current, and the motor and the driven machine can withstand the sudden starting shock. Examples of this are machines that start unloaded, as well as those that require little torque; or machines may be equipped with some form of unloading device to reduce starting torque, as in the use of an unloader valve in a compressor. A clutch may be inserted between a machine and motor so that the motor may be started unloaded. When the motor is up to speed the clutch is engaged. Clutches are sometimes used on large machines so that maximum horsepower can be exerted during breakaway without serious power system disturbance. Use of clutches also permits using motors with lower torque and locked-rotor currents. In most instances, up-to-date installations use solid-state motor controllers to better advantage. Many of the older types of starters are still in use and will continue to provide good service for many more years. As they deteriorate, they are usually replaced by a solid-state type of starter so that the clutch arrangements are unnecessary.

Figure 12-38 shows the general-purpose enclosure for a full-voltage starter. This type of starter is designed for full-voltage starting of polyphase squirrel-cage motors and primary control of slip-ring motors. This type of starter may be operated by remote control with pushbuttons, float switches, thermostats, pressure switches, snap switches, limit switches, or any other suitable two- or three-wire pilot device.

## STARTING SEQUENCE

If full-voltage starting produces excessive current demands on the distribution system, motors should be started individually or in blocks of permissible size by using some method of time delay, such as motor driver, pneumatic, or mercury plunger timing relays. When large and small motors are to be started on a common power system, best results are obtained by starting the largest sizes first. This gives larger motors the advantage of full-line capacity. If synchronous

**Fig. 12-34** Elementary drawing of the control circuitry for a consequent-pole starter. (Allen-Bradley)

**Fig. 12-35** *Connections for different sequences and methods of operation.* (Allen-Bradley)

**Fig. 12-36** *Typical wiring diagrams for two-speed consequent-pole starter.* (Allen-Bradley)

motors are on the system with other types of ac motors, the synchronous units should always be started first since they provide voltage stability for starting the induction motors.

## PROTECTION AGAINST LOW VOLTAGE

Low-voltage protection is needed while the motors are running even though systematic starting permits all motors to be started without excessive line voltage drop. When three-wire control circuits are used, a severe dip in line voltage or a momentary complete outage breaks the control-sealing circuits, and the controller drops out and stops the motor. This provides low-voltage protection and prevents simultaneous acceleration of all motors to full speed after being slowed down by a voltage dip. However, all motors are disconnected from the line during the voltage dip, and each must be restarted.

## TIME-DELAY PROTECTION

It is possible to wire the circuitry so that a time-delay undervoltage arrangement can be used. This permits dropout of the controllers on low-voltage dips but allows restarting automatically if normal voltage is restored within a preset time delay. The usual time delay is 2 seconds or less.

Time-delay undervoltage protection on controllers will prevent some complete shutdowns but should be applied with caution. If used on all motor controllers, restoration of voltage within the time-delay setting after a voltage dip causes each motor to attempt to accelerate simultaneously, thus producing excessive currents that may operate backup protection and starter overload devices and disconnect the motors.

Pilot devices such as pressure, float, or temperature switches automatically start and stop motors as the demand arises. On severe voltage dips or voltage failure, motor controllers drop open even though the

**Fig. 12-37** *Elementary diagram of a four-speed, two-winding controller and the possible arrangements for motor connections.*
(Allen-Bradley)

demand switch is closed. Upon restoration of full voltage all units attempt to restart at the same time. This operating hazard can be overcome by adding a time delay in the starting circuit of each motor and timing the demand for starting at slightly different intervals. Time delays of various units can then be staggered so that at the restoration of voltage only one unit at a time will be started.

**Time-Delay Protection** 219

| SPEED | L1 | L2 | L3 | OPEN | TOGETHER |
|---|---|---|---|---|---|
| LOW | T1 | T2 | T3 | ALL OTHERS | T4, T5, T6, T7 |
| 2ND | T6 | T4 | T5, T7 | ALL OTHERS | |
| 3RD | T11 | T12 | T13 | ALL OTHERS | T14, T15, T16, T17 |
| HIGH | T16 | T14 | T15, T17 | ALL OTHERS | |

(A)

| SPEED | L1 | L2 | L3 | OPEN | TOGETHER |
|---|---|---|---|---|---|
| LOW | T1 | T2 | T3 | ALL OTHERS | T4, T5, T6, T7 |
| 2ND | T11 | T12 | T13 | ALL OTHERS | T14, T15, T16, T17 |
| 3RD | T6 | T4 | T5, T7 | ALL OTHERS | |
| HIGH | T16 | T14 | T15, T17 | ALL OTHERS | |

(B)

## CONSTANT TORQUE

| SPEED | L1 | L2 | L3 | OPEN | TOGETHER |
|---|---|---|---|---|---|
| LOW | T1 | T2 | T3, T7 | ALL OTHERS | |
| 2ND | T6 | T4 | T5 | ALL OTHERS | T1, T2, T3, T7 |
| 3RD | T11 | T12 | T13, T17 | ALL OTHERS | |
| HIGH | T16 | T14 | T15 | ALL OTHERS | T11, T12, T13, T17 |

(C)

| SPEED | L1 | L2 | L3 | OPEN | TOGETHER |
|---|---|---|---|---|---|
| LOW | T1 | T2 | T3, T7 | ALL OTHERS | |
| 2ND | T11 | T12 | T13, T17 | ALL OTHERS | |
| 3RD | T6 | T4 | T5 | ALL OTHERS | T1, T2, T3, T7 |
| HIGH | T16 | T14 | T15 | ALL OTHERS | T11, T12, T13, T17 |

(D)

| SPEED | L1 | L2 | L3 | OPEN | TOGETHER |
|---|---|---|---|---|---|
| LOW | T1 | T2 | T3 | ALL OTHERS | |
| 2ND | T6 | T4 | T5 | ALL OTHERS | T1, T2, T3 |
| 3RD | T11 | T12 | T13 | ALL OTHERS | |
| HIGH | T16 | T14 | T15 | ALL OTHERS | T11, T12, T13 |

(E)

## VARIABLE TORQUE

| SPEED | L1 | L2 | L3 | OPEN | TOGETHER |
|---|---|---|---|---|---|
| LOW | T1 | T2 | T3 | ALL OTHERS | |
| 2ND | T11 | T12 | T13 | ALL OTHERS | |
| 3RD | T6 | T4 | T5 | ALL OTHERS | T1, T2, T3 |
| HIGH | T16 | T14 | T15 | ALL OTHERS | T11, T12, T13 |

(F)

**Fig. 12-37**  *(Continued)*

**Fig. 12-38** *Full-voltage starters (NEMA), open type, without enclosure: (A) size 3; (B) size 5. (Allen-Bradley)*

## REVIEW QUESTIONS

1. What is voltage spread?
2. What is the purpose of a centrifugal switch on a single-phase motor?
3. How can the direction of the rotation be reversed on a split-phase motor?
4. What type of motor uses pushrods and a wound armature?
5. Where are capacitor-start motors used?
6. How are capacitor-start motors reversed when standing still?
7. What advantage does the permanent split-capacitor motor have?
8. What are shaded-pole motors most likely to be used for?
9. What is needed to get a split-phase motor to run?
10. How much current does an across-the-line motor draws when it starts?
11. What is the advantage of reduced-voltage motor starting?
12. What is another name for primary resistor starters?
13. What is the major disadvantage of the autotransformer starter?
14. What type of starting does part-winding starters provide?
15. What is the least expensive method of motor starting?
16. Where are wye-delta starters typically used?
17. Why are wye-delta starters used with delta-wound squirrel-cage motors?
18. Why are compelling relays needed?
19. What happens to motor speed when more poles are added?
20. How do consequence pole motors obtain two speeds?

# Chapter 13

# Solid-State Reduced-Voltage Starters

## PERFORMANCE OBJECTIVES

After studying this chapter, you will be able to:

1. Define thyristor operation.
2. Explain what gating does.
3. Explain solid-state stepless acceleration.
4. Describe the operation of a diac in a control circuit.
5. Describe the operation of a triac in a control circuit.
6. Explain how surge supressors are installed on magnetic devices.
7. Describe lightning surge protection.

## ELECTROMECHANICAL DEVICES

The electromechanical devices used for years are still reliable and working in many installations. They are used to provide sequencing and interlocking tasks. They are simple in construction, flexible in use, and have many contact combinations. They can also handle large currents and break the circuit as required.

Solid-state devices have no moving parts and no contacts to clean, replace, or adjust. They use transistors, triacs, diacs, and SCRs to do the switching. These logic elements can perform the same functions in a solid-state system as relays do in the electromechanical systems (Fig. 13-1).

***Fig. 13-1*** *Solid-state reduced-voltage controller. (Square D)*

The solid-state control device has many advantages that make it desirable for the various environments in which it has to operate. It has no contacts to become dirty or malfunction when needed to control a critical sequence of operations. The solid-state control devices are more reliable than electromechanical devices. They come in sealed-in modules that can be plugged into a rack and replaced as a unit if anything goes wrong with the circuitry.

## REDUCED-VOLTAGE STARTING

Reduced-voltage starting can be accomplished in a number of ways. However, in solid-state circuitry it is somewhat simpler than described previously. The exact details of the circuit functions are somewhat more complex than those of the electromechanical system; however, a complete understanding of solid-state physics and/or electronics is not necessary in order to grasp the workings of the simple devices utilized to perform the operations of solid-state switching and control.

## SILICON-CONTROLLED RECTIFIERS

The silicon-controlled rectifier (SCR) is the device used most often to control electric motors. The proper name for an SCR is thyristor. However, popular use of the term SCR has made it part of the literature and accepted by everyone working in the field. It is a specialized type of semiconductor used for control of electrical circuits.

An SCR conducts current in a forward direction only. The symbol for an SCR is shown in Fig. 13-2. Current flows through an SCR from the cathode (C) to the anode (A). The illustration indicates the SCR also has a gate (G).

The function of the SCR is shown in the circuit diagram in Fig. 13-3. The most typical use of an SCR is for a controlled circuit. Examples include a light dimmer or a speed control for a motor. This type of circuit is

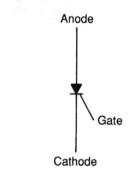

***Fig. 13-2*** *Symbol for SCR.*

***Fig. 13-3*** *Schematic of SCR-controlled circuit.*

**224** Solid-State Reduced-Voltage Starters

illustrated in Fig. 13-3. The resistor in this circuit, R, is a rheostat, or adjustable resistor. This is used to control the amount of voltage delivered to the gate of the SCR. The more voltage delivered, the greater the flow. Thus, adjusting the rheostat can serve to control the circuit. If the circuit illuminates a lamp, lowering the voltage to the rheostat dims the bulb. If the load is a motor, its speed is slowed. Figures 13-4, 13-5, and 13-6 show what typical SCRs look like with their leads identified according to cathode, gate, and anode connections.

**Fig. 13-4** *Drawing of a typical SCR.*

**Fig. 13-5** *Drawing of a typical SCR.*

**Fig. 13-6** *Larger currents require larger SCRs.*

One of the main reasons for using semiconductor devices for motor control is the device's ability to start a motor under reduced-voltage conditions and thus allow the motor to accelerate to full speed at a lower torque level. By reducing the high current inrush the mechanical shock to the driven equipment is reduced.

A reduced-voltage solid-state motor starter uses SCRs for power control. Inasmuch as an SCR conducts in the direction of the arrow in the symbol, it means that current flows only one way in an SCR. To use an SCR to its advantage on ac it is necessary to use two of them in reverse parallel (Fig. 13-7). SCRs have to be turned on in order to conduct current through them; that is, they need a gate pulse to turn them on. Once an SCR is turned on or gated, it does not stop forward current flow. Full wave control uses two SCRs in each phase. Three-phase operation must utilize six diodes, connected as shown in Fig. 13-8.

**Fig. 13-7** *Parallel SCRs for one phase.*

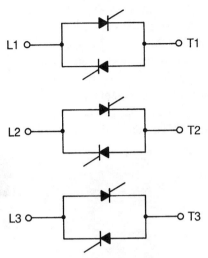

**Fig. 13-8** *Three-phase SCR arrangement.*

The current through an SCR can be controlled by gating the SCR at different times within the half cycle. This also controls the acceleration time of the motor. If the gate pulse is applied early in the half cycle, the output is high. If the gate pulse is applied late in the half-cycle, only a small part of the waveform is passed through and the output is low. So by controlling the SCR's output voltage the motor acceleration characteristics can be controlled. (Fig. 13-9).

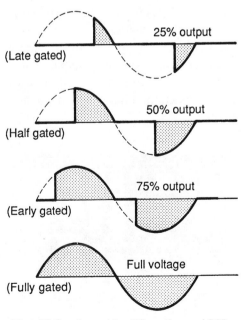

**Fig. 13-9** *Outputs for differently gated SCRs.*

## SOLID-STATE STEPLESS ACCELERATION

The class 8660 solid-state reduced-voltage controller provides smooth, stepless acceleration of a three-phase induction motor. The controller offers several standard and optional features to control, monitor, and protect the motor during the start and run modes of operation. Modular construction of the controller adds flexibility and ease of maintenance (Fig. 13-10). Soft start is accomplished by gradually turning on six silicon-controlled rectifiers. Two SCRs are connected in a back-to-back or reverse-parallel arrangement and mounted on a heat sink to make up a power pole. The power pole also contains a printed circuit board and a thermal sensor.

*Fig. 13-10*  Power pole. (Square D)

Firing of the SCRs is controlled by the modules on the logic rack. These modules also check for correct startup and running conditions and provide a visual indication of controller status through the use of light-emitting diodes (LEDs). Each module has a specific location and function. Figure 13-11 shows a logic module rack.

*Fig. 13-11*  Logic module rack. (Square D)

## LOGIC RACK

The logic rack is located on the lower part of the controller and has sockets for eight plug-in modules (Fig. 13-12). Each module has a specific location and performs a specific function in the operation of the controller. The module in the first position is internal to the controller and provides wiring connections between the power pole and the logic modules. The modules in positions 2 through 8 control the firing of the SCRs, check for correct startup and running conditions, and provide a visual indication of the controller status through the use of LEDs. The B-2 module goes in the second position, one of the B-3 modules goes in the third position and so on. The specific module functions are noted below.

*Fig. 13-12*  Module position in logic rack. (Square D)

### B-2 Module

This module provides logic voltages and checks for correct starting conditions. The control can be started if the control power LED is *on* and the start inhibit LED is *off*.

### B-3 Module

A three-phase, temperature-compensated, solid-state overload relay is supplied as an integral part of the controller. It provides class 10, inverse-time trip characteristics that protect against harmful motor overloads. There is a different B-3 module for each of the four controller current ratings of 200, 320, 500, and 720 A. Motor full-load current settings are adjustable by the use of potentiometers on the B-3 module. An overload condition will automatically de-energize the controller, close the alarm contact, and light the *trip-* and *start-inhibit* LEDs. An overload test feature on the logic rack assembly provides a check for operation of the solid-state overload circuitry. Overload trip time is a function of the current-limit setting. The lower the current-limit setting, the longer the trip time. Trip times for three current-limit settings are shown in Table 13-1. Longer trip times for high-inertial loads can be provided on the other types of controllers. Form Z72 provides class 30, inverse-time trip characteristics by using a special B-3 module and power poles with higher current ratings. Trip times for class 30 overloads are shown in Table 13-1.

*Table 13-1* Overload Trip Times

| Current Limit % of MFLC | Trip Time (Seconds) | |
|---|---|---|
| | Standard Class 10 | Form Z72 Class 30 |
| 150 | 90 | 250 |
| 300 | 30 | 90 |
| 425 | 5 | 40 |

## B-4 Module

The starting method that is used is determined by the B-4 module. Current limit starting is standard and is adjustable by the use of a potentiometer on the logic rack assembly. Optional starting methods are available. A description of each of the starting methods follows.

**Current Limit (B-4A Module)** The current-limit feature will limit the motor current to a preset level at all times during start and run conditions. Current limit is adjustable between 150 and 425% of motor full-load current by way of a potentiometer located on the logic rack. If a shorting contactor is used, this feature will be present only in the start condition (Fig. 13-13).

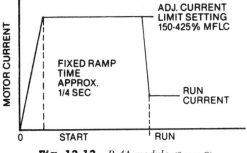

*Fig. 13-13* B-4A module. (Square D)

**Linear Timed Acceleration (B-4B) Tachometer Feedback** This option allows the motor speed to be increased linearly with the time until the motor reaches full speed (Fig. 13-14). Start time is adjustable from 3 to 30 seconds and does not fluctuate with motor loading. This method gives the smoothest acceleration but requires a tachometer input. Motor current is limited to the current-limit setting.

*Fig. 13-14* B-4B module. (Square D)

**Voltage Ramp (B-4C Module)** This option allows the applied motor voltage to increase linearly from 0 to 100% over an adjustable period of 3 to 30 seconds. The motor current is limited to the current-limit setting. This method provides acceleration that is approximately linear from zero to full speed but does not require a tachometer. The actual acceleration time depends on the motor and load (Fig. 13-15).

*Fig. 13-15* B-4C module. (Square D)

**Current Ramp (B-4D Module)** This option supplies a breakaway currrent to the motor at start and then linearly ramps the current up to the current-limit setting. Breakaway current is adjustable from 0% to 150% of motor full-load current. Ramp time is adjustable from 0 to 7 seconds. This method provides the greatest control of starting current (Fig. 13-16).

**Fig. 13-16** B-4D module. (Square D)

**Accel/Decel (B-4E Module)** This option provides both a soft start and a soft stop. Starting characteristics are identical to the voltage ramp start B-4C This option also allows the applied motor voltage to decrease linearly from 50% to 0% over an adjustable period of 3 to 30 seconds to provide a soft stop. Provisions for an emergency stop are included (Fig. 13-17).

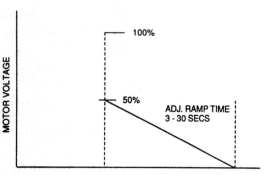

**Fig. 13-17** B-4E module. (Square D)

If any one of these occurs, the controller will shut off and the appropriate LEDs will be lighted. The phase unbalance function is activated whenever three-phase power is present at the controller line terminals but is disabled during starting. A fault condition occurs when voltage unbalance is greater than the unbalance setting. The voltage unbalance setting is adjustable from 5% to 14% as defined by NEMA standards.

The phase-reversal function is activated whenever three-phase power is present at the controller line terminals. A fault condition occurs if the three phases are not in correct sequence. Without the B-7 module, the controller is phase insensitive and will operate with any phase sequence.

The underload function is activated after the motor is "up to speed." A fault condition occurs when the motor drops below the underload setting, which is adjustable from 0% to 90% of motor full-load current. This can be disabled by adjusting the setting to zero.

## B-5 Module

The B-5 module determines the correct firing sequence of the SCRs.

## B-6 Module

The B-6 module provides the firing phase angles of the SCRs, which determines the percent of conduction for each SCR.

## B-7 Voltage Monitor Module

This optional module provides three separate functions:

- Phase unbalance
- Phase reversal
- Underload

## B-8 Energy-Saving Module

The energy-saving module will automatically adjust the voltage to the motor when load fluctuations occur. The motor will maintain full speed and required torque but draw less kVA when the load decreases. If the load increases, the module will respond by increasing the kVA so that the motor and load do not slow down in speed. This feature cannot be used on controllers with shorting contactors.

## SHORTED SCR SWITCH

If an SCR shorts, the short is detected and the shorted SCR switch will flip to the *yes* position. This switch will also trip the shunt trip circuit breaker (if used) ahead of the controller. If there is an open circuit between the controller and the motor, the shorted SCR circuitry will trip. This can occur if there is an open disconnect switch between the controller and motor. Isolation contactors should be placed ahead of the controller. A motor load must be connected to the controller to prevent nuisance tripping of the shorted SCR circuitry.

## ELEMENTARY WIRING DIAGRAMS FOR SOLID STATE

The solid-state reduced-voltage controller with an isolation contactor is shown in Fig. 13-18. Keep in mind that the M, SR2, OT, alarm, shorted SCR, and

**NOTES:**

1. M, SR2, OT, ALARM, SHORTED SCR, AND UP-TO-SPEED RELAYS ARE MOUNTED ON THE CONTROLLER AND WIRED INTERNALLY.
2. M DENOTES THE COIL FUNCTION OF THE SOLID STATE REDUCED VOLTAGE CONTROLLER.
3. THE SR2 RELAY CONTROLS THE START AND STOP SEQUENCE, AND ALSO HAS CONTACTS THAT MAY BE USED AS ELECTRICAL INTERLOCKS.
4. OT IS AN OVER TEMPERATURE SWITCH THAT OPENS WHEN THAT CONDITION EXISTS.
5. OL IS THE OVERLOAD RELAY CONTACT. IT OPENS WHEN: AN OVERLOAD IS DETECTED; L1, L2 OR L3 VOLTAGE IS NOT PRESENT; OR THE 120V CONTROL VOLTAGE IS MISSING.
6. THE ALARM CONTACT CLOSES WHEN AN OVERLOAD IS DETECTED.
7. THE SHORTED SCR CONTACT CLOSES WHEN THAT CONDITION EXISTS. IT IS USED WITH A CIRCUIT BREAKER OR DISCONNECTING SWITCH WITH A SHUNT TRIP COIL.
8. THE UP-TO-SPEED CONTACT CLOSES WHEN THE SCR'S ARE IN FULL CONDUCTION. IT IS USED WITH A SHORTING CONTACTOR.

**Fig. 13-18** *Solid-state reduced-voltage controller with an isolation contactor.* (Square D)

up-to-speed relays are mounted on the controller and wired internally. Figure 13-19 shows the solid-state reduced-voltage controller with a shorting contactor and Fig. 13-20 shows the controller with a shorting contactor and an isolation contactor.

**Fig. 13-19** *Solid-state reduced-voltage controller with a shorting contactor.* (Square D)

**Fig. 13-20** *Solid-state reduced-voltage controller with a shorting contactor and an isolation contactor.* (Square D)

## DIAC

The diac is basically a two-terminal device. It has a parallel-inverse combination of semiconductor layers. This combination of layers permits the triggering of the device in either direction (Fig. 13-21). As you remember, the SCR allowed triggering in only one direction. Thus the diac has the ability to conduct in

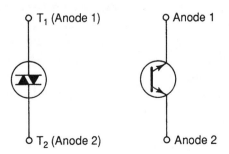

**Fig. 13-21** *Symbols for a diac.*

both directions when an ac signal voltage is applied across its terminals. There are a number of applications for such a device. One of them is in the control of ac electric motors. They may also be used in proximity detectors.

Note in the symbol that the diac does not have a gate or control element. It can be used as a bidirectional trigger diode (Fig. 13-23B). Current can flow either way when enough voltage is supplied for breakover. Typically, the firing potential is about 30 V in either direction. The diac is in its *off* state until the voltage across terminals T1 and T2 exceeds the breakover voltage. In power control circuits a diac can be used for more effective control of the turn-on point for the gate electrode of either a triac or an SCR.

## TRIAC

The triac is basically a diac with a gate terminal. The gate terminal controls the turn-on conditions of this bilateral device. The gate current can control the action of the device in either direction. This is similar to that of the SCR. However, the characteristics of the triac are somewhat different from those of the diac. Figure 13-22 shows the symbol and the location of the gate terminal.

**Fig. 13-23** (A) Triac demonstration circuit; (B) triac using a diac to trigger the gate.

**Fig. 13-22** Symbol for a triac.

By placing the triac in a circuit it is possible to indicate how it works (Fig. 13-23). In this arrangement the switch is used to select various conditions for the triac. The load can be either a light bulb or an ac motor. When the switch is in position 1 there is no gate connection. The triac does not conduct. The motor does not run. There is no trigger voltage applied to the gate. In position 2, a diode is placed in the circuit and with its polarity so arranged to allow a trigger voltage applied to the gate on the positive pulse of the ac applied to the circuit. The triac conducts, but only one-half of the ac sine wave.

This means that only about one-half of the normal current is applied to the motor. This is the same arrangement as with an SCR. An ac motor may have a problem with this type of pulsating dc voltage. When the switch is moved to position 3, the full ac sine-wave voltage is applied to the gate, with, of course, a reduction in value caused by the resistor R. Now that both halves of the ac sine wave are applied to the gate, the triac conducts full time and the full value of ac is applied to the ac motor. The motor then runs at full speed. R can be made a variable type and its value would then control the amount of ac current that passes through the triac and to the motor.

Another arrangement for the triac is shown in Fig. 13-23B. Here a diac is used to trigger the triac. The trigger voltage is controlled by the variable resistor. This allows for better regulation of the motor.

Triacs are packaged in the same types of cases as SCRs, so it is difficult or impossible to tell by a visual inspection which type is in the package. The numbers on the package indicate whether it is an SCR or a triac. There are triacs available today that can handle in excess of 10-kW loads.

## LIGHT-EMITTING DIODES

Light-emitting diodes (LEDs) are used as indicator lights on the module panels for solid-state controllers. They are small, give enough light for the purpose, and draw very little current. They are available in red, green, and amber (Fig. 13-24).

**Fig. 13-24** *Light-emitting diode (LED); symbol for LED.*

LEDs are made of gallium-arsenide junctions, a semiconductor material. Creation of electron-hole pairs is a reversible process. Energy is released when an electron recombines with a hole. In gallium-arsenide, an electron drops directly into a hole and a photon of energy is emitted. The gallium-arsenide junctions provide the best conditions for the generation of radiation in the visible range. Some are made for infrared radiation.

LEDs are used as indicator lamps. In most instances, they must be used in series with a resistor. They are also used as logic indicators for computer circuits. When reverse biased, the LED is nonconducting. This means that you have to have the proper polarity connections to the cathode and anode in order for it to glow. It is capable of conducting current when it is forward biased. It emits light when conducting with a forward bias current. An LED usually operates on 1 to 3 V. Excessive current will destroy an LED, and this calls for a series resistor in most circuits.

## USING SOLID-STATE CONTROL AND ELECTROMAGNETIC DEVICES

When solid-state controls are utilized in circuits that have electromagnetic devices, there are problems with the "dirty" power source. The buildup and collapse of a magnetic field whenever a coil of wire or inductor is energized and de-energized produces spikes and other types of electrical noise. These spikes can cause problems with solid-state devices since they are susceptible to voltage surges and spikes that are commonplace with the energizing of relay coils and the turning on and off of electric motors.

## SURGE SUPPRESSORS

Surge suppressors are installed on magnetic device coils, such as relays, contactors, and motor starters. A voltage-surge suppressor may have its leads connected to the coil terminals. The purpose of the suppressor is to limit voltage noise and overvoltage spikes produced by the starter coil when the coil circuit is opened.

The surge suppressor shown in Fig. 13-25 is made to be easily mountable directly across the coil terminals of contactors and starters with 120 and 240 V ac coils. The purpose of the suppressor is to limit voltage transients for applications requiring interface with solid-state components. One suppressor is required for each coil.

**Fig. 13-25** *Surge suppressor for mounting across coil terminals. (Allen-Bradley)*

Figure 13-26 shows two types of surge suppressors used to reduce the high transient voltages generated when the coil circuit is opened. These suppressors are used with relay coils and other electromechanical devices.

**Fig. 13-26** *Surge suppressor: (A) for mounting under a relay; (B) for mounting on coil terminals. (Allen-Bradley)*

**Fig. 13-27** *Resistor-capacitor combination surge suppressor.*
(Allen-Bradley)

Figure 13-27 is a surge suppressor used to protect solid-state devices against electrical transients that can result whenever electromechanical devices are operated.

Suppressors are for use with relays, timers, ac contactors, and starters. This suppressor consists of a resistor-capacitor combination sealed in epoxy.

## LIGHTNING PROTECTION

Lightning can also present some high-voltage surges. Secondary surge arrestors can be installed to prevent problems associated with lightning (Fig. 13-28). The arrestor shown in Fig. 13-28A is for single-phase, two- or three-wire grounded service. Two of them may be installed to provide protection on 208Y/ 120-V ac, three-phase, four-wire services. This suppressor will handle 1500 A at 940 V, 5000 A at 1600 V, 10,000 A at 2200 V, and 20,000 A at 3250 V.

A suppressor for use on 650-V ac phase-to-ground maximum is shown in Fig. 13-28B. It is used for three- or four-wire grounded service such as single phase three-wire, three-phase three-wire, or three phase four-wire systems. This suppressor will handle 1500 A at 2200 V, 5000 A at 2900 V, and 10,000 A at 3400 V, and

**Fig. 13-28** *(A) Secondary surge arrestors used in lightning protection for electrical systems; 175-V ac phase-to-ground maximum. (B) Secondary surge arrestors for lightning protection used for electrical systems; 650-V ac phase-to-ground maximum.*
(Square D)

20,000 A at 4000 V. These are maximum discharge voltages that appear across the arrestor during the passage of the discharge current. Discharge current is the current at the arrester during sparkover.

## REVIEW QUESTIONS

1. What does SCR stand for?
2. What is another name for the SCR?
3. What does phase unbalance detection mean?
4. What does phase reversal mean?
5. How are LEDs used on solid-state devices?
6. What is a "dirty" power source?
7. What is the purpose of a surge suppressor?
8. How are solid-state devices protected from lightning?
9. What is a diac used for?
10. What is a triac?

# CHAPTER 14

# Speed Control and Monitoring

## PERFORMANCE OBJECTIVES

After studying this chapter, you will be able to:

1. Describe how the speed of an electric motor is controlled.
2. Explain the difference between a synchronous motor and an induction motor.
3. Describe how field excitation of a synchronous motor is obtained.
4. Explain how the speed of a synchronous motor is determined.
5. Describe an amortisseur.
6. Explain various starting methods for synchronous motors.
7. Describe Korndorfer starting.
8. Explain how speed is regulated by resistance.
9. List types of speed control for wound-rotor motors.
10. Explain how secondary resistances are used for wound-rotor induction motor control.
11. List the reason for using solid-state adjustable-speed controllers.
12. Describe how frequency changing is used to change motor speed.

## MOTOR SPEED CONTROL

Controlling the speed of an electric motor is possible in some instances. However, there is always some price to pay for speed control. A motor is usually designed to operate at a given speed, and any deviation from that speed causes it to have less starting torque or to run hot.

Industrial processes call for motors that can be varied in their speed. This means that methods must be designed to handle the variable-speed requirement. The first place to look is at the characteristics of the motors themselves as to what they will or will not tolerate in terms of speed variations.

## SQUIRREL-CAGE MOTORS

The speed of a squirrel-cage induction motor (Fig. 14-1) is nearly constant under normal load and voltage conditions but is dependent on the number of poles and the frequency of the ac source. This type of motor slows down, however, when loaded by an amount that is just sufficient to produce the increased current needed to meet the required torque.

**Fig. 14-1** *Squirrel-cage electric motor.*

The difference in speed for any given load between synchronous and load speed is called the slip of the motor. Slip is usually expressed as a percentage of the synchronous speed.

Synchronous speed equals the speed of the rotating field. Since the amount of slip is dependent on the load, the greater the load, the greater the slip will be, that is, the slower the motor will run. This slowing of the motor, however, is very slight, even at full load, and amounts from 1 to 4% of synchronous speed. Thus the squirrel-cage type is considered a constant-speed motor.

This type of motor is not suitable for industrial applications where a great amount of speed regulation is required, because the speed can be controlled only by a change in frequency, number of poles, or slip. Speed control by changing the frequency is becoming very popular. The number of poles is sometimes changed either by using two or more distinct windings or by reconnecting the same winding for a different number of poles.

## SYNCHRONOUS MOTORS

A synchronous motor is by definition one that is in unison or in step with the phase of the alternating current that operates it. This condition is only approximated in practice because there is always a slight phase difference. Any single-phase or polyphase alternator will operate as a synchronous motor when supplied with current at the same potential, frequency, and wave shape that it produces as an alternator, the essential condition in the case of an alternator being that it must be speeded up to synchronism before being put into the circuit.

A synchronous motor may have either a revolving armature or a revolving field. Most synchronous motors are of the revolving field type. The stationary armature is attached to the stator frame, while the field magnets are attached to a frame that revolves with the shaft.

The field coils are excited by direct currents, either from a small dc generator (usually mounted on the same shaft as the motor and called an exciter) or from some other source. Figure 14-2 shows a directly connected exciter.

In most industrial applications today, solid-state electronics provide the dc needed for operation of this type of motor. Solid-state electronics also provides the change in frequency needed to control the speed of the motor.

**Fig. 14-2** *Synchronous motor with a directly connected exciter.*

## Excitation

Field excitation for a synchronous motor is obtained from a separate exciter set driven by an induction motor, from a direct-connected or belted exciter, or from a constant dc voltage supply such as a station bus. Standard excitation voltage is either 125 or 250 V, but the motor field winding is designed for an excitation voltage approximately 10% below this, to allow for voltage drop in the line.

## Speed

The speed of a synchronous motor is determined by the frequency of the supply current and the number of poles of the motor. This means that the operating speed is constant for a given frequency and number of poles. The equation for the determination of motor speed is

$$\text{rpm} = \frac{\text{frequency} \times 120}{P}$$

where P is the number of poles of the motor.

All motors are built with an even number of poles, so the available speeds on 60 Hz range from 3600 rpm for a two-pole machine down to 80 rpm for a machine containing 90 poles. This allows the motor to be directly connected to its load, even at lower speeds, where induction motors cannot be used advantageously because of low operating efficiency and power factor.

Some motors are required to operate at more than one speed but are constant-speed machines at a particular operating speed. For example, when a speed ratio of 2:1 is required, a single-frame, two-speed synchronous motor may be suitable. Four-speed motors are used when two speeds that are not in the ratio of 2:1 are desired.

The single-frame, two-speed motor is usually of the salient-pole type of construction, with the number of poles corresponding to the low speed. High speed is obtained by regrouping the poles so as to obtain two adjacent poles of the same polarity, followed by two poles of opposite polarity. This gives the effect of reducing the number of poles on the rotor by one-half for high-speed operation. Corresponding changes in the stator connections are also made. This switching is usually accomplished automatically by means of magnetic starters, by manually operated pole-changing equipment. Figure 14-3 shows a synchronous motor and exciter with the exciter-field rheostat, field switch, and exciter-field meters.

**Fig. 14-3** *Synchronous motor and exciter with rheostat, switch, and meters.*

## Starting

To make a synchronous motor self-starting, a squirrel-cage winding is usually placed on the rotor. After the motor reaches a speed slightly below synchronous, the rotor is energized. When synchronous motors are started, their dc fields are not excited until the rotor has practically reached full synchronous speed. The starting torque required to bring the rotor up to this speed is produced by induction.

In addition to a dc winding on the field, synchronous motors are generally provided with a damper or amortisseur winding. It consists of short-circuited bars of brass or copper embedded in slots in the pole faces and joined together at either end by means of end rings. This winding usually termed a squirrel cage winding, enables the motor to obtain sufficient starting torque for the motor to start under load.

The starting torque necessary to bring the motor up to synchronous speed is termed the pull-in torque. The maximum torque that the motor will develop without pulling out of step is termed the pull-out torque.

When the stator winding in the synchronous motor is being excited by the ac line connection, it immediately sets up a rotating magnetic field. The rotating flux of this field cuts across the damper winding of the rotor and induces secondary currents in the bars of this winding. The reaction between the flux of these secondary currents and that of the rotating stator field produces the torque necessary to start the rotor and to bring it up to speed.

When the rotor has been brought up to nearly synchronous speed (as an induction motor because of the damper winding), the dc field poles are excited and the strong flux of these poles causes them to be drawn into step or full synchronous speed with the poles of the rotating magnetic field of the stator. During normal operation, the rotor continues to revolve at synchronous speed as if the dc poles were locked to the poles of the rotating magnetic field of the stator. Because a synchronous motor has no slip after the rotor is brought up to full speed, no secondary currents are induced in the bars of the damped windings during normal operation.

## Starting Methods

In starting a synchronous motor as an induction motor, the voltage impressed on the motor should be reduced in starting and while coming up to speed. This reduced starting voltage is usually obtained from a starting compensator (autotransformer) similar to that used in the starting of an induction motor.

**Autotransformer Starting** With the autotransformer method of starting, the usual practice is to close the starting contactor first. This connects the stator to the reduced voltage. When a speed near synchronization has been reached, the starting contactor is opened and the running contactor is closed, thus connecting the motor to the full-line voltage. After synchronous speed has been reached, the field switch is closed through a moderate amount of resistance. The field current is now adjusted in order to make the motor operate at the desired power factor.

**Full-Line Voltage Starting** If the motor is to operate at high starting torque, it is common practice to use a full-line starting voltage in connection with a time-delay over-current relay. The relay will operate before the surge of starting current can damage the motor windings. Figure 14-4 shows the typical automatic across-the-line synchronous motor starter. Across-the-line starting and reduced-voltage starting have already been discussed. However, the reduced-voltage diagram is shown in Fig. 14-5. Note how the autotransformers are switched into the circuit. This method of reduced-voltage starting uses three switches or contactors. The two starting switches or contactors are connected to each side of the autotransformer or compensator and are both closed at the same time. After the motor has attained nearly synchronous speed, the starting switches are opened and the running switch is closed. This sounds great in theory but is not always successful. The motor may not become synchronized when it is supposed to, so you may have to start over again and try to get it to synchronous speed or use an automatic method.

*Fig. 14-4* Elementary diagram of an across-the-line motor starter.

**Fig. 14-5** *Diagram of a reduced-voltage starter for a synchronous motor.*

**Reactance Starting** Reactance starting is similar to the reduced-voltage starting methods, except that the first step is obtained by reactance in series with the motor armature instead of autotransformers. In the reactance method of starting, more current is required from the line for the same torque on the first step than when compensators are used. It has one advantage. No circuit opening is required when the motor is transferred to running voltage. The transfer is accomplished by short-circuiting the reactance.

**Resistance Starting** A typical circuit using the resistance method of reduced-voltage starting is shown in Fig. 14-6. Switch 1 is closed first. This connects the motor to the line through the entire resistance. Switches 2, 3, and 4 are then closed, with a time interval between each closing. Each switch, in turn, short circuits a part of the resistance. This method of starting is sometimes used when power company rulings require several progressive steps of starting current.

**Korndorfer Starting** The reactance and resistance methods are similar to starting the motor using the Korndorfer method. It permits the motor to be started

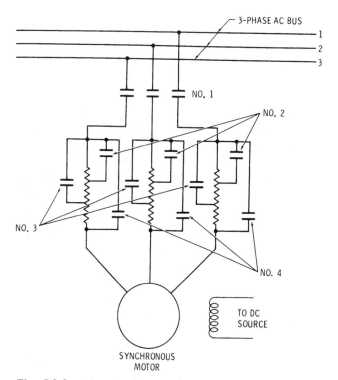

**Fig. 14-6** *Schematic diagram of a resistance-type synchronous motor starter.*

Synchronous Motors 237

without opening the motor circuit. The motor is first connected through suitable taps of a compensator, and then started by connecting the compensator to the line. Full voltage is connected by first opening the neutral of the starting compensator. This allows the motor to run with part of the compensator winding in series with the motor. Then the entire compensator winding is short circuited (Fig. 14-7).

**Fig. 14-7** *Schematic diagram of a Korndorfer-type synchronous motor starter.*

Switch 1 is closed first. This connects one of the compensator windings to the line. Then switch 2 is closed, completing the motor circuit at reduced voltage. As the motor increases its speed, a timing relay, operated by switch 2, opens the circuit of 3. This, in turn, opens the transformer neutral. Switch 4 is closed next. This connects the motor to full-line voltage by shorting the compensator sections. By opening switch 2, the reduced-voltage taps of the compensator are disconnected and the permanent running connection to the motor is completed.

**Other Methods of Starting** An auxiliary prime mover, usually an induction motor, may be used as a starter. This method of starting is applied to the motors that have no squirrel-cage winding, or it is used with alternators converted to motor use. This type of motor cannot start under load.

## Uses for Synchronous Motors

Synchronous motors may be used for power factor correction; for constant-speed, constant-load drives; and for voltage regulation. Because of the higher efficiency possible with synchronous motors, they can be used advantageously on most loads where constant speed is required. Typical applications are compressors, fans, blowers, line shafts, centrifugal pumps, rubber and paper mills, and to drive dc generators.

## WOUND-ROTOR MOTORS

The wound-rotor motor differs from the squirrel-cage type. It has wire-coil windings in its rotor instead of a series of conducting bars in the rotor. Inserting external resistance in the motor circuit when starting will develop a high torque with a comparatively low starting current. As the motor comes up to speed, the resistance is gradually removed until, at full speed, the rotor is short-circuited. Speed can be regulated, within limits, by varying the amount of resistance in the rotor circuit (Fig. 14-8).

**Fig. 14-8** *Wiring diagram with resistor connections.*

## Speed Regulation by Resistance

Resistors can be used to regulate the speed if they are of the proper size to prevent overheating from constant use. The resistors used in starting are used only for a short time, but those used for continuous motor speed reduction are in use for longer periods of time. This means that a resistor must be selected for its intended purpose.

DC motors produce the most effective variable speed outputs. However, wound-rotor motors, because of their adjustable rotor resistance, are one of the few means of speed control available for ac motors.

Wound-rotor motors are just that—they have a wound rotor. They are insulated coils of wire that are not permanently short circuited, as in the squirrel cage motor, but are connected in regular succession to form a definite polar area having the same number of poles as the stator. The ends of these rotor windings are brought out to collector rings, usually referred to as slip rings.

Currents induced in the rotor are carried by means of slip rings (and carbon brushes riding on the slip rings) to an externally mounted resistance (Fig. 14-9). These resistances can then be regulated or changed according to the needs of the start sequence. By changing the resistance in the rotor circuit it is possible to change the speed of the motor. Once it has come up to synchronous speed the resistors are then short circuited and the motor runs with characteristics similar to a squirrel-cage type.

However, some resistance can be left in the circuit to aid in speed control, that is, of course, if the size (wattage rating) of the resistors is such as to withstand the constant current flow through them. By placing a high resistance in the rotor circuit, it is possible to start the motor and have it produce high starting torque with low starting current.

## Types of Speed Control

The wound-rotor motor can be used where the speed range is small, where the speeds desired do not coincide with a synchronous speed of the line frequency, and where the speed must be gradually or frequently changed from one value to another. This includes compressors, pulverizers, stokers, and conveyors.

A smooth, no-jerk start can be obtained by using the wound-rotor motor. It is simply a matter of supplying the right control equipment.

## Multiswitch Starters

Figure 14-10 shows how a typical multiswitch starter for a wound rotor is wired into the rotor circuit. This type of starter is used in the secondary circuits of large wound-rotor induction motors up to 2000 hp with rotor currents up to 1000 A. Contact levers are of the double-pole type and are mechanically arranged in such a manner that they must be closed in a predetermined sequence, and only one at a time. Since the switches are designed for hand-over-hand operation, a desirable time element is introduced that prevents too-rapid acceleration of the motor. When the final switch has been closed, it is held in place by a magnetic coil, and because of the mechanical interlocking feature, all

**Fig. 14-9** *Starter-controller for a wound rotor induction motor.*

**Fig. 14-10** *Diagram of a typical multi-switch starter for a wound rotor motor.*

**Wound-Rotor Motors** 239

other switches remain closed. This type of starter is just that—a starter, it is not useful as a speed regulator.

## Drum Controllers

Drum controllers can be used for starting and for speed control of the wound-rotor motor (Fig. 14-11). Drum controllers are made to handle both stator and rotor circuits. The cylinder mounting the contact segments are built in two insulated sections. When they are built to handle the rotor circuit, only the stator circuit is controlled by a circuit breaker or line starter. In addition to starting and regulating, speed-regulation drum collectors are commonly used for speed-reversing duty as well.

Motor-driven controllers are used in certain drives requiring close automatic speed regulation such as in large air-conditioning plants, blowers, stokers, and similar applications. Some of these installations have been in use for a number of years and are gradually being replaced by more modern motor control methods.

## Magnetic Starters

Magnetic starters are built to regulate motor speed, start the motor, and to set the speed of the motor. They consist of a magnetic contactor for connecting the stator circuit to the line, and one or more accelerating contactors to commutate the resistance in the rotor circuit. The number of secondary accelerating contactors varies with the rating, a sufficient number being used to assure smooth acceleration and to keep the inrush current within practical limits. The operation of the accelerating contactors is controlled by a timing device, which provides definite time acceleration. For high-voltage service, the primary contactor is usually of the oil-immersed type. The diagram of a typical magnetic starter for use with a wound-rotor induction motor is shown in Fig. 14-12.

## Resistors

Generally, the secondary resistors for wound-rotor induction motors are designed for star connection. Resistors for most manual controllers may be connected with all three secondary phases closed or with one secondary phase open on the first point of the controller. Resistors for magnetic controllers are connected with all three phases closed in the secondary on the first point. The torque obtained with a resistor of a given class number varies with the connection used on the first point of the controller.

Keep in mind that wound-rotor motors can be started with a load and without drawing too much

**Fig. 14-11** *Non-reversing drum controller for a wound-rotor motor with a three-phase secondary.*

**Fig. 14-12** *Magnetic starter contactor for use with a wound rotor motor.*

current. They can be used for such loads as those with back pressures set up by fluids and gases, as in reciprocating pumps and compressors. They are also used in elevators and cranes. They do, however, have some disadvantages.

Disadvantages are the initial cost since they do require a wound rotor. The slip rings and brushes do need maintenance from time to time. Resistors and the switching arrangements require periodic inspection and maintenance.

## Solid-State Adjustable-Speed Controllers

Solid-state adjustable-speed controllers are available to produce smooth starts and energy savings. They are, in most instances, maintenance free and are easy to operate and make it simple to train new operators.

The reason for using solid-state adjustable-speed controllers is because they provide step-less, smooth adjustable-speed control of the ac wound-rotor motor. This means the elimination of resistors, liquid rheostats, and reactors as well as magnetic clutches. They all consume energy that is not the case with the solid-state controller. The solid-state circuitry is used to provide the excitation to the rotor. By controlling the rotor current it is possible to control the motor speed and thereby its torque.

## FREQUENCY SPEED CONTROL

Solid-state ac motor control is accomplished by changing the frequency of the power source. Westinghouse's Accutrol line is an adjustable-speed ac drive package in ratings from 1 to 5 hp at 230 V and 3 to 250 hp at 460 V, three-phase, 60 Hz (Fig. 14-13).

Motor speed is adjusted by controlling the output voltage and frequency of the unit. This is accomplished

**Fig. 14-13** *Solid state ac motor control.* (Westinghouse)

**Frequency Speed Control** 241

by rectifying the incoming ac supply voltage and changing it to dc. The dc voltage is inverted by a three-phase inverter section to an adjustable frequency output whose voltage is adjusted proportionately to the frequency to provide constant volts per hertz excitation to the motor terminals up to 60 Hz. Above 60 Hz, the voltage may remain constant at rated volts. In this way energy-efficient low-loss speed control is obtained in the range 2 to 120 Hz.

This type of speed control does have advantages over dc machines inasmuch as the dc motors are hard to maintain and have problems in environments that are wet, corrosive, or explosive. These controls are found in food-packing plants, dairies, chemical plants, sand and gravel plants, paper mills, and cement plants. Centrifugal pumps and blowers are particularly suited for use with this type of control, as considerable reduction in energy consumption can be achieved by varying the speed to control the flow of gas or fluids instead of using throttling devices such as valves, dampers, or fluid recirculators.

## MULTISPEED STARTERS

Multispeed starters are designed for the automatic control of two-speed squirrel-cage motors of either the consequent-pole or separate-winding types (Fig. 14-14). These starters are available for constant horsepower, constant-torque or variable-torque, three-phase motors (Fig. 14-15). Multi-speed motor starters are commonly used on machine tools, fans, flowers, refrigeration compressors, and many other types of equipment.

**Fig. 14-14** *Open-type, two-speed, separate-winding, multi-speed starter.* (Square D)

**Fig. 14-15** *Starter diagram for a consequent-pole motor with constant torque or variable torque.* (Allen-Bradley)

## SPEED MONITORING

Speed sensing a switch can be used to sequence conveyors where it is necessary for one conveyor to be running at nearly full speed before a second conveyor is started. The switch can also be used to indicate which direction material on a conveyor is moving from the rotation of a suitable driven shaft.

The electronic speed switch is a rugged, self-contained, rotary shaft-speed detector (Fig. 14-16). If the shaft speed exceeds or falls below an adjustable, preset value, the speed switch detects the change and actuates external relays, audible alarms, or warning lamps.

**Fig. 14-16** *Electronic speed switch.* (Reliance)

Output power is switched by a triac, solid-state switch. This model is available in the pictured foot-mounted model or the flange-mounted model. Table 14-1 shows the available speed ranges for the speed switch.

Table 14-1  Speed Ranges[a]

| Range Dial Setting | Speed Range | |
|---|---|---|
| | 5-5000 rpm (Standard) | 0.7–700 rpm (Option D) |
| 1 | 5–15 | 0.7–2 |
| 2 | 15–50 | 2–7 |
| 3 | 50–100 | 7–20 |
| 4 | 150–500 | 20–70 |
| 5 | 500–1500 | 70–100 |
| 6 | 1500–5000 | 200–700 |

[a]Speed switch range is field adjustable by dial setting to the range limits shown. After the desired general speed range limits shown. After the desired general speed range is selected, specific speed is set by turning an adjustable potentiometer.

A tachometer generator allows accurate monitoring of machine operating speeds. When this is tied into a closed-loop speed regulator, the tachometer generator can be used to control the machine speed (Fig. 14-17).

**Fig. 14-17**  Tachometer generator. (Reliance)

## REVIEW QUESTIONS

1. How can the speed of a squirrel-cage motor be changed?
2. What is an exciter? Where is it useful?
3. What is the rpm range of motors?
4. What is a damper winding on a motor?
5. What is the Korndorfer method of starting a motor?
6. What is another function of a synchronous motor?
7. What type of motors produce the most effective variable-speed outputs?
8. What types of motors use multi-switch starters?
9. What two purposes do drum controllers serve on wound-rotor motors?
10. What are the disadvantages of wound-rotor motors?
11. What is the main reason for using solid-state adjustable-speed controllers?
12. What does an inverter do?
13. How is a tachometer generator useful in regard to machine speeds?
14. Define a synchronous motor.
15. How is field excitation for a synchronous motor obtained?
16. What determines the speed of a synchronous motor?
17. When are four-speed motors used?
18. How are synchronous motors made self-starting?
19. What is another name for a damper winding?
20. What is another name for a starting compensator?

# 15
CHAPTER

# Motor Control and Protection

## PERFORMANCE OBJECTIVES

After studying this chaper you will be able to:

1. Describe manual starter operation.
2. Describe the "soft start" method of motor control.
3. Explain how sequence control is accomplished.
4. Explain how automatic sequence control is accomplished.
5. Describe how jogging is accomplished.
6. List advantages and disadvantages of plugging a motor.
7. Define antiplugging.
8. Describe electronic motor braking.
9. List the disadvantages of electronic motor braking.
10. Describe how mechanical braking works.
11. Understand how thruster brakes are used.
12. List the advantages of hydraulic brakes and magnetic brakes.
13. Explain what causes chattering brakes.
14. Describe overload protection for a motor.
15. List the advantages of a line-voltage monitor.
16. Explain how programmable motor protection works.
17. Explain how a remote temperature detector module works.

## MOTOR CONTROL

Controlling a motor can be a simple task or it can be complicated, depending on the needs of the machine that is powered by the motor. Motors are incorporated into many machines. This makes it necessary for the motor to be controlled according to the needs of the machine it powers. Then, of course, as the motor is used, it is also abused. This means that it must be protected if it is to run properly and for long periods of time as needed for any given purpose. Keep in mind that not all motors are meant to be variable-speed motors. Some are made to run at a constant speed.

Motors must be started, stopped, reversed, and the speed must be controlled. They are also jogged, plugged, and in some instances stopped rather quickly. All these operations require equipment and circuitry to accomplish the job correctly without damage to the motor.

This chapter covers these operations and the equipment needed to cause proper operation of an electric motor under varying load conditions.

## MANUAL STARTERS

Manual starters (Fig. 15-1) provide full-line voltage starting, reliable thermal overload protection, as well as undervoltage protection. Typical applications are on woodworking machinery, metal sawing machines, and many other machine tools where undervoltage protection is needed to meet safety standards. By removing jumper A, a remote emergency stop operator may be wired to the vacated terminals. Note how the pilot light is wired in the circuit so that the light is on when the motor is energized. Three-phase operation using the same type of starter is shown in Fig. 15-2. The pushbutton wiring is shown in Fig. 15-3. Note how the wiring diagram and the elementary diagram differ but contain the same information. The elementary diagram shows how the pilot device is energized. The M shown in the diagram represents the contactor coil that will close the contacts in the three-phase lines to the motor when energized. A complete circuit for energizing the coil is provided by pressing the start button.

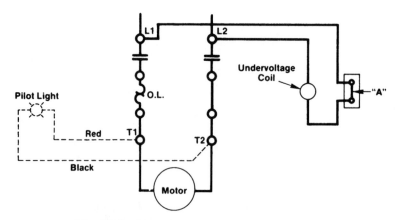

*Fig. 15-1* Manual starter, single-phase. (Allen-Bradley)

246   Motor Control and Protection

**Fig. 15-2** *Manual starter, three-phase.* (Allen-Bradley)

**Fig. 15-3** *Three-phase starters.* (Allen-Bradley)

**Fig. 15-4** *Single-phase motor using standard three-phase starter.* (Allen-Bradley)

A variation on the standard start-stop starter is shown in Fig. 15-4, where the three-phase starter is used with a single-phase motor. Figure 15-5 is a variation that is used with a three-phase motor. It provides more than one start-stop station. The elementary diagram shows how the start buttons are in parallel with the contacts of the starter M coil. This is a useful arrangement when a motor must be started and stopped from any of several widely separated locations. Using this particular circuitry, it is also possible to use only one start-stop station and have several *stop* buttons at different locations to serve as emergency stops.

Standard-duty start-stop stations are provided with connections A shown in Fig. 15-5. This connection must be removed from all but one of the start-stop stations used.

Low-voltage protection is a method of protecting an operator from injury from automatic restart of a machine upon resumption of voltage after a power failure. This is normally accomplished with a magnetic starter with three-wire control. This protection can be had with standard manual starters (Fig. 15-6). The low-voltage feature is accomplished by a continuous-duty solenoid assembly built into the overload relay mechanism. When

**Manual Starters** 247

**Fig. 15-5** *Variations with start-stop stations.* (Allen-Bradley)

**Fig. 15-6** *Manual starter wiring diagram with low-voltage coil.*
(Square D)

a power failure occurs or the line voltage is disconnected, the solenoid will de-energize and mechanically open the starter contacts. When power is restored, the starter must be manually reset before the contacts can be closed and normal operation resumed.

Low-voltage protection is required by OSHA 1910.213b3 and 1910.217b8iii on certain woodworking machines and all mechanical power presses. NFPA 79 Section 130-21 requires it on certain metalworking machines. Some local safety regulations have extended it to other applications, such as mixers, conveyors, or wherever operator safety could be in jeopardy.

# SOLID-STATE MOTOR CONTROLLER

It is possible to obtain the advantages of solid-state electronics in motor controllers. In fact, an entire chapter is dedicated to the programmable controller and its advantages and uses. In this chapter, however, the smaller microcomputer-controlled starters for standard squirrel-cage induction motors are highlighted.

Allen-Bradley makes the smart motor controller (SMC), which has a soft-start, current limiting, and full-voltage starting. Three modes are possible with the same controller: the soft start, current limit, and full voltage (Fig. 15-7).

**Fig. 15-7** *Smart motor controller (SMC), microprocessor-controlled starting.* (Allen-Bradley)

**Soft Start** This method has the most general application. The motor voltage gradually increases during the acceleration ramp period. The ramp period can be adjusted from 2 to 30 seconds. Then the user sets it for the best starting performance over the required load range.

**Current Limit** This starting mode is used when it is necessary to limit the maximum starting current. The current limit is adjusted according to the starting current restriction. This can be adjusted for 200 to 450% of full-load amperes.

**Full Voltage** For applications requiring a full-load start, the acceleration ramp time is set to a minimum of 0.25 second. This, in effect, allows the controller to start the load across-the-line.

Solid-state electronics have been utilized successfully in the production of motor controls. In time, they will probably replace the electromechanical devices. However, the cost of the electronic devices will continue to decrease with better circuitry and more competition. Older machines will be retrofitted with the newer electronics. Some packages are already available to be attached to existing setups.

# SEQUENCE CONTROL

In some applications it is necessary to make sure that one starter is not operational until the other has been energized. This type of use is found in equipment that may have the need for high-pressure lubrication or with hydraulic pumps.

These auxiliary devices have to be operational before the machine is turned on. Fig. 15-8 shows the starters arranged for sequence control of a conveyor system. The two starters are wired so that M2 cannot be started until M1 is running. This is necessary if M1 is driving a conveyor fed by another conveyor driven by M2. Material from M2 conveyor would pile up if the M1 conveyor could not move and carry it away.

If a series of conveyors is involved, the control circuits of the additional starters can be interlocked in the

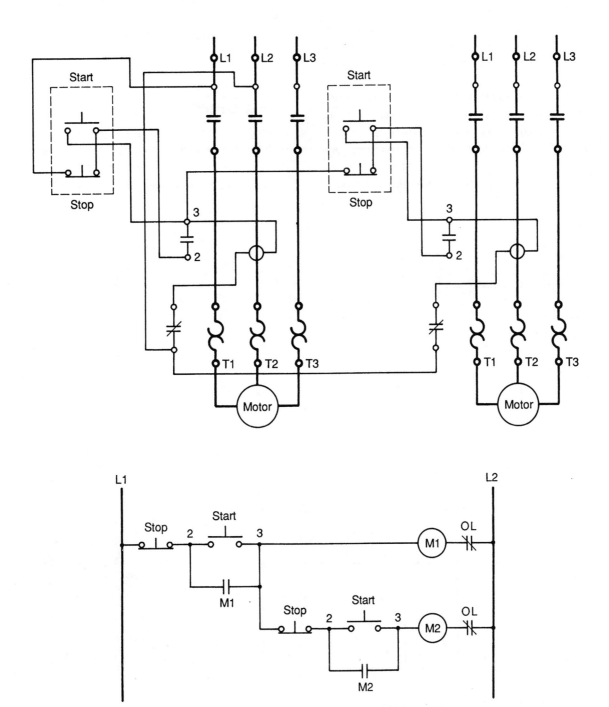

NOTE: Control circuit is connected only to the lines of Motor 1.

**Fig. 15-8** *Sequence control diagrams.* (Allen-Bradley)

same way. That is, M3 would be connected to M2 in the same step arrangement that M2 is now connected to M1, and so on.

The M1 button, *stop* button, or an overload on M1 will stop both conveyors. The M2 *stop* button or an overload on M2 will stop only M2.

## WHEN STARTING ANY ONE REQUIRES ANOTHER

Several motors can be run independently of each other with some of the starters actuated by two-wire and some by three-wire pilot devices. Whenever any one of these motors is running, a pump or fan motor must also run (Fig. 15-9).

A master start-stop pushbutton station with a control relay is used to shut down the entire system in an emergency. Control relay (CR) provides three-wire control for M1, which is controlled by a two-wire control device such as a pressure switch. Motors M2 and M3 are controlled by start-stop pushbutton stations.

Auxiliary contacts on M1, M2, and M3 control M4. These auxiliary contacts are all wired in parallel so that any one of them may start M4. On some starters, auxiliary contacts have been added to M2 and M3 for this purpose. The standard hold-in contact on M1 may be used as an auxiliary if wire Y is removed. Hold-in contacts are not required when a two-wire control device is used.

When this system is used, the phase connections on all of the starters must be the same. That is, L1 of each starter must be connected to the same incoming phase line; L2 and L3 of each starter must be phased out similarly.

## Automatic Sequence Control

Having automatic sequence control is also possible with the arrangement shown in Fig. 15-10. In this system it is desired to have a second motor start automatically when the first one is stopped. The second motor is to

**Fig. 15-9** *Sequence control diagrams.* (Allen-Bradley)

250   Motor Control and Protection

**Fig. 15-10** *Sequence control diagrams.* (Allen-Bradley)

**When Starting any One Requires Another**

run only for a given length of time. A good application of this might be found where the second motor is needed to run a cooling fan or a pump after the first motor has stopped.

To accomplish this, an off-delay timer (TR) is used. When the *start* button is pressed, it energizes both M1 and TR. This operation of TR closes its time-delay contact, but the circuit to M2 is kept open by the opening of the instantaneous contact. As soon as the *stop* button is pressed, both M1 and TR are dropped out. This closes the instantaneous contact on TR and starts M2. M2 will continue to run until TR times out and the time-delay contact opens.

# JOGGING

Jogging, or inching, is defined by NEMA as the momentary operation of a motor from rest for the purpose of accomplishing small movements of the driven machine. One method of jogging is shown in Fig. 15-11. The selector switch disconnects the holding circuit interlock and jogging may be accomplished by pressing the *start* button.

There are several means of accomplishing the jogging operation. Figure 15-12 shows how jogging is done using a control relay. Pressing the *start* button energizes the control relay that in turn energizes the

**Fig. 15-11** *Jogging with a selector switch.*

**Fig. 15-12** *Jogging with a control relay.*

**252** Motor Control and Protection

**Fig. 15-13** *Jogging using a selector switch pushbutton.*

starter coil. The normally-open starter interlock and relay contact then form a holding circuit around the *start* button. However, pressing the *jog* button energizes the starter coil independent of the control relay and no holding circuit forms. Then jogging can be obtained simply by pushing the *jog* button and releasing it independent of the *start* button.

Jogging can also be accomplished by using a selector pushbutton. The use of a selector pushbutton to obtain jogging is shown in Fig. 15-13. In the *run* position the selector pushbutton gives normal three-wire control. In the *jog* position, the holding circuit is broken and the jogging is accomplished by pressing the button.

## Forward or Reverse Jogging

Jogging in the forward or reverse direction is possible if the wiring shown in Fig. 15-14 is followed. This control scheme permits jogging the motor either in the forward or reverse direction, whether the motor is

**Fig. 15-14** *Jogging using a control relay for reversing starter.*

Jogging 253

at a standstill or is rotating in either direction. Pressing the *start-forward* or *start-reverse* buttons energizes the corresponding starter coil, which in turn closes the circuit to the control relay. The relay picks up and completes the holding circuit around the *start* button. As long as the relay is energized, either the forward or reverse contactor will remain energized. Pressing either *jog* button will de-energize the relay, releasing the closed contactor. Further pressing of the *jog* button permits jogging in the desired direction.

## PLUGGING

Plugging is defined by the NEMA as a system of braking in which the motor connections are reversed so that the motor develops a countertorque. Thus it exerts a retarding force. In the scheme shown in Fig. 15-15 the motor is run in one direction only and must come to a complete stop when the *stop* button is pressed. The reverse contactor of the reversing switching is used only for plug stopping and not for running in reverse. The lockout solenoid is built into some of the speed switches and its function is to guard against an accidental turn of the motor shaft, closing the speed switch contacts and starting the motor. This protective feature is optional and the speed switch can be furnished without lockout solenoid if desired.

## Plugging a Motor to Stop from Either Direction

With the system shown in Fig. 15-16, the motor can be started in either direction by pressing the proper button. Pressing the *stop* button will plug the motor to stop from either direction. A standard reversing switch is used for this purpose.

The lockout solenoid is a built-in part of the speed switch and it guards against an accidental turn of the motor shaft closing the speed switch contacts and starting the motor. The control relay and the pushbutton station are standard parts.

## Antiplugging

Antiplugging protection is defined by NEMA as the effect of a device that operates to prevent application of countertorque by the motor until the motor speed has been reduced to an acceptable value. With the motor operating

**Fig. 15-15** *Plugging diagrams.* (Allen-Bradley)

**Fig. 15-16** *Plugging diagrams. (Allen-Bradley)*

NOTE: CR must be located within the starter enclosure.

in one direction, as shown in Fig. 15-17, a contact on the antiplugging switch opens the control circuit of the contactor used for the opposite direction. This contact will not close until the motor has slowed down, after which the other contactor can be energized. In this schematic the motor can be reversed, but it must not be plugged.

## BRAKING

Electric motors can be brought to a stop or braked both electrically and mechanically. In some instances it is necessary to use a combination of both. This usually happens when the motor is connected to a load that is not easily stopped or cannot be disconnected easily.

### Electronic Motor Brake

The electronic motor brake made by Square D provides a simple, effective means of braking an ac squirrel-cage motor (Fig. 15-18). It can be used for woodworking machines such as saws and sanders, and for machine tools such as lathes and drills, as well as for conveyor systems, textile machinery, and centrifuges. Heating, venting, and air-conditioning fans and many other machines in varied industries may also use this type of braking.

The major advantages of the electronic methods versus the mechanical brake system are:

- No friction, wear, or maintenance
- Adjustable soft-stop capability

Braking 255

**Fig. 15-17** Antiplugging diagrams. (Allen-Bradley)

**Fig. 15-18** Electronic motor brake. (Square D)

**256  Motor Control and Protection**

- No mechanical connection to the motor shaft
- Multimotor braking capability
- Easily wired to a new or existing machinery
- Unaffected by hostile motor environment

Electronic braking is commonly known as dynamic braking. Dynamic braking of an ac induction motor is generally accomplished by exciting its stator windings with dc current. The amount of braking torque is directly proportional to the dc current passing through the stator windings of the motor (Fig. 15-19).

The electronic motor brake is designed such that the braking contactor closes before the thyristor (SCR) switches the braking current on. The contactor will not open until after the braking current has been switched off. This allows the braking contactor to be rated for current-carrying capacity only and not for the higher make-and-break duty.

An additional circuit detects when the motor has come to a halt, switches off the braking current, and permits the motor to restart. No braking time adjustment is required. The maximum braking time is factory preset at 10 seconds. Braking torque is adjustable by use of a single potentiometer. This is an ideal braking system for jobs where there is a variable load and for multispeed three-phase motors.

## Mechanical Braking

Electric motors can also be stopped when necessary by using a mechanical means. These are similar to what is used with automobiles. Some of them rely on an electric current to energize the solenoid to cause the brake shoes to tighten around the motor shaft and stop it.

Reliance makes an electromechanical brake for motors up to 10 hp and 3600 rpm (Fig. 15-20). It has friction disks that are self-resetting with a manual release lever. The magnet coils are encapsulated to protect them from dirt and moisture. Antirattle springs are incorporated to reduce vibration and noise.

**Fig. 15-19** *Wiring diagram for electronic motor brake.* (Square D)

**Fig. 15-20** *Electromechanical brakes.* (Reliance)

Dynamic braking of a motor may cause threaded fasteners connected to the motor shaft to loosen, due to the reverse torque applied. Use positive-locking fasteners or fastening compound to prevent such loosening.

Note: Electronic motor brakes will not stop the motor if power is lost or disconnected.

This type of electronic motor brake can be used to stop a load and signal a mechanical brake system to hold it. In addition, the brake will interface with either jogging, reversing, multispeed, or reduced-voltage motor starter applications.

In some instances it is necessary to have a mechanical brake since dynamic braking is not sufficient to stop the motor rotation completely after power is removed. These brakes may be actuated whenever power is removed from the motor circuit. An electromagnet holds brake shoes away from the motor shaft whenever the motor is energized. Once power has been removed the brake is automatically applied by spring action (Fig. 15-21). This type of braking is very useful in elevators and similar installations.

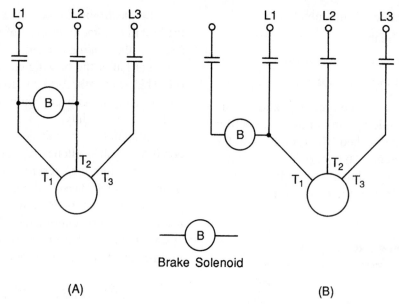

**Fig. 15-21** *AC brake coil hook ups for across-the-line starting.*

## Thruster Brakes

Thruster brakes are used with ac or dc motors and provide a smoothly applied fixed torque for hold or for stopping (Fig. 15-22). They are used on crane travel drives, lift bridges, conveyors, and similar applications to reduce load sway, and affect loading to motors and the mechanical system. These brakes are released by a thruster mechanism. This self-contained mechanism contains an ac square-cage motor and hydraulic pump. When de-energized the brake sets smoothly as the pumping action ceases.

## Magnetic Brakes

Brakes are selected by the amount of torque required for the particular application. Generally, the full-load torque of the motor is used as a basis for determining the brake torque required. This can be calculated by using the following formula for both ac and dc motors:

$$\text{Torque} = \frac{\text{rated hp} \times 5252}{\text{rated rpm}}$$

Depending on the characteristics of the drive, the braking torque required may be more or less than the full-load torque of the motor. In addition to being selected to meet the torque requirements of the particular application, the magnetic brake used for stopping must be selected to prevent overheating of the brake wheel when operated on the anticipated duty cycle.

**Fig. 15-22** *Thruster brakes.* (Square D)

## Hydraulic Brakes

Hydraulic brakes are used with ac or dc motors to provide an operator-controlled infinitely adjustable torque for slowing and stopping. These are used on crane travel drives, mill machines, conveyors, and similar jobs. They are spring released, hydraulically applied shoe-type friction brakes designed to meet American Iron and Steel Engineers (AISE) standards for mounting. The standard brake includes corrosion-resistant hardware and grease fittings (Fig. 15-23). Figure 15-24 shows the typical piping diagram for one brake.

**Fig. 15-23** *Hydraulic brake. (Square D)*

## MOTOR PROTECTION

Motor protection can take many forms, inasmuch as various types of motors are used to drive many types of machines. One of the most commonly used motor applications is to drive pumps. Surge protection and backspin are both present in pumps and must be considered in the circuitry design of the starters.

## Surge Protection and Backspin

Surge protection and "backspin" are two of the factors to be considered when protecting motor used to power pumps. Surge protection is often necessary when the pump is turned off and the long column of water is stopped by a check valve. The force of the sudden stop may cause surges that operate the pressure switch contacts, subjecting the starter to chattering.

Figure 15-25 shows how the system provides protection on both starting and stopping. Backspin is included automatically. Two-timing relays are used here, one to provide surge protection on starting and one to provide surge protection on stopping and backspin protection. TR1 is an on-delay tuner used for surge protection on starting. When the pressure switch contact closes, relay CR, the starter and two timers are energized. The instantaneous contact on TR1 closes, bypassing the pressure switch contact and preventing the pump motor starter from dropping out even though starting surges open the pressure contact. After the timing period, the time-delay contact TR1 opens the bypass and pressure switch (PS) can then stop the pump at the proper pressure. TR2 is an off-delay timer for surge protection on stopping and backspin protection. Once turned off the system cannot be operated again until timer TR2 has timed out and its normally closed contact is closed.

**Fig. 15-24** *Typical piping diagram for one brake. (Square D)*

**Fig. 15-25** *Pump operation with surge protection on starting and stopping.* (Allen-Bradley)

Another system that provides backspin protection and surge protection on stopping is shown in Fig. 15-26. It also has time delay between pressure switch closing and motor starting. The pressure switch energizes the timer (TR), but the motor cannot start until the time-delay contact has closed. The timer can thus be set for a time long enough to allow all surges and backspin to stop.

The dashed lines show how a selector switch can be added to bypass the pressure switch if necessary. This is often used for motor testing purposes. It does not eliminate the time delay, however. If the selector switch is added, wire A must be removed.

## Overload Protection

The overload relay is a manual reset, eutectic alloy, thermal-type overload device (Fig. 15-27). When coordinated with the proper short-circuit protection, the overload relay is intended to protect the motor, motor controller, and power wiring against overheating due to excessive overcurrents.

**Fig. 15-26** *Pump operation with backspin protection and surge protection.* (Allen-Bradley)

**Fig. 15-27** *Overload relay, eutectic alloy, thermal type.* (Allen-Bradley)

The ratchet stud assembly is heated by current flowing through the heater element. Relay operation occurs when the temperature of the ratchet and stud reaches the melting point of the eutectic alloy, freeing the ratchet stud and opening the NC contact.

To reset the overload relay contact, it is necessary to press and release the reset operator after the eutectic alloy has solidified. Approximately 2 minutes are required for the alloy to solidify.

## Automatic Reset Overload Relay

An indirectly heated, automatic reset noncompensated thermal relay is shown in Fig. 15-28. The temperature-sensitive unit of this relay is a bimetallic U-shaped strip that is mechanically coupled to a precision snap-action switch. The bimetallic strip is heated by the motor current that flows through a heater element affixed in close proximity. As the bimetal is heated, a deflection is produced in it by the different rates of expansion of the

**Fig. 15-28** *Overload relay-automatic reset. (Allen-Bradley)*

**Fig. 15-30** *Line-voltage monitor. (Allen-Bradley)*

two metals. A sustained current, greater than the rating of the heater element, will develop sufficient deflection to open the contact of the snap switch.

## Inverse Time Current Relay

A magnetically operated inverse time current overload relay that can be used in the protection of ac or dc motors is shown in Fig. 15-29. Both the tripping current and tripping time are easily adjustable. The relay is usually supplied with normally-closed contacts and automatic reset. To prevent relay damage, current through the relay coil should be interrupted after the relay trips.

**Line-Voltage Monitor** The line-voltage monitor detects phase failure, voltage imbalance, phase reversal, and undervoltage (Fig. 15-31A). It provides prestart and running protection on the line side of the point of connection and connects directly to three-phase lines with a 600-V maximum. It can be used with potential transformers where line voltage exceeds 600 V. It has automatic reset and an LED indicates normal voltage conditions and energized output relay CR.

**Fig. 15-29** *Inverse time current relay. (Allen-Bradley)*

## Solid-State, Line-Voltage and Line-Current Monitor Relays

The line-voltage monitor and line-current monitor are solid-state devices designed for use in three-phase systems to protect motors and other loads against abnormal voltage/current conditions. In general, the line-voltage monitor is applied where prestart protection and line-side protection are important, whereas the line-current monitor is applied where line- and load-side protection is important (Fig. 15-30).

**Fig. 15-31** *(A) Line-voltage monitor in the circuit; (B) line-current monitor in the circuit. (Allen-Bradley)*

**Motor Protection**

**Line-Current Monitor** The line-current monitor detects phase failure, current imbalance, and phase reversal. It also provides running protection on the line and load side of the point of connection when used in a single single-motor branch circuit. Inputs connect to standard current transformers (5 A secondary). An LED indicates that supply voltage is present. The manual reset device also has an LED to indicate normal current conditions and an energized output relay CR (Fig. 15-31B). Current imbalance protection is effective during the motor running period only. Phase failure and reversal protection is provided during both starting and running periods.

## Programmable Motor Protection

Allen-Bradley's programmable motor protector combines sophisticated, comprehensive, and coordinated motor protection into a modular system (Fig. 15-32).

**Fig. 15-32** *Programmable motor protector: (A) remote RTD module; (B) programmer monitor.* (Allen-Bradley)

It is intended for protection of large expensive motors that are often critical to a system or process. Typically, these motors are medium-voltage (2300 to 7200 V) or large low-voltage (200 hp or larger) motors.

By processing incoming data and looking for trends in various motor parameters, the device can provide a high degree of coordinated motor protection. It annunciates an abnormal condition and provides an output that trips the motor off-line. Alarm and trip contacts can also be used to initiate an orderly shutdown of the process.

## Protective Module

The protective module is the main module in the system. It has inputs, an annunciator panel, and outputs. This module receives inputs from external potential transformers (for three-phase line voltage) and current transformers (for three-phase line current). These inputs represent a portion of real-time motor data coming into the device for processing (Fig. 15-33). The remaining inputs to the protective module consist of the outputs from the programmer/monitor and the remote resistance temperature detector (RTD) module.

The protective module examines incoming motor data and compares them to user-preprogrammed limits. The module then determines if the motor should be taken off-line. It provides alarm and trip signals, and visually indicates the abnormality.

## Remote Temperature Detector Module

The remote RTD module serves as a remote temperature gathering panel. It is a microprocessor-based device that performs temperature data acquisition and coordination on demand from the protective module. The temperature data are used by the protective module to help construct a thermal model of the motor copper and iron.

This remote RTD module scans eight RTDs (two bearing and six winding). It takes an analog signal, digitizes it, performs a linearization function (to compensate for nonlinear RTD characteristic), scales it, and communicates the value to the protective module on demand. For motors not having RTDs embedded in the stator windings and bearings, motor protection can still be provided using just the protective module.

An optional instrumentation card allows the programmable motor protection (PMP) to provide a variety of metering functions (Fig. 15-34).

The following information can be provided:

- Elapsed time running: hours
- Total energy consumption: MWh
- Power factor: lead/lag
- Power: kW

**Fig. 15-33** *System configuration with communications option.* (Allen-Bradley)

**Fig. 15-34** *Instrumentation card for PMP.* (Allen-Bradley)

## REVIEW QUESTIONS

1. What do manual starters provide for a motor?
2. What is meant by low-voltage protection?
3. How is low-voltage protection provided?
4. What is meant by soft start?
5. What is meant by current limit?
6. What is jogging? How is it done?
7. What is plugging? How is it done?
8. How are motors braked or stopped?
9. What are six advantages of the electronic braking method?
10. What is another name for electronic braking?
11. What is backspin protection?
12. What causes chattering in a starter?
13. When are line monitors used?
14. What does a line voltage monitor do?
15. What is meant by annunciates?
16. Where are manual starters used?
17. What does the symbol Ⓜ represent?
18. Who requires low-voltage protection?
19. What are now replacing electromechanical devices?
20. Why is a master start-stop button used with a control relay?

# 16
CHAPTER

# Three-Phase Controllers

## PERFORMANCE OBJECTIVES

After studying this chapter you will be able to:

1. List factors to be considered in the operation and control of any motor.
2. List the types of three-phase starters.
3. Draw a ladder diagram for the operation of start-stop jog on a three-phase motor.
4. Describe how to reverse a three-phase motor.
5. List the advantages of solid-state controllers for three-phase motors.

## THE THREE-PHASE MOTOR

The squirrel-cage three-phase motor is highly reliable and efficient at essentially constant speed and requires little or no maintenance. Depending on construction, it may be classified as normal torque, normal starting current; normal torque, low starting current; high torque, low starting current; high slip; low starting torque, normal starting current; low torque, low starting current.

The three-phase squirrel-cage motor can be used in many different locations and for various applications, including rotary compressors, machine tools, large fans, light conveyors, milling machines, agitators, elevators, hoists, punch presses, centrifugal pumps, and blowers. It is made in the range 1/2 to 400 hp.

The wound-rotor motor is used where limited speed control and speed adjustments under fluctuating load are required. This type of motor is made in the range of half to several thousand horsepower. It can be found driving conveyors, fans, lift bridges, cranes, hoists, and metal rolling mills.

The synchronous motor is made in the range of 20 to several thousand horsepower. It is used for power factor correction and for exact slow-speed drives and maximum efficiency on continuous loads above 75 hp.

These motors can be started with comparative ease with the equipment available today. However, there are some uses that require special consideration. A controller is something more than just a starter. A starter causes the motor to start and stop. The controller not only starts the motor, but it also has the ability (according to its intended design) to reverse the motor and to vary its speed and torque as needed. It is this facet of motor control that requires special consideration and slightly different use of electrical circuitry and electro-mechanical devices. More solid-state devices such as programmable controllers are being introduced into the field of control and will require a better grasp of electronics to be fully understood.

The following factors are to be considered in the operation and control of any motor. These conditions can cause the motor to be damaged or destroyed.

- *Low voltage.* The motor tries to do the work, but the low voltage causes excessive currents that overheat the motor.
- *Heavy-duty load cycles.* Constant starting, stopping, jogging, inching, or plugging overloads the motor.
- *Excessive loads.* The load is too big for the motor.
- *Locked rotor.* Motor jams or cannot get running.
- *High-inertia load.* Motor takes abnormally long time to accelerate.
- *High ambient temperature.* The surrounding temperature is high, which heats the motor. Thus the motor may not be able to handle as big a load safely as in a lower ambient temperature.
- *Single phasing.* One of the power lines fails. The motor will draw higher line current. Two overload relays with properly sized elements will protect the motor except in rare cases such as when the motor is fed by an ungrounded wye-delta supply transformer.
- *Poor ventilation.* Motor cannot get cool air to take away some of the heat (cannot be protected by overload (OL) relays).

## STARTERS
### Simplest Type

The simplest type of three-phase starter is shown in Fig. 16-1. The wiring diagram shown in Fig. 16-2 makes it easy to visualize the wiring of the starter. The armature and crossbar or the overload reset mechanism is not shown in the wiring diagram since these parts need not be considered from the wiring standpoint. Note the location of L1, L2, and L3 at the top of the starter and T1, T2, and T3 at the bottom. As can be seen here, overload protection is built into the device.

***Fig. 16-1*** *Starter, size 1. (Allen-Bradley)*

**266** Three-Phase Controllers

**Fig. 16-2** *Diagram for starter.* (Allen-Bradley)

## Full-Voltage Reversing Starter

Full-voltage starting with the capability of reversing the motor calls for a little more complex wiring (Fig. 16-3). Remember, that to reverse the direction of the rotation of a three-phase motor it is necessary to change two of the incoming lines. Note in Fig. 16-4 that L1 and L3 are reversed when the contacts on the left are closed. It also means that the contacts in L1, L2, and L3 are opened when the reversing contacts close. L2 is not switched, so look at the diagram closely to see how the reversing contacts are directly across the L2 contacts, indicating that there is no change in this line. Also notice how the starters are electrically interlocked to avoid both contactors being closed simultaneously. The contacts are also mechanically interlocked so that both cannot be closed at the same time. Normally, L1 and L2 are switched to reverse direction.

**Fig. 16-3** *Full-voltage reversing starter.* (Allen-Bradley)

## Forward-Reverse-Stop

Standard wiring with the *forward-reverse-stop* push-button station is shown in Fig. 16-5. The *stop* button must be pressed before changing direction.

**Fig. 16-4** *Wiring diagram for the full-voltage reversing starter.* (Allen-Bradley)

**Fig. 16-5** *Forward-reverse-stop three-phase starter.* (Allen-Bradley)

A mechanical interlock and electrical interlocks are supplied as standard on all reversing starters. Limit switches can be added to stop the motor at a certain point in either direction. Connections A and B must be removed when the limit switches are used.

## Start-Stop Jog

The purpose of jogging is to have the motor operate only as long as the *jog* button is held down (Fig. 16-6). The starter must not lock-in during jogging. That is why the jog relay (CR) is used.

Pushing the *start* button operates the jog relay. This causes the starter to lock in through one of the relay contacts. When the *jog* button is pressed, the starter operates, but this time the relay is not energized and thus the starter will not lock in.

**Starters** 267

shown in Fig. 16-9. The motor can be started in either *high* or *low* speed. The change from low to high can be made without first pressing the *stop* button. However, when changing from high to low, the *stop* button must be pressed between speeds.

**Fig. 16-8** Two-speed manual starter. *(Allen-Bradley)*

**Fig. 16-6** Start-stop jog *three-phase starter. (Allen-Bradley)*

## Reversing Starter

Figure 16-7 shows how a manually operated reversing starter operates. Note the absence of a low-voltage control circuitry. This means that the contacts are closed, either to start or to reverse by manually (by hand) setting the contacts. L1 is changed with L3 and L2 just shorted in the reverse positions with the contacts shorting or paralleling the start contacts.

**Fig. 16-7** Manual reversing starter. *(Allen-Bradley)*

## Two-Speed Starter

A two-speed starter for a motor with separate windings for low and high speeds resembles that shown in Fig. 16-8. As the situation becomes more demanding, it is necessary to wire up motors to do various things under different conditions. A typical connection for a two-speed separate winding motor is

**Fig. 16-9** *Starter for a two-speed, separate-winding motor starter. (Allen-Bradley)*

**268** Three-Phase Controllers

## DUPLEX MOTOR CONTROLLERS

The controller operates first one motor and then the other on each successive closing of pilot device A. (Fig. 16-10). When pilot device B closes, both motors are energized. Typical applications include pump motors, where a second pump is required for peak demand periods. For this application, both pilot devices may be float or pressure switches and B is set to operate after A and only if both pumps are required.

*Fig. 16-11*  Medium-voltage controllers. (Square D)

*Fig. 16-10*  AC duplex motor controllers. (Square D)

Both pilot devices must be two-pole, but B can be omitted if only alternation is required and both motors are never required to run simultaneously. If one motor is running and its disconnect switch is opened, an overload relay trips, or the starter is de-energized for any reason, the other motor will automatically be started.

## MEDIUM-VOLTAGE CONTROLLERS

Medium voltage usually refers to the range from 2200 to 7200 V. This higher voltage range calls for protection for those who work around the equipment. Larger cabinets are needed to contain all the devices utilized in the control of these higher-horsepower motors. Figure 16-11 shows the housing for two of the controllers. Part A is used for squirrel-cage, reduced-voltage, non-reversing reactor controller, and part B is used for synchronous reduced-voltage, non-reversing autotransformer and reactor controllers.

Full-voltage controllers are used when full starting torque and resulting inrush current are not objectionable. One-high construction provides complete isolation for each controller and permits space for adding optional power and control devices. Figure 16-12 shows how the squirrel-cage, full-voltage, non-reversing motor is controlled. Note the three fuses used for current limiting and the three current transformers plus the bridge circuit made up of diodes.

Figure 16-13 shows a feeder disconnect circuit. Feeder disconnect controllers use mechanically held contactors that remain closed on loss of power. They are opened by use of a manual-trip pushbutton or an optional electrical solenoid release. Feeder disconnect controllers are available with either vacuum or air brake contactors of the bolted or draw-out design. These controllers are used frequently to disconnect transformers and in transfer schemes in place of metal-clad circuit breakers or disconnect switches.

A full-voltage, squirrel-cage, reversing controller is shown in Fig. 16-14. These controllers are used to control motors being operated in forward and reverse directions, where full starting torque and resulting inrush current are not objectionable to the motor. Reversing controllers are available with either vacuum or air-break contactors.

The squirrel-cage, reduced-voltage motor using an autotransformer is shown in its controller configuration in Fig. 16-15. These controllers provide maximum torque with a minimum of line current while providing taps to permit torque and line current to be varied. Vacuum or air-break contactors are available.

The reduced-voltage, primary reactor, squirrel-cage motor controller permits the starting of motors without the high inrush currents and voltage variations

**Fig. 16-12** *Controller for squirrel-cage, full-voltage, non-reversing motors.* (Square D)

**Fig. 16-13** *Controller for feeder disconnect.* (Square D)

**Fig. 16-14** *Controller for squirrel-cage, full-voltage, reversing motors.* (Square D)

**Fig. 16-15** *Controller for squirrel-cage, reduced voltage, auto-transformer motors.* (Square D)

**Medium-Voltage Controllers**

associated with full-voltage starting (Fig. 16-16). Synchronous, full-voltage, nonreversing motor controllers are used with motors where constant speed and plant power factor correction are desired. Some typical industrial applications are pulp and paper mills, lumber mills, rubber mills, metal rolling mills, gas compressors, centrifugal fans, blowers, generators, crushers, and grinders (Fig. 16-17).

**Fig. 16-16** *Controller for square-cage, reduced voltage, nonreversing motors.* (Square D)

**Fig. 16-17** *Controller for synchronous, full-voltage, nonreversing motors.* (Square D)

The brushless synchronous, full-voltage motor controller is shown in Fig. 16-18. It is used for synchronous motors required in explosive atmospheres. Brushless motors have the same advantages as regular synchronous motors, and since brushes are not used, less maintenance is required.

The controller shown has brushless field control with an incomplete sequence relay, dc power supply for the exciter field with a thyrite protector, a powerstat for field adjustment, loss of excitation protection, and pull-out protection.

The synchronous, reduced-voltage, autotransformer controller shown in Fig. 16-19 provides maximum torque with a minimum of line current while providing taps to permit torque and line current to be varied. Further, these controllers are used where constant speed and plant power-factor correction are desired.

**Fig. 16-18** *Controller for synchronous, brushless, full-voltage motors.* (Square D)

**Fig. 16-19** *Controller for synchronous, reduced-voltage, autotransformer motors.* (Square D)

**Medium-Voltage Controllers**

**Fig. 16-20** *Controller for synchronous, reduced-voltage, primary reactor motors. (Square D)*

The primary reactor synchronous motor controllers are used for drives where maximum efficiencies are required and when full starting torque and resulting inrush current are objectionable to the system. Further, these controllers are used where constant speed and plant power-factor correction are desired (Fig. 16-20).

## SOLID-STATE MOTOR CONTROLLER

The Bulletin 2050 controller made by Allen-Bradley provides controlled current (torque) starting of squirrel-cage motors from 30 to 1200 hp rated at 208 through 575 V ac at 50/60 Hz. These controllers are used on applications such as conveyors, pumps, compressors, and various other loads where minimum shock starting and smooth step-less acceleration is required (Fig. 16-21). Three acceleration modes are provided; current ramp, constant current, and linear timed.

**Current Ramp** This method has the most general application. During acceleration, a low initial current is gradually increased to a limiting-start current value. Smooth starting and optimum performance are achieved by adjusting the acceleration ramp time (rate of current increase). In most cases no other adjustments are required.

**Fig. 16-21** *Solid-state motor controller: (A) separate 200-hp power module and logic module with mounting plate; (B) 200-hp logic module mounted on power module. (Allen-Bradley)*

**Constant Current** This is a variation of the current ramp method, used where the principal requirement is to limit the starting current. With the acceleration ramp time set at minimum, the start current limit is adjusted in accordance with starting current restrictions.

**Linear Timed Acceleration** For applications requiring a controlled acceleration time or linear rate of speed increase, the controller can be set up for linear timed acceleration. A tachometer is installed on the motor and provides a feedback signal that is used by the controller

**274** Three-Phase Controllers

to increase the motor speed linearly with time, according to the acceleration ramp potentiometer setting.

## Modules

The controller consists of modules. The power module contains three power pole assemblies, each having back-to-back SCRs with single-bolt clamping. The power poles share a single heat sink/base and cooling fan. See Fig. 16-22 for a block diagram of the combination controller with some of the options marked with an asterisk.

The logic module can be mounted on the front of the power module or it can be mounted separately. External features include diagnostic LEDs, a terminal strip connecting the control devices, and the acceleration ramp adjustment potentiometer.

The energy-saver module is an option. It can reduce operating costs by reducing the motor power losses. It is recommended for certain applications, such as where motors run unloaded for long periods. It can be installed within the logic module.

## Solid-State Advantages

There are a few advantages claimed for solid-state controllers as opposed to electromechanical types. The main advantage is maintenance. Inasmuch as there are no moving parts such as contacts, there is little call for cleaning and adjusting them. It is possible to incorporate voltage regulation, transient suppression, and snubber circuits for protection against changes in temperature and changes in voltage. It is also possible to incorporate into the circuitry shorted SCR and open-phase protection. Phase-reversal detection is possible as well as protection against startup in an incorrect sequence. Over-current protection can also be designed for protection in various operating modes.

## MOTOR CONTROL CENTERS

As technology becomes more complex, integrated control equipment is becoming more in demand. There is a decided advantage to integrating all control and power requirements into one centralized package. It can be pre-engineered, pre-wired, and fully tested before it is delivered, thereby saving time not only in installation but in testing.

Motor control centers are used in a wide variety of industrial and commercial applications, such as pulp and paper mills, sawmills, building products, food processing, can plants, wastewater treatment plants, coal and bulk handling, chemical plants, and oil and gas production, to name but a few. In other words, it can be used wherever three-phase motors are used (Fig. 16-23).

**Fig. 16-22** *Block diagram of combination controller with options designated with asterisk.* (Allen-Bradley)

**Fig. 16-23** *Motor control center.* (Allen Bradley)

The control center shown in Fig. 16-23 houses a three-phase 600-V rated bus network that distributes power to various vertical sections. The main horizontal bus is located in the center of the section. This provides better heat dissipation and power distribution and makes the main bus accessible from floor level for easier and safer maintenance. The center-fed 300-A rated vertical bus supplies power to individual units above and below the horizontal bus for effective 600-A capacity, providing virtually unrestricted unit arrangement.

Different sizes and ratings of units with varying degrees of complexity are available depending on the horsepower and voltage of the motor and its particular control requirements. Standard units are available to handle basic starting-stopping-reversing, multi-speed, and reduced-voltage starting applications. Motor control centers are not only used to house basic control devices; they are used to package solid-state technology.

Programmable controllers, adjustable-frequency drives, solid-state reduced-voltage starters, and solid-state protective devices represent the kind of technology that is integrated into today's motor control centers.

## REVIEW QUESTIONS

1. Where are wound-rotor motors used?
2. What is the range of available horsepower for synchronous motors?
3. Why shouldn't a motor lock-in during jogging?
4. When are full-voltage controllers used?
5. Where are primary reactor synchronous motor controllers used?
6. What is an acceleration ramp?
7. When is an energy saver recommended?
8. What are motor control centers used for?
9. What is the simplest type of three-phase starter?
10. What wires are switched in order to reverse a motor?

# 17
CHAPTER

# DRIVES

# PERFORMANCES OBJECTIVES

After studying this chapter, you will be able to:

1. Describe adjustable-frequency ac drives.
2. Identify three major types of inverters.
3. Figure rpm when frequency and number of poles are known.
4. Describe a variable-voltage inverter.
5. Explain how pulse-width-modulated inverters work.
6. Describe the operation of eddy current drives.
7. Explain how open-loop controls operate.
8. Explain how closed-loop controls operate.
9. List the advantages and disadvantages of dc drives.
10. Describe solid-state digital ac drive advantages.

Industry is constantly striving to find controls that will increase productivity and reduce energy costs. This striving has about ceased, inasmuch as electronics has taken over and produced the desired results. However, many electromechanical devices are still in use and will be for many years. It is both of these worlds—electromechanical and electronic—that we discuss in this chapter.

The ac induction motor is the mainstay for energy conversion in the United States. It is found in industry, in commerce, and in the home. Our lifestyle would be unimaginable without it. It is the major converter of electrical energy into another usable form. For this purpose, about two-thirds of the electrical energy produced is fed to motors.

## FANS, BLOWERS, AND PUMPS

Fans, blowers, and pumps consume much of the electrical energy to operate the ac motors that power them. It has been estimated that approximately 50% of the motors in use today are attached to these types of loads. Fans, blowers, and pumps are particularly attractive to look at for energy savings. Several other methods of control for fans and pumps have been advanced recently that show energy savings over traditional methods.

Fans are designed for their maximum load. However, they do not always operate at maximum capacity. This means that there is a possibility of energy savings if the speed of the motor can be changed according to the load demand. In most instances the outlet dampers for fans and throttling valves for pumps were used to control their outputs. However, neither of these controls improved the efficiency of the pump or fan. What is needed is a control method to adapt fans and pumps to varying demands that do not decrease the efficiency of the system as much. Newer methods include direct variable-speed control of the fan or pump. This method produces a more efficient means of flow control than the existing methods.

In addition, adjustable-frequency drives offer a distinct advantage over other forms of variable-speed control. As can be seen in Fig. 17-1, by changing the speed or actual rpm of the fan, the performance of the fan changes producing a different airflow.

**Fig. 17-1** Variable-speed drives change fan rpm. (Allen-Bradley)

## ADJUSTABLE-SPEED DRIVES

There are several types of adjustable-speed drives that can be used on fans. These include variable-pitch belt drives, eddy current drives, dc drives, and adjustable frequency drives.

## ADJUSTABLE-FREQUENCY AC DRIVES

Adjustable-frequency drives are commonly called inverters. They are available in a range of horsepower from fractional to 1000. They are designed to operate standard induction motors. This allows them to be added easily to an existing system. The inverters are often sold separately because the motor may already be in place (Fig. 17-2).

The basic drive consists of the inverter itself, which converts the 60-Hz incoming power to a variable frequency and a variable voltage. The variable frequency is

**Fig. 17-2** Industrial ac drive. (Allen-Bradley)

the actual requirement that will control the motor speed. Three major types of inverter designs are in use today: current source inverters (CSI), variable-voltage inverters (VVI), and pulse-width-modulated inverters (PWM).

Keep in mind that ac motor speeds are a function of frequency and the number of poles:

$$\text{rpm} = \frac{120 \times \text{frequency (Hz)}}{\text{number of poles}}$$

Synchronous motors will run at the synchronous speed as determined by the formula. Their speed will not change with load changes within the pull-out torque capacity of the motor. An induction motor's speed-torque characteristics are shown in Fig. 17-3. Motor output requirements are dictated by the load. Friction-type loads such as conveyors require constant torque from the drive motor. Certain machine tool applications require constant horsepower. Most fans and blowers require variable torque. Generally, standard ac motors can produce constant torque throughout the speed range with a constant volts/Hz supply. The converter is adjusted to provide this supply (Fig. 17-4).

The converter and inverter are two different devices. The converter changes from ac to dc and the inverter changes dc to ac, whereas in actual practice the inverter changes the ac to dc and back to ac again at a different frequency. It may also be a reshaped waveform when changed back to ac again for frequency control purposes.

## CSI INVERTERS

The CSI inverter controls the current output to the motor. The actual speed of the motor is sensed by the use of other circuits. This is then compared to the reference speed and an error is used to generate a demand for more or less current to the motor. The output switching devices, usually SCRs, are switched at the desired frequency to "steer" the current to the motor. Current source inverters are available in a wide range of horsepower but most often are found in the range of 50 hp and above.

This type of inverter is used on a standard induction motor and is readily available, reliable, and easy to repair. If the inverter fails, the motor can be operated directly across the incoming line for continued operation. The inverter can adapt its operation to prevent overloads caused by accelerating the high-inertia loads found in some applications. The current control limits fault currents that will minimize damage on a major fault or overload condition. This inverter may require tachometer feedback for speed regulation. A tack generator must be added and is not a standard option for induction motors. If the tack feedback signal is lost during operation, the drive may run away to full speed. The inverter has to be matched to the motor's electrical characteristics. The inverter is sensitive to those characteristics, and improper operation may occur if the motor is replaced with a different type or size.

The inverter design requires the motor to be connected to operate at all. The inverter cannot be run or tested without the motor. The inverter uses a phase-controlled rectifier for current control. This method produces low power factor at low speeds.

The size of the major components usually causes these inverters to be the largest of the drives in overall size. All the power delivered to the system may go through a conversion within the inverter. Large power devices must be used to handle this.

## VARIABLE-VOLTAGE INVERTERS

The variable-voltage inverter controls the voltage and frequency to the motor to produce variable speed operation. The distinguishing characteristic between this type of inverter and the PWM inverter is the scheme used to control the voltage. VVI inverters control the voltage in a separate section from the output section used for frequency generation. Usually, the voltage control is done using a phase-controlled input bridge rectifier circuit at the input of the inverter. The frequency

*Fig. 17-3*  Synchronous motor speed-torque curve.

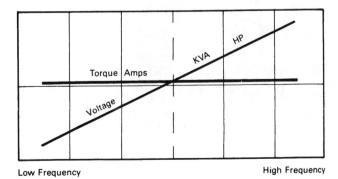

*Fig. 17-4*  AC motor operation curve.

control is accomplished by an output bridge circuit that switches the variable voltage to the motor at the desired frequency. These drives are available from fractional horsepower to about 500 hp.

The standard induction motor is readily available, reliable, and easy to repair and can be used with the VVI inverter. The inverter can achieve efficiencies of 90% at full speed and full load. If the inverter fails, the motor can still be operated directly across the incoming line for continued operation. The inverter can adapt its operation to prevent overloads caused by accelerating the high-inertia loads found in some applications.

Installation of this type of inverter is simple. Just three power leads to the motor are used. No tach feedback is required, and the drive can be located large distances from the motor being controlled. The drive can be tested and operated without requiring a motor to be connected. More than one motor can be operated from the same inverter. Also, the inverter is not sensitive to changing the combination of motors operated as long as the total load current does not exceed the inverter's rated current.

There are some drawbacks to the VVI inverter. The initial cost of the inverter system is high. The total power delivered to the motor must be converted by the inverter. This requires high-power components within the inverter. The inverter has a large portion of sophisticated circuits that require skilled technicians for service.

## PULSE-WIDTH-MODULATED INVERTERS

These inverters accomplish both frequency and voltage control at the output section of the drive. The output voltage is always a constant amplitude and by chopping or pulse-width modulation, the average voltage is

**Fig. 17-5** *PWM wave forms.* (Allen-Bradley)

controlled (Fig. 17-5). These drives are available from 1 to 1000 hp.

Some of the features of this type of inverter are its use on induction motors. It has good efficiency, up to 90% at full speed and full load. A diode bridge rectifier is used to rectify the incoming power. This permits a good power factor throughout the full operating speed range of the inverter (Fig. 17-6).

Control and logic functions are integrated onto a single LSI chip. This chip contains approximately 6300 transistors to perform the complex logic and control function. The use of this chip actually simplifies

**Fig. 17-6** *PWM drive.* (Allen Bradley)

drive construction by reducing the number of electrical connections needed, thus providing more reliability and quality.

If the inverter fails, the motor can be operated directly across the incoming line for continued operation. The inverter can adapt its operation to prevent overloads caused by accelerating the high inertial loads found in some applications. No tack is needed, so there are only three wires to connect. The control can be operated large distances from the motor.

This type of drive can also be tested without requiring a motor to be connected. More than one motor can be operated from the same inverter. Also, the inverter is not sensitive to changing the combination of motors operated as long as the total current is within the rated current limits of the inverter.

Some of the less desirable features of this type of inverter are its high cost initially and the fact that the total power delivered to the motor must be converted by the inverter. This requires high-power components within the inverter.

It does have some sophisticated circuits that require skilled technicians to repair or service. However, the use of large-scale-integrated circuits and microprocessor circuits permit self-diagnostics that aids in troubleshooting. Printed circuit board substitution can be done by unskilled service workers.

The eddy current drive consists of two distinct parts. One part, the mechanical unit, consists of the eddy current clutch and an induction motor. The motor runs at constant speed and provides a source of energy for the clutch. By controlling the excitation to the clutch, the amount of slip between the motor and the output shaft can be regulated and the output speed varied. If the excitation is high, the output speed increases toward the motor speed. If the excitation is lowered, the speed decreases toward zero speed. A given speed is maintained by balancing the excitation of the clutch to the load requirement.

Eddy currents are small currents created in the core when magnetic fields are changed. Eddy currents generate heat. They flow in the opposite direction to the current that induced them. The effect is to resist the flow of current in the core. Eddy currents can be minimized by using laminating techniques. Laminating is the building of an object through use of several layers of material. When this method is used, each lamination is varnished. Varnishing insulates the layers from each other, which increases resistance to eddy currents. Eddy currents in motors and transformers must be reduced as much as possible inasmuch as they represent power consumption for no work accomplished. By varying the excitation to the clutch in the eddy current drive it is possible to vary the eddy currents accordingly.

The clutch excitation is controlled by the eddy current controller. The controller uses high-gain amplifiers and a closed-loop speed control circuit to sense the need for clutch excitation. The mechanical unit has a tachometer generator mounted on the output shaft to provide speed feedback for the controller. The drive cannot regulate speed without this type of feedback.

Eddy current drives are available as integral units with the induction motor mounted or as just the eddy current coupling alone, which must be connected to an induction motor. There are also special designs available for use on vertical pumps. The drives are available from integral horsepower ratings all the way up to clutches large enough to handle several thousand horsepower.

Some of the advantages and good features of the eddy current drives are that the first costs are usually smaller than for adjustable-frequency drives, the controllers are much smaller than for other drives, and they need to handle 10% or less of the total power being delivered to the system. The control circuitry is less sophisticated and complex than that found in other systems. The eddy current coupling and the control respond well without overloading when operating high-inertia loads.

However, there are some disadvantages to this type of drive. Eddy current clutches are not common devices. On-site repair or even local repair may not be available. The special clutch does not permit the motor to operate the load directly. If the mechanical unit or the controller needs repair, the system is down.

The efficiency of the system decreases with speed. This is due to the output speed being controlled by slip within the clutch. Tach feedback is required to maintain speed control. If speed feedback is lost, the drive will go to full speed. Eddy current clutch brand names include Louis-Allis's Adjusto Speed and GE's Kinatrol.

## VARIABLE-PITCH DRIVES

The variable-pitch drive is a method of speed control that uses the mechanical means of belts and variable pitch sheaves or pulleys to change speed. The power source is a standard induction motor. Often these units are enclosed and have a gear reduction built in for reduced speed ranges. The horsepower range is generally limited from 5 to 50 hp, with not much available outside that range.

However, the cost is an important factor in this type of speed control. These systems are among the lowest-cost methods of achieving variable speed. The principle of operation is well known and easy to understand, and construction is simple.

The disadvantages are headed by remote control. This is not an inherent feature of the variable pitch drive.

This is because the drive uses mechanical means to vary the speed. Electrical control signals must be adapted to existing mechanical controls. The stress of variable-speed operation on belts requires periodic checks and the replacement of belts from time to time. High-inertia loads may cause problems. This may require over sizing the drive or custom motors. Special shutdown and startup procedures may be required to prevent overloading the motor. Running at constant speed for extended periods of time may cause grooving in the sheaves. This degrades speed control and decreases belt life.

## WOUND-ROTOR AC MOTOR DRIVES

Wound-rotor motor drives use a specially constructed motor to accomplish speed control. The motor rotor is constructed with windings that are brought out of the motor through slip rings on the motor shaft. These windings are connected to a controller that places variable resistors in series with the windings. The torque performance of the motor can be controlled using these variable resistors. Wound rotor motors are most common in the range of 300 hp and above.

There are a few advantages to this type of drive. The initial cost is moderate for the high-horsepower units. Not all the power need be controlled. That results in a moderate size and simple controller. The simple construction of the motor and control lends itself to maintenance without the need for a high level of training. High-inertia loads work well with this type of motor.

However, there are some disadvantages that should be taken into consideration for this type of drive. The motor has slip rings and is not readily available. Efficiency suffers at low speeds. The drive usually is limited to a speed range of 2:1. The speed regulation is also poor on the fan-type loads.

## DC DRIVES

The dc drive technology is the oldest form of electrical speed control. The drive system consists of a dc motor and a controller. The motor is constructed with armature and field windings. Both of these windings require dc excitation for motor operation. Usually, the field winding is excited with a constant level voltage from the controller.

## REGULATED SPEED DRIVES

There are various speed control methods used on dc motors. One of the simplest is the rheostat. The rheostat has an effective speed control range of about 4:1 with poor regulation of motor speed against changes in load and torque and line voltage. This control method is very inefficient because of the power dissipated in the rheostat.

Another method, the variable transformer with rectifiers, can run dc motors over a wider speed range (up to 10:1) at an improved regulation compared to the rheostat control. It is also more efficient than the rheostat.

The SCR (thyristor) controls have half-wave operation and characteristics that are similar to the variable transformer control. However, SCR systems with full-wave rectification can achieve a 20:1 speed range when used with IR compensation techniques, that is, pseudo-closed-loop current sensing techniques.

If a tachometer is used for feedback, true closed loop feedback control is possible with SCR circuits. A full-wave or three-phase SCR control may achieve speed ranges of up to 100:1. Due to the pulsating nature of SCR control techniques, the speed stability of motors below 1 hp may not always be good in the lower speed range, due to the low moment of inertia of the motor and due to the load. However, in integral-horsepower machines, excellent speed range can be achieved.

Transistor controllers can handle closed-loop speed control with a speed range of over 1000:1 and with regulation better than 1 or 2% based on set speed. Transistor controls have their best operational advantage for motors up to 1 hp, with some servomotors up to 5 hp. For smaller motor sizes, a continuous control method is used, but for motors above 1/3 to 1/2-hp a PWM technique is the most efficient control method.

## OPEN-LOOP CONTROLS (TRADITIONAL APPROACH)

DC motor speed controls have been around a long time. Perhaps the oldest and most widely used control for small dc motors is the series resistor speed control, shown in Fig. 17-7. The variable resistance is inserted in series with the armature and field circuit. The control has good starting characteristics with large torque available at low speed but has a runaway speed tendency at small-load torque conditions. This makes the control useful only for control applications with somewhat fixed friction conditions.

The shunt motor variable resistor speed control is shown in Fig. 17-8. Here a series resistor is inserted in the armature circuit.

The field winding is excited with a constant voltage. When more resistance is inserted in the armature circuit, speed regulation degenerates. This type of control works well for constant load torque rather than for a widely varying torque situation.

Figure 17-9 shows a control that uses a variable transformer to control the armature voltage of a shunt

**Fig. 17-7** *Electrical diagram and speed-torque curve for series motor resistor speed.*

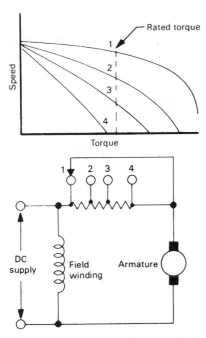

**Fig. 17-8** *Electrical diagram and speed-torque curve for shunt motor resistor speed.*

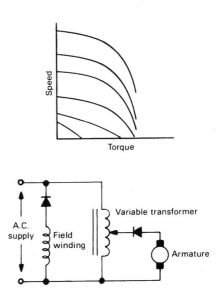

**Fig. 17-9** *Shunt motor with variable transformer to control armature voltage.*

motor with constant field excitation. The resulting speed-torque characteristics are much improved over variable resistor control characteristics, with a more uniform speed regulation over a wider speed range.

Figure 17-10 shows how a shunt motor connected to a variable field resistor operates. The action of this control circuit is unique in that the motor speed is variable only above the speed it would have without the field resistor control. This has an undesirable effect in that the torque constant of the motor will decrease with

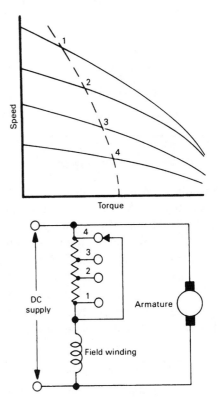

**Fig. 17-10** *Shunt motor with variable field resistor control.*

**Open-Loop Controls (Traditional Approach)**

increasing resistance insertion (field weakening). The net effect or result is that the armature current for a given torque will increase with higher speed. The motor can easily be overloaded. This control circuit is used only in unique cases where load conditions are both predictable and well controlled.

In some sophisticated open-loop control methods, such as the motor-generator armature control shown in Fig. 17-11, a constant-speed motor drives a generator with an adjustable control field voltage. The generator will produce an adjustable voltage that is delivered to the armature of the motor. The resulting torque-speed characteristics are improved over those shown in Fig. 17-8. This is because the regulation is essentially independent of the speed setting.

**Fig. 17-11** *Open-loop control using an armature control method.*

This results in superior motor speed control performance over any of the methods shown previously. However, due to the cost of the motor-generator set and associated field control, this method has not been practical for the small motor speed control applications most commonly used in home and industry. Therefore, the motor-generator control method has been confined primarily to industrial uses of large motor speed control, that is, for motors with more than 1 hp.

## CLOSED-LOOP CONTROLS

Many of the open-loop controls are adequate, but the trend toward better speed regulation, such as in meeting servo systems requirements, has necessitated closed-loop control (Fig. 17-12). In the basic or elementary form, a closed-loop control consists of an actuator (the motor), a comparator, an amplifier, and a sensor (the generator). The more sophisticated controls use this basic circuitry and then build on it to get motors to produce the torque and speed required when needed.

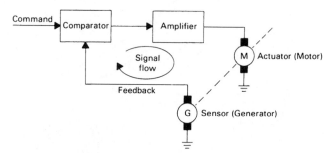

**Fig. 17-12** *Closed-loop speed control system.*

## DC DRIVES: ADVANTAGES AND DISADVANTAGES

Among the advantages for dc drives is the fact that the dc drive technology is simpler than the ac drive technology. It has been in existence for a long time and is well known. DC drives have good efficiency through the speed range. DC controllers are smaller than adjustable-frequency drives, but the motors are larger than induction motors.

Disadvantages of the dc drives are the fact that the dc motor is not always available or is not considered a shelf item. A tack generator for good speed regulation is a requirement. If tach loss occurs, the drive may run away to full speed. The power factor decreases with speed. Bypass is not possible because of the construction of the dc motor. Full power conversion of all power supplied to the motor is required by the controller. Larger power devices are required.

## DC DRIVES AND SYSTEMS

Custom-engineered drives and coordinated drive systems are available in ratings from 4 to 1500 hp. A wide variety of packaging and functional modifications provide flexibility and performance levels for demanding applications as dictated by machine characteristics and process requirements.

Figure 17-13 is a programmable logic controller with 75-hp Speedpak drive for a pull-through-slitter application. It is mounted in a standard enclosure. System-engineered drives serve complex manufacturing processes. In many cases they are integrated with programmable controllers and a variety of standard industrial control products to provide the user with a total control system.

**Fig. 17-13** *Programmable logic controller for 75-hp motor.* (Allen-Bradley)

**Fig. 17-14** *Digital ac drive.* (Allen-Bradley)

## SOLID-STATE DIGITAL AC DRIVES

The Bulletin 1352 is a versatile package capable of controlling the speed of a standard induction motor. It has the latest microprocessor and power semiconductor technology for controlling ac induction motors. The control panel is used to set up, operate, and troubleshoot the drive—all at the touch of a button. Take a look at Fig. 17-14. All parameters are entered in numerical format and stored in an electrically erasable programmable read-only memory chip (an EEPROM). There are no potentiometers to adjust and no other equipment required to calibrate the drive.

The control panel (Fig. 17-15) serves as a fully functional operator station, and a full complement of

To set-up the drive simply:

1. Push [PARAM] to select the parameter mode

2. Push [+] or [–] to select the parameter to be changed

3. Push [DISPLAY] to show the current value of the parameter

4. Push [+] or [–] to change the value of the parameter

5. Push [RESET] to enter the new value

In the normal operation mode, pushing the display button allows the operator to monitor:
☐ Drive frequency reference(s)
☐ Output frequency
☐ Motor load
☐ Output voltage

In the event of a fault, the corresponding diagnostic code is displayed to indicate one of the following conditions:
☐ Overcurrent
☐ Short circuit
☐ Undervoltage
☐ Over voltage
☐ Overtemperature
☐ Semiconductor fault
☐ Processor fault

**Fig. 17-15** *Control panel operation for the Bulletin 1352 digital ac drive.* (Allen-Bradley) inputs and outputs are provided for hand-wiring to external devices or interfacing to other equipment. Faults are displayed on the control panel in an easy-to understand digital format. In addition, all control logic values and drive input-output (I/O) points can be monitored to further enhance drive troubleshooting.

## REVIEW QUESTIONS

1. What are two types of adjustable-speed drives that can be used on fans?
2. What is another name for adjustable-frequency drives?
3. What are the three major types of inverter designs used today?
4. What dictates motor output requirements?
5. What is the range of horsepower on which CSI inverters operate best?
6. What does VVI mean?
7. What are some of the drawbacks to using VVIs?
8. What does PWM mean?
9. What are the two parts of an eddy current drive?
10. How are eddy current minimized?
11. What is the main advantage to an eddy current drive?
12. What is the common horsepower range of wound-rotor motors?
13. Why is rheostat control inefficient for dc motors?
14. What is the main disadvantage of open-loop controls?
15. What is used to make up a closed-loop control?
16. What is an EEPROM?
17. What is the purpose of a control panel for a digital ac drive?
18. List three advantages and three disadvantages of dc drives.
19. What is an SCR control?
20. List three advantages and three disadvantages for ac motor drives.

# 18
CHAPTER

# Transformers

## PERFORMANCE OBJECTIVES

After studying this chapter, you will be able to:

1. Explain how a transformer operates.
2. Describe autotransformer operation.
3. List transformer losses.
4. Describe how transformers are made environmentally safe.
5. Define polychlorinated biphenyls (PCBs).
6. Explain why three-phase transformers are chosen for motor control.
7. Define third harmonics.
8. Draw the schematic for a buck and boost transformer.
9. Define an askarel.
10. List three dry-type transformers.
11. Troubleshoot and maintain oil-filled transformers.
12. Locate proper current demands for various horsepower single-phase and three-phase motors
13. Draw wye-to-wye connections.
14. Draw wye-to-delta connections.
15. Draw delta-to-delta connections.
16. Draw delta-to-wye connections.

**Fig. 18-1** *The primary of the coil is coupled to the secondary by the magnetic flux lines.*

## TRANSFORMERS

Transformers make it possible to utilize the high voltages generated at power plants. They can be used to step up voltages to allow electrical power to be transported from the generator site to the user. Transformers make it possible to do many things with electricity. This is a relatively silent device that is hidden in enclosures above and below ground. It has no moving parts (except ventilation fans in some cases) and is over 99% efficient. Transformers can range in size from tiny assemblies of the size of a pea to behemoths weighing over 500 tons. However, the principles that govern the function of electrical transformers are the same, no matter what the size or the use to which it is put.

A transformer functions with no physical connection between the input and output coils. The principle used is mutual inductance. Current flows into the transformer primary coil. This current creates a magnetic flux. The magnetic flux, in effect, couples the primary coil with the secondary coil. Voltage is induced in the secondary coil. The induced voltage may be varied by increasing or decreasing the magnetic field. The result of a transformer's operation is the induction of electromotive force (EMF) in the secondary coil (Fig. 18-1).

## IRON-CORE TRANSFORMER

Iron-core transformers use the mutual inductance principle to transfer power between primary and secondary windings. An iron core provides a low-reluctance path for the magnetic flux. The iron core is represented by two straight lines between the primary and secondary coils.

Alternating current (ac) changes its magnetic field constantly so that a transformer can use this changing magnetic flux field to transfer ac power from the primary to the secondary (Fig. 18-2). Note that the two windings are not connected in this type of transformer (autotransformers will be examined later—they have only one winding). The only means of transferring energy from the primary winding to the secondary winding is the magnetic field.

### Turns Ratio

Transformers are either step-up or step-down. If the input voltage is higher than the output voltage, the device is a step-down transformer. In a step-up transformer, the output voltage is higher than the input. The input-output voltage relationship depends on the turns ratio, which describes the relationship between the windings of the primary and secondary coils. The first number in a given ratio is for the secondary. The second number is for the primary. The ratio of the primary

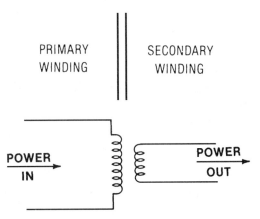

**Fig. 18-2** *Electric power is transferred from the primary to the secondary by the mutual coupling of the magnetic fields.*

to secondary voltage is the same as the primary-to-secondary turns ratio:

$$\frac{E_s}{E_p} = \frac{N_s}{N_p}$$

We can also state that the power in the primary equals the power in the secondary, minus any losses.

A moderate overload causes the temperature of the transformer to be increased. Therefore, the current available is actually a function of the amount of heating permissible in the transformer itself.

## AUTOTRANSFORMERS

Autotransformers provide a simple means of transforming voltage and impedance, but they do not provide the isolation between the primary and the secondary given by a regular transformer. The secondary load is really a part of the primary circuit whether the autotransformer is being used to step up or step down the voltage. A good example of autotransformer use in the industry is the buck and boost transformer. It is used to increase or decrease the voltage as needed (Fig. 18-3).

Autotransformers are restricted in their applications because it is not always possible to have one end grounded as required by the National Electrical Code (NEC) in some installations.

## Turns Ratio

The voltage ratio for an autotransformer is directly related to the number of turns between the tap and the common terminal and over the entire winding. Voltages and impedance can be calculated for an autotransformer in the same way that they are calculated for a transformer with separate windings. The voltage is directly proportional to the number of turns in the

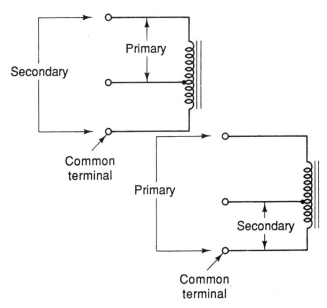

**Fig. 18-3** *Primary and secondary are both part of a single winding on an autotransformer.*

winding. The impedance ratio is equal to the square of the turns ratio:

$$\frac{N_p}{N_s} = \sqrt{\frac{Z_p}{Z_s}} \quad \text{or} \quad \frac{Z_p}{Z_s} = \left(\frac{N_p}{N_s}\right)^2$$

## TRANSFORMER LOSSES

Most transformers operate warm, and some are quite hot if used for a long period of time. This heat represents a power loss and means an efficiency of less than 100%.

### Eddy Current Losses

The greatest loss comes from eddy currents. They are currents induced into the core material. Since the core material has low electrical resistance, there are many closed loops for high current. This current in the core material generates heat and is wasted power.

Eddy currents are not desired. They are produced by the magnetic flux that links the primary and secondary windings. The flux lines are necessary for the transfer of electrical energy from the primary to the secondary. However, they also induce voltages into the core material (Fig. 18-4A). Flux produced by the eddy currents opposes the desired flux that links the primary and secondary windings. This means that a greater power demand is made on the source, and the primary current is higher than the amount required to supply the secondary load.

**Fig. 18-4** *(A) Solid core with high eddy current losses; (B) laminated core with eddy current losses practically eliminated.*

**Fig. 18-5** *Functional drawings showing hysteresis loss.*

Eddy current losses are reduced to a minimum by using a laminated core (Fig. 18-4B). With a laminated core, the eddy current paths in the core are parallel to that for the current in the windings. These thin layers of material create a high-resistance electrical path and reduce the circulation of eddy currents. At the same time, the iron laminations provide a low-reluctance magnetic path, and the structure has no adverse effect on transformer action.

## Hysteresis Losses

Hysteresis also plays a role in the loss presented by transformer operation. These losses are due to the properties of the iron core. Iron is slow to change polarity with changes in current and magnetic field polarity. The delay is known as hysteresis, or slowness to change properties (from north to south or south to north as the current changes polarity) (Fig. 18-5).

Note how the north-south poles are aligned in the magnetized piece and scattered in the nonmagnetized piece in between the change from north to south pole. This change from a north-south orientation to a south-north orientation consumes energy inasmuch as the metal structure is actually changed with the change in polarity. Hysteresis losses in transformers are minimized by using silicon steel inasmuch as it will change its polarity with a minimum of effort.

## Copper Losses

Copper losses are due to the resistance of the copper wire in the primary and secondary coils. Large-size wire helps to minimize these losses.

## CORE CONSTRUCTION

There are at least two usable core designs for power transformers (Fig. 18-6). The core form has the low-voltage and high-voltage windings at different locations

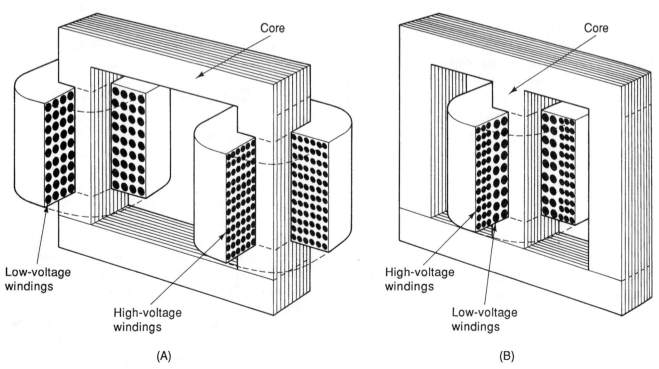

**Fig. 18-6** *(A) Core-form and (B) shell-form transformer designs.*

290  Transformers

on the frame or core (Fig. 18-6A). The shell-form construction makes a more efficient type of transformer inasmuch as the secondary is wound on top of the primary (Fig. 18-6B). This allows for maximum flux to cut the secondary windings. The magnetic field is also concentrated by the middle leg and contained within the core by the other two legs of the core. The laminations are E-shaped and inserted within the coil first in one direction and then the other. Then I-shaped laminations are used to complete the flux path, and the E-I combination makes a complete path for the magnetic flux.

In a three-phase transformer, the primary is wound and then the secondary on top of it, but there are three coil sets, one on each leg of the transformer core. Figure 18-7 shows how this is done with a cutaway view of a three-phase transformer. Note the tube coolers, where the hot oil moves up and then through the tubes to be cooled and to reenter the tank through the bottom.

## TRANSFORMER OIL

In order to make a better transformer, the industry discovered in 1929 that polychlorinated biphenyls (PCBs) made a good insulating oil and also served to keep the coil from overheating since it carried the heat to an outside cooling arrangement (Fig. 18-7). However, PCBs

**Fig. 18-7** *Typical transformer, cutaway view.*

have been determined to be toxic and to cause cancer in human beings and animals. The Environmental Protection Agency (EPA) keeps close tabs on transformers that utilize this cooling oil:

1. *PCB transformers:* All transformers that contain over 500 ppm (parts per million) of PCB. These cannot be rebuilt and must be labeled with the EPA label.
2. *PCB-contaminated transformer:* All transformers not in category 1 must be assumed to be contaminated with between 50 and 500 ppm, unless proven otherwise by testing. They may be rebuilt, but PCB fluid must be burned in an EPA-approved incinerator.
3. *Non-PCB transformers:* They have less than 50 ppm PCB.

The historical significance of this development in the transformer-manufacturing process makes interesting reading, inasmuch as it also shows the development of the technology for the past 60 years.

## EPA Highlights[1]

1929—Polychlorinated biphenyls (PCBs) invented by Swann Chemical Company.

1933—General Electric Company patented the application for use as a dielectric fluid in transformers and placed in service.

1977—US Environmental Protection Agency (EPA) published the final rule for "Polychlorinated Biphenyls (PCBs) Manufacturing, Processing, Distribution in Commerce and Use Prohibitions."

1982—Electrical Use Rule required periodic inspection and records of PCB leaks and also established new classification for PCB transformers, which pose an exposure risk to food plants and animal feeds.

1985 (July)—Fire Hazard Rule imposed new restrictions due to PCB transformers involved in fires, requiring removal of high-risk transformers or installation of fault protection by October 1, 1990.

1985 (August 16)—Fire Hazard Rule requires the reporting of any PCB transformer involved in a fire that caused tank to rupture and details cleanup procedures.

1985 (October 1)—Deadline for removal of PCB transformers in food and feed plants. Also prohibited the installation of used PCB transformers into commercial buildings. Commercial buildings are defined as public assembly properties, educational properties, institutions, stores, office buildings, and transportation centers, such as airports and bus or train stations.

1985 (December 1)—Fire Hazard Rule further requires the following, regarding all PCB transformers:
1. Removal of all combustibles in and around the transformer.
2. Notify local fire department of location of PCB transformer.
3. Access ways to PCB transformers that must be marked and identified for firefighters.
4. Owners of PCB transformers in or near commercial buildings must register the transformer with the building owner in writing with a complete description of the transformer and location. ("Near" means within 30 m.)

1990 (October 1)—All PCB transformers in or near commercial buildings must be removed, retrofitted, or equipped with new fault protection based on their classification.
1. Transformer secondary voltage less than 480Y/277 V requires high fault current protection.
2. A radial transformer with a secondary voltage of 480Y/277 V or higher must be equipped with both high and low fault current protection.
3. Use of network PCB transformers is prohibited as of this date.

After reviewing the requirements above, it should be evident that the best permanent solution is a retrofit or transformer replacement, providing a new non-PCB transformer.

## AUTOTRANSFORMERS: THREE- AND SINGLE-PHASE

The three-phase autotransformer is chosen for its economy and energy considerations. They are used by those who have a 480Y/277- or 208Y/120-V, three-phase, four-wire distributing system in the building (Fig. 18-8).

An autotransformer cannot be used on a 480- or 240-V, three-phase, three-wire delta system. A grounded neutral phase conductor must be available in accordance with the NEC Article 210-9, Exception 1.

Other than this article, autotransformers are installed under the same requirements as any other dry-type transformer. Figure 18-9 shows a typical wiring diagram for the autotransformer. Table 18-1 lists some of the characteristics of these transformers at 30 to 300 kVA (kilovolt-amperes).

---

[1]Information courtesy of Square D Company.

TAPS—Four 2.5% 2 FCAN and 2 FCBN

Wiring diagram for a three phase autotransformer

**Fig. 18-8** *Autotransformer. (Square D)*

**Fig. 18-9** *Typical wiring diagram for autotransformer. (Square D)*

Since the source and load share a common winding, there is no isolation of the load from the source. Where this is a requirement, it is recommended that the shielded isolating transformer be used. This is especially true for computers and other sensitive loads.

## Impedances

Impedances of autotransformers are quite a bit lower than comparable two-winding transformers. Although this provides better voltage regulation, it must be considered when making short-circuit studies for proper protective devices. Note the *X/R* and percentage *Z* values in Table 18-1. Table 18-2 shows impedances at 170°C.

*Table 18-2  Impedance at 170°C*

| | Impedance (%) | |
|---|---|---|
| kVA | Autotransformers | Insulating Transformers |
| 30 | 2.1 | 5.5 |
| 45 | 3.3 | 5.7 |
| 75 | 3.7 | 5.2 |
| 112 | 2.4 | 6.9 |
| 150 | 3.5 | 6.7 |
| 225 | 2.6 | 6.6 |
| 300 | 3.5 | 3.7 |

Source: Courtesy of Square D.

## Third Harmonics

Third harmonics are always present in wye-wye connections, but are kept at a minimum by using a three-legged core construction. For general use, this presents no problem. Sound levels are indicated in Table 18-3.

*Table 18-1  Autotransformers*

| | | Approx. Dimensions | | | | Average | |
| kVA | Temp. Rise (°C) | Height (in.) | Width (lb.) | Depth (in.) | Weight (lb.) | X/R | % Z |
|---|---|---|---|---|---|---|---|
| 30 | 150 | 23 | 22 1/4 | 15 | 250 | 1.4 | 2.1 |
| 45 | 150 | 23 | 22 1/4 | 15 | 275 | 1.0 | 3.3 |
| 75 | 150 | 30 | 30 | 20 | 425 | 1.2 | 3.7 |
| 112.5 | 150 | 30 | 30 | 20 | 605 | 1.0 | 2.4 |
| 150 | 150 | 42 | 36 | 24 | 750 | 1.5 | 3.5 |
| 225 | 150 | 42 | 36 | 24 | 1065 | 1.1 | 2.6 |
| 300 | 150 | 42 | 36 | 24 | 1375 | 2.6 | 3.5 |

Source: Courtesy of Square D.

Table 18-3  Sound Levels for Transformers

| kVA | Sound Level (dB)[a] | | |
| --- | --- | --- | --- |
| | Design Levels | | |
| | Auto | Insulating | NEMA Standard |
| 30 | 43 | 43 | 45 |
| 45 | 44 | 44 | 45 |
| 75 | 44 | 47 | 50 |
| 112 | 44 | 49 | 50 |
| 150 | 50 | 50 | 50 |
| 225 | 50 | 51 | 55 |
| 300 | 52 | 54 | 55 |

Source: Courtesy of Square D.
[a]dB, decibel. A 1-dB change in sound is probably the smallest change the trained ear can detect. Mathematically, it is equal to $10 \times \log_{10}$.

## Grounding

Article 250-26 of the NEC deals with grounding of separately derived ac systems. It is best to study the requirements of Article 250-26, 250-79(c), and Section 250-5 before connecting the transformers. Figure 18-10 shows the typical drawing for grounding a two-winding transformer, and Fig. 18-11 shows how the common grounded neutral is installed for autotransformers as per the NEC 210-9.

**Fig. 18-10**  Two-winding transformers. (Square D)

In the case of the autotransformer, the grounded conductor of the supply, whether it be a 480Y/277- or 208Y/120-V system, is brought into this transformer to the common HO-XO terminal and the ground is established to satisfy the NEC.

Requirement. Running the fourth wire is usually a negligible expense, since in most cases, the transformer is very close to its supply. The autotransformer connection is shown in Fig. 18-11.

**Fig. 18-11**  Autotransformer with common grounded neutral. (Square D)

## BUCK AND BOOST TRANSFORMERS

Buck and boost transformers are used to increase or decrease voltage. They are insulating transformers that have 120 × 240 V primaries and either 12/24- or 16/32-V secondaries. When used as isolating transformers, they carry the rated kilovolt-amperes stated on the nameplate. Their prime use and value, however, lies in the fact that the primary and secondary of a buck and boost transformer can be interconnected and the unit used as an autotransformer. By varying the manner in which the two primary and two secondary windings are connected, numerous ratios and current ratings can be obtained.

There are many applications where a slight adjustment in voltage (either upward or downward) is desirable or necessary. The use of a buck and boost transformer is one of the most economical and compact means of accomplishing adjustments of this type. When used as an autotransformer, a buck and boost unit will carry loads in excess of its nameplate rating. The increased ampacity is dependent on the ratio and voltage to which the transformer is subjected.

The application of buck and boost transformers as autotransformers has long been a source of confusion to users. Key to this confusion undoubtedly lies in the fact that when used as an autotransformer, the buck and boost unit carries loads in excess of its nameplate rating.

The nameplate rating shows the load that can be carried continuously by the unit when used as an isolation transformer, with a line voltage of 120 or 240 V (Fig. 18-12). There is no direct electrical connection between the primary windings (letters H) and the secondary windings (letters X). If used in this way, each winding must carry full nameplate load.

Figure 18-13 shows a buck and boost transformer connected as an autotransformer. Note that in this case, the primary windings (letters H) and the secondary windings (letters X) are in direct electrical contact at H1-X1 junction. Because the primary and secondary windings are interconnected, they share the load, rather than each having to carry the full load. Consequently, the transformer can carry a load greater than on the nameplate. The winding section from H1 to H4 carries only the difference of the primary and secondary currents, thus permitting the higher-load kilovolt-ampere ratings.

### Calculating the Load

Increasing or decreasing line voltage by small percentages can be accomplished easily and economically by the proper use of buck and boost units as autotransformers. Depending on the percentage of change desired and the

**Fig. 18-12** *Dashed lines indicate customer wiring; 120-V line can be transformed to 24-V load.*
*(Square D)*

**Fig. 18-13** *Dashed lines indicate customer wiring; 208-V line can be transformed to handle a 230-V load.* *(Square D)*

base voltage, the proper connections must be used; see Fig. 18-14, schematics 1 to 4.

The percentage of voltage change is given in the following:

Boosting:

$$\text{Change (\%)} = \frac{\text{High voltage} - \text{low voltage}}{\text{low voltage}} \times 100$$

Bucking:

$$\text{Change (\%)} = \frac{\text{High voltage} - \text{low voltage}}{\text{high voltage}} \times 100$$

Example:

Boosting 207 V to 230 V:

$$\text{Change (\%)} = \frac{230 - 207}{207} \times 100$$

$$= \frac{23}{207} \times \frac{100}{1} = 11.1\%$$

## Calculating Single-Phase Kilovolt-Amperes

The single-phase kilovolt-amperes is given by

$$\frac{\text{Volts} \times \text{load amperes}}{1000}$$

Buck and Boost Transformers

**Fig. 18-14** *Wiring the buck and boost transformer.* (Square D)

## Three-Phase Buck and Boost Transformers

So far the discussion has centered on single-phase buck and boost transformers. Figure 18-14, schematics 5 to 10, show various means of connecting the transformers in three-phase arrangements.

## Calculating Three-Phase Kilovolt-Amperes

The three-phase kilovolt-amperes is given by

$$\frac{\text{Volts} \times \text{load amperes} \times 1.73}{1000}$$

Single-phase transformers use one transformer unit. Three-phase transformers use two or three transformers to boost or buck. Three-phase loads can be served by using two single-phase units connected in open delta form. Three-phase loads can also be served by using three single-phase units connected in wye.

Before using autotransformers, check local codes for any restrictions to their use.

## DRY-TYPE TRANSFORMERS

Newer makes of oil-filled transformers have eliminated the use of PCBs in their tanks. Dry-type transformers do have some advantages and are used to replace the older PCB types. Dry-type transformers increase the efficiency of electrical systems considerably by permitting voltages greater than 600 V to be conducted, as near as physically possible, to the electrical center of the load. This reduces to the minimum money spent for line losses and for larger secondary systems. It is not practical to do this with oil-filled transformers, which must be installed outdoors at a safe distance from a building, or in a fireproof vault, which cannot be usually located for the most efficient power distribution. Although recent NEC changes permit vaults with lower fire ratings if they are sprinkler protected, many engineers and users object to installing any water pipes in electrical equipment rooms. Some electrical equipment rooms also contain an emergency distribution system, including the generator. A water leak or sprinkler head malfunction could disable the entire electrical system.

A second type of application, which is quite common, is the use of dry-type transformers on a roof. Since they are usually lighter than a comparable oil-filled transformer, a lighter roof structure is required. Also, they do not require any special provisions for containing the oil in case of a leak or fire.

Another major area of use has been in high-rise buildings, permitting the economical distribution of power to the various floors using voltages up to 15 kV, and then stepping down to the utilization voltage.

Since the use of askarel transformers was prohibited because of the environmental concerns raised by PCBs, new liquids with suitable cooling and insulating qualities have been developed and applied to transformer use.

The NEC classifies them as "less flammable." This means that they can burn but are less combustible than mineral oil, or "nonflammable," which means that they do not have a flash point or fire point. Because this is a new application for most of these liquids, there are few historical field data to determine if there are some unknown, undesirable characteristics that might develop in some of these liquids over a period of years. Figure 18-15 shows dry-type transformers with and without a cabinet.

## Cooling

The addition of cooling fans increases the capacity of a transformer by 33.33%. This is an economical way to handle short-time load peaks or emergency overloads. Fans can be provided at the time the transformer is purchased, or provision can be made for the future addition of fans in the field. Fan cooling is generally economical to provide on transformers 300 kVA and larger. Automatic fan cooling systems have heat-sensing thermistors in each phase of a three-phase or single-phase transformer.

## Insulation and Temperature Rise

Most dry-type transformer manufacturers use UL-component-recognized 220°C insulation systems for 30 kVA and above. The 220°C represents the ultimate temperature of the winding and is the sum of the

**Fig. 18-15** *Dry-type transformers: (A) open cabinet; (B) closed cabinet. (Square D)*

permissible rise (by resistance measurement) of the windings (150°C), the hot differential allowance (30°C), and the ambient allowance (40°C). This 220°C insulation system was formerly called the H Insulation System by the National Electrical Manufacturers Association (NEMA) and the American National Standards Institute (ANSI). The letter designation H was dropped due to the confusion caused by UL calling this same 220°C system a C Insulating System.

## Taps

Taps are made standard with two 2.5% above and below normal primary voltage. Four 2.5% taps below normal primary voltage can be obtained from most manufacturers.

### SPECIAL TRANSFORMERS

Outdoor transformers of the ventilated dry type are tamperproof with special baffled louvers, and are constructed of heavy-gage sheet metal that has been cleaned and painted. Periodic inspections are recommended, but there is no oil or filter to check. The transformers are available in all voltages up to 15 kV and sizes through 1500 kVA (Fig. 18-16).

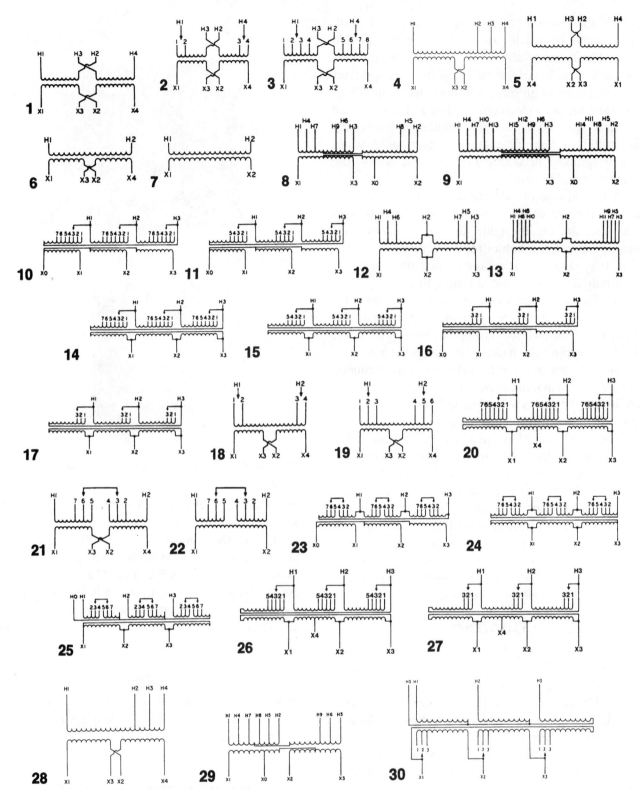

**Fig. 18-16** Transformer wiring diagrams. (Square D)

*Pad-mounted transformers* (dry type) are ventilated and have a 4160-V delta primary and 208Y/120-V secondary with two 2.5% taps (Fig. 18-17). This type of housing may also contain a primary hook-stick-operated switch, transformer, and a limited amount of secondary breakers and switches. All outdoor transformers over 600 V have lightning arresters as standard equipment for extra protection against lightning or switching surges.

**Fig. 18-17** *Pad-mounted dry-type transformer.* (Square D)

*Epoxy units* are made for substations and are particularly suited for applications requiring a dry-type transformer with good performance characteristics. The windings are completely impregnated with epoxy resin, which together with fiberglass cloth and tape forms the solid dielectric system. The solid dielectric system protects the windings from its environment with respect to moisture and airborne contaminants. This solid dielectric system also provides the ability to withstand thermal shock and the mechanical forces of a short circuit. They are ideal replacements for PCB units (Fig. 18-18).

**Fig. 18-18** *Three-phase cast-epoxy unit substation transformer.* (Square D)

## LIQUID-FILLED TRANSFORMERS

The transformer coils are immersed in an oil or silicone to increase the efficiency of the transformer. The liquid has the ability to transfer heat from the coil to the outside of the case efficiently. The liquid also has insulating qualities that are very desirable at high voltages. Now that PCBs are eliminated from the manufacture of transformers, it is necessary to make sure that the replacements are environmentally safe. So far, the oils used are as fire resistant as the PCB oil.

### Safety Notice

Keep in mind that to work with transformers, adequate training is needed. Extremely hazardous voltages are present in energized units. Adequate training and safe working procedures for this and related high-voltage equipment are needed to make sure that you can work with the equipment safely.

Working with and inspecting the transformer tank and its contents requires special instructions on its contents and operation. Some of these are mentioned here to inform those who may wish to obtain further training before work on the equipment.

## SUBMERSIBLE TRANSFORMER

Some manufacturers use gray paint for the exterior of the transformer and others use green (Fig. 18-19). These three-phase transformers consist of a core and coil assembly designed to reduce operating losses, and to provide adequate mechanical strength should a system fault occur. The 75- to 5000-kVA core-coil assemblies are of the five-legged design with wound cores (Fig. 18-20). By using the five-legged design, the possibility of tank heating due to stray flux paths under short-circuit conditions is eliminated. Computer design techniques produce a core-coil assembly with high efficiency and low losses for the most economical operating characteristics. Strip aluminum secondary windings offer high axial short-circuit strength and fast coil heat dissipation.

### Maintenance

Oil leaks are rare but, if detected, must be repaired at once to avoid the liquid level dropping below energized parts, creating a possibility of flashover or tank overheating. If required, the transformer must be refilled to its proper operating level. Small pinhole leaks in the exterior metal, weld seam, or other locations, resulting in slow dripping, can be repaired with

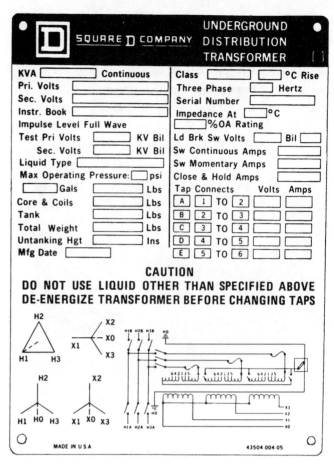

**Fig. 18-19** *Three-phase liquid-filled submersible small power transformer.* (Square D)

**Fig. 18-20** *Pad-mounted liquid-filled compartmental transformer.* (Square D)

a lifetime durable epoxy patch kit. The transformer must first be de-energized. Then a temporary oil stick is applied. Next the epoxy is applied. This usually eliminates the need for a vacuum pump to stop the oil leak while the epoxy is being cured. Larger leaks may require the use of a vacuum pump. The main reason for maintaining the level and preventing the oil to leak out is the entry of moisture into the tank. The moisture can cause problems at high voltages and can also cause the oil to become contaminated, necessitating its removal and replacement. Figure 18-21 shows all the parts of a liquid-filled substation transformer. Note the location of the pressure relief device, fill plug, liquid-level gage, dial-type thermometer, nameplate, and drain valve. The fins on the back are used to cool the liquid as it passes through the tubes from top to bottom. Hot oil rises and enters the tubes at the top of the fins. As it trickles down, it is cooled by the outside air. Once cooled, it reenters the tank at the bottom to once again pick up excess heat and conduct it to the outside of the tank and away from the coils inside.

**Fig. 18-21** *Three-phase, liquid-filled substation transformer.*
(Square D)

# TRANSFORMER APPLICATIONS

Transformers are used in business, industry, and commerce as well as for schools and homes. They come in many sizes and shapes, as has been seen so far in this chapter. Six typical installations and applications for industrial work are discussed in the next part of this chapter.

## Distributing Power at High Voltages

Transformers are used for distributing power at 480 or 600 V and stepping down voltage at the point of use (for 240-V motors, or 120-V equipment and lights). This results in better regulation of voltage and minimized line loss and reduces wiring costs. See Fig. 18-22 for a diagram illustrating these points.

## Double-Wiring Elimination

Transformers are used to eliminate double wiring. For maximum safety, 120-V lighting and control circuits may be obtained from 240-, 480-, or 600-V power circuits by installing dry-type transformers at the most convenient location to the load. This eliminates separate circuits and independent metering for power and light and often results in large savings. See Fig. 18-23 for examples of this.

## Operating 120/240-V Equipment from Power Circuits

Transformers are used to operate portable tools, electrical control devices, alarms, relays, soldering irons,

**Fig. 18-22** *120/240-V single phase from a three-phase high-voltage source.*

Transformer Applications

**Fig. 18-23** *120-V lighting and control circuits from a three-phase source.*

heating pots, small heat-treating furnaces, bench welders, and other high-current devices more economically from 120- or 240-V circuits supplied through a dry-type transformer from a high-voltage power circuit (Fig. 18-24).

**Fig. 18-24** *Operating 120/240-V single-phase from 240/ 480-V, or 600-V three-phase.*

## Isolating Circuits

Transformers are used because of their ability to isolate one circuit from another. They can be air-cooled transformers and used as a means of subdividing circuits to accommodate independent demand.

They can be connected to a three-phase, 480-V circuit, dry-type transformer to provide 120/240-V, three-wire, single-phase power for lighting loads. They can provide 120-V single-phase lighting loads or 240-V single-phase lighting or power loads. Transformers permit grounding of each low-voltage circuit (Fig. 18-25).

## Changing from a Four-Wire to Three-Wire Circuit

Transformers can be used to produce a three-wire, 120/240-V, single-phase circuit from the 120-V, two-wire circuit of a four-wire, 208Y/120-V, three-phase source. A three-phase primary of 240, 480, or 600 V may be used to provide a three-wire, 120/240-V, single-phase circuit (Fig. 18-26).

## Stepping-Up or Stepping-Down Voltage

Transformers can be used to step up or step down voltages simply by virtue of the turns ratio in the primary as compared to the secondary (Fig. 18-27). Whenever the voltage source is lower or higher than the nominal required by the equipment load, a buck and boost transformer may be used.

### MOTOR TRANSFORMERS

Transformers arse very necessary to obtain the proper voltage for the operation of motors. Table 18-4 shows transformer ratings required for the operation of standard induction motors at standard voltages and at 60 Hz. Table 18-5 shows how to convert kilowatts to horsepower or horsepower to kilowatts, and Table 18-6 shows what can happen with improper voltage levels in a plant.

**Fig. 18-25**  *Subdividing circuits.*

**Fig. 18-26**  *Three-wire secondary circuit from a four-wire, three-phase source.*

Motor Transformers

**Fig. 18-27** Buck and boost transformers.

**Table 18-4** Transformer kVA Rating Required for Operation of Standard Induction Motors at Standard Voltages

| AC Motors | Horse-Power | Full-Load Amperes | | | | Minimum Wire Size[a] | | | | Minimum Transformer (kVA)[b] |
|---|---|---|---|---|---|---|---|---|---|---|
| | | 115 V | 220 V | 230 V | 440 V | 115 V | 230 V | 230 V | 440 V | |
| Single-phase | $\frac{1}{6}$ | 4.4 | | 2.2 | | 14 | | 14 | | 0.53 |
| | $\frac{1}{4}$ | 5.8 | | 2.9 | | 14 | | 14 | | 0.70 |
| | $\frac{1}{3}$ | 7.2 | | 3.6 | | 14 | | 14 | | 0.87 |
| | $\frac{1}{2}$ | 9.8 | | 4.9 | | 14 | | 14 | | 1.18 |
| | $\frac{3}{4}$ | 13.8 | | 6.9 | | 14 | | 14 | | 1.66 |
| | 1 | 16 | | 8 | | 12 | | 14 | | 1.92 |
| | $1\frac{1}{2}$ | 20 | | 10 | | 12 | | 14 | | 2.4 |
| | 2 | 24 | | 12 | | 10 | | 14 | | 2.88 |
| | 3 | 34 | | 17 | | 8 | | 12 | | 4.1 |
| | 5 | 56 | | 28 | | 6 | | 10 | | 6.72 |
| | $7\frac{1}{2}$ | 60 | | 40 | | 3 | | 8 | | 9.6 |
| | 10 | 100 | | 50 | | 1 | | 6 | | 12 |
| Three-phase | $\frac{1}{2}$ | | | 2 | 1 | | 14 | | 14 | 0.9 |
| | $\frac{3}{4}$ | | | 2.8 | 1.4 | | 14 | | 14 | 1.2 |
| | 1 | | | 3.5 | 1.8 | | 14 | | 14 | 1.5 |
| | $1\frac{1}{2}$ | | | 5 | 2.5 | | 14 | | 14 | 2.1 |
| | 2 | | | 6.5 | 3.3 | | 14 | | 14 | 2.7 |
| | 3 | | | 9 | 4.5 | | 14 | | 14 | 3.8 |
| | 5 | | | 15 | 7.5 | | 14 | | 14 | 6.3 |
| | $7\frac{1}{2}$ | | | 22 | 11 | | 10 | | 14 | 9.2 |
| | 10 | | | 27 | 14 | | 10 | | 14 | 11.2 |
| | 15 | | | 40 | 20 | | 8 | | 12 | 16.6 |
| | 20 | | | 52 | 26 | | 6 | | 10 | 21.6 |
| | 25 | | | 64 | 32 | | 4 | | 8 | 26.6 |
| | 30 | | | 78 | 39 | | 3 | | 8 | 32.4 |
| | 40 | | | 104 | 52 | | 1 | | 6 | 43.2 |
| | 50 | | | 125 | 63 | | 0 | | 4 | 52.0 |

Source: Courtesy of Square D.
[a]Not more than three conductors in cable or raceway.
[b]Allow 20% additional kVA if motors are started more than once per hour. Data above computed from standard motor data as listed in National Electric Code®. For estimating only. For OEM application, check exact requirements with factory.

*Table 18-5* Conversion Tables for Horsepower to Kilowatts and Kilowatts to Horsepower for Electric Motors

| Kilowatts to Horsepower | | | | Horsepower to Kilowatts | | | |
|---|---|---|---|---|---|---|---|
| kW | hp | kW | hp | hp | kW | hp | kW |
| 1 | 1.341 | 55 | 73.733 | 1 | .746 | 55 | 41.03 |
| 2 | 2.681 | 60 | 80.436 | 2 | 1.492 | 60 | 4.76 |
| 3 | 4.022 | 65 | 87.139 | 3 | 2.238 | 65 | 48.49 |
| 4 | 5.363 | 70 | 93.842 | 4 | 2.984 | 70 | 52.22 |
| 5 | 6.703 | 75 | 100.545 | 5 | 3.730 | 75 | 55.95 |
| 6 | 8.044 | 80 | 107.248 | 6 | 4.476 | 80 | 59.68 |
| 7 | 9.384 | 85 | 113.951 | 7 | 5.222 | 85 | 63.41 |
| 8 | 10.725 | 90 | 120.654 | 8 | 5.968 | 90 | 67.14 |
| 9 | 12.065 | 95 | 127.357 | 9 | 6.714 | 95 | 70.87 |
| 10 | 13.406 | 100 | 134.048 | 10 | 7.460 | 100 | 74.60 |
| 11 | 14.747 | 110 | 147.47 | 11 | 8.206 | 110 | 82.06 |
| 12 | 16.087 | 120 | 160.87 | 12 | 8.952 | 120 | 89.52 |
| 13 | 17.428 | 130 | 174.28 | 13 | 9.698 | 130 | 96.98 |
| 14 | 17.768 | 140 | 187.68 | 14 | 10.444 | 140 | 104.44 |
| 15 | 20.109 | 150 | 201.09 | 15 | 11.190 | 150 | 111.90 |
| 16 | 21.450 | 160 | 214.50 | 16 | 11.936 | 160 | 119.36 |
| 17 | 22.790 | 170 | 227.90 | 17 | 12.682 | 170 | 126.82 |
| 18 | 24.131 | 180 | 241.31 | 18 | 13.428 | 180 | 134.28 |
| 19 | 25.471 | 190 | 254.71 | 19 | 14.174 | 190 | 141.74 |
| 20 | 26.812 | 200 | 268.12 | 20 | 14.920 | 200 | 149.20 |
| 22 | 29.493 | 220 | 294.93 | 22 | 16.412 | 220 | 164.12 |
| 24 | 32.174 | 240 | 321.74 | 24 | 17.904 | 240 | 179.04 |
| 26 | 34.856 | 260 | 348.56 | 26 | 19.396 | 260 | 193.96 |
| 28 | 37.537 | 280 | 375.37 | 28 | 20.888 | 280 | 208.88 |
| 30 | 40.218 | 300 | 402.18 | 30 | 22.380 | 300 | 233.80 |
| 32 | 42.899 | 325 | 435.69 | 32 | 23.872 | 325 | 242.45 |
| 34 | 45.580 | 350 | 469.21 | 34 | 25.364 | 350 | 261.1 |
| 36 | 48.261 | 400 | 436.24 | 36 | 26.856 | 400 | 298.4 |
| 38 | 50.943 | 450 | 603.27 | 38 | 28.348 | 450 | 335.7 |
| 40 | 53.624 | 500 | 670.30 | 40 | 29.840 | 500 | 373.0 |
| 42 | 56.305 | 600 | 804.36 | 42 | 21.332 | 600 | 447.6 |
| 44 | 58.986 | 700 | 938.42 | 44 | 32.824 | 700 | 522.2 |
| 46 | 61.667 | 800 | 1072.48 | 46 | 34.316 | 800 | 596.8 |
| 48 | 64.349 | 900 | 1206.54 | 48 | 35.808 | 900 | 671.4 |
| 50 | 67.030 | 1000 | 1340.60 | 50 | 37.300 | 1000 | 746.0 |

Source: Courtesy of Square D.
Note: Conversions not given may be obtained by adding appropriate values given.

*Table 18-6* Improper Voltage Levels Affect Performance of Plant Equipment

| | | |
|---|---|---|
| Motors | Torque reduced 19% by 10% undervoltage, and motor temperature increases and shortens life of motor. | Starting current (inrush current caushing voltage dips) is increased 12% and power factor decreased 5% when 10% overvoltage exists. |
| Rectifier loads: Electroplaters, battery chargers, static dc supplies for cranes, dc motor drives, magnetic chucks, precipitators | With a 10% undervoltage electroplating deposition rate drops 10 to 20%; battery charging rate falls 15 to 25%; precipitator cleaning power drops 20% magnetic chuck holding power is reduced 19%. | Metallic rectifiers withstand 50% less transient surge when operated at 10% overvoltage. |
| Magnetic devices: Solenoids for clamping and ejecting, vibrating feeders, magnetic brakes, solenoid valves, motor starter contactors, ac relays | Solenoids take longer to open a valve, eject a part, close a relay, or close a starter. Holding power of relays varies as the square of the voltage, and at 10% undervoltage is so reduced that vibration or minor voltage that vibration or minor voltage dips will drop out contactors. | Wear and distortion of solenoid surfaces is substantially greater. Saturation of solenoid cores with associated drastic increase of operating current and heating takes place. |

Source: Courtesy of Square D.

## REVIEW QUESTIONS

1. How efficient are transformers?
2. What is a buck and boost transformer?
3. Why are eddy currents undesirable in a transformer?
4. What is hysteresis?
5. How are hysteresis losses minimized?
6. How are transformer copper losses minimized?
7. Why were PCBs used in transformers?
8. What took place on October 1, 1990 that affected transformers?
9. What is a third harmonic?
10. Why are dry-type transformers better in some cases than oil-filled ones?
11. How can a transformer's capacity be increased by 33%?
12. How are epoxy transformers used?
13. Which is the primary of a transformer?
14. What is induced voltage?
15. What is mutual inductance?
16. If you know the secondary voltage, primary voltage, and the number of turns in the primary, how do you find the number of turns in the secondary of a transformer?
17. What is an autotransformer?
18. What limits the use of autotransformers?
19. What generates the greatest losses in a transformer?
20. What circulates through tube coolers?

## REVIEW PROBLEMS

Transformers are a very important part of any electrical system. They are utilized to make sure that the correct voltage and current are available where needed. A quick review of the turns ratio and some of the factors related to transformers will be useful.

1. When connected to a 60-Hz, 120-V circuit, a 24-V transformer delivers up to 4 A to a 24-V solenoid. How much current does the transformer demand from the primary circuit when the full 4 A flows in its secondary?
2. How much current flows in the primary of a transformer whose primary is connected to a source of 120 V and whose secondary provides 12 A at 12 V? Assume 98% efficiency for the transformer.
3. If the turns ratio of a step-up transformer is 1:5. what is the voltage of the secondary if the primary is connected to a source of 120 V, 60 Hz?
4. A 16-V transformer delivers 1.0 A to a chime. If the primary voltage is 120 V, what is the current through the primary?
5. How much voltage do you get from a 64-VA transformer if the current rating of the secondary is 4 A?
6. A step-up transformer has 600-V output at 250 mA. The input voltage is 120 V. What is the primary current needed to produce the 250 mA in the secondary?

# 19
CHAPTER

# Power Generation

## PERFORMANCE OBJECTIVES

After studying this chapter, you will be able to:

1. Draw a sine wave.
2. Identify parts of an ac generator.
3. Draw output waveforms for single-phase, two-phase, and three-phase generators.
4. Explain why exciters are needed.
5. Draw wye-to-wye connections.
6. Describe how output frequency is determined for an alternator.
7. Draw a transfer switching circuit.
8. Describe how an automatic transfer switch operates.
9. Explain the difference between the two types of UPS systems.
10. Describe systems for paralleling multiple power sources.
11. Define cogeneration.
12. Explain peak-load shaving.

## BASIC PRINCIPLE

The basic principle for the generation of an EMF in an ac generator or alternator is the same as in a dc generator. The generation of an EMF in an armature conductor depends solely on a relative motion between the conductor and the magnetic field. Two constructions are possible. The magnetic field may be stationary, and the armature may rotate. In this case, the magnetic field is called the stator and the armature is called the rotor. Or the magnetic field may rotate and the armature may be stationary, in which case the magnetic field is called the rotor and the armature is called the stator.

In almost all dc generators, the field is stationary and the armature is rotated. But in almost all ac generators, the armature is stationary and the field is rotated. The latter type of construction provides some advantages. A rotating armature requires slip rings to carry current to the external load. Such rings are difficult to insulate. They are frequent sources of trouble often causing open and short circuits. A stationary armature needs no slip rings. Thus armature leads can be continuously insulated conductors from the armature coils to the bus bars. It is more difficult to insulate conductors in a rotating armature than in a stationary armature because of the centrifugal force that results from rotation. Also, the stationary armature allows alternators to operate with higher voltages than those in dc generators.

Inasmuch as an ac motor will not operate without a source of alternating current, it is important that a closer examination of the device that provides the power for the operation of all ac motors be made. Figure 19-1

**Fig. 19-1** Loop of wire rotating in a magnetic field. This basic alternator produces the wave form shown in Fig. 19-2.

shows a simplified drawing of an ac generator. Certain features of this generator are basic to the design of all ac generators. Note how the loop of wire rotating in a magnetic field creates the sine-wave output.

## Sine Wave

The voltage produced by this generator is an alternating voltage. One complete revolution of the coil produces 1 Hz of voltage. That is, the voltage builds up from zero to a maximum, then drops to zero, then builds up again in the opposite direction to a maximum. Finally, the voltage drops to zero to complete the hertz or cycle. Such a hertz of alternating current or voltage is represented by a sine wave (Fig. 19-2).

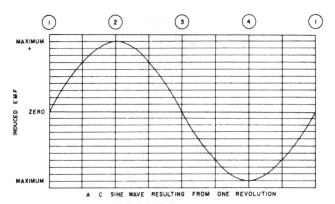

*Fig. 19-2* Sine-wave output of a single phase alternator.

## COMPONENTS OF THE ALTERNATOR

### Rotors

The rotating part of the generator is called the rotor. The field of the ac generator is placed on the rotor. It is either a salient-pole type or a turbo type (Fig. 19-3).

*Fig. 19-3* Salient-pole rotor for an alternator.

*Fig. 19-4* Turbo-type rotor for an alternator.

Compare the salient-pole rotor and Fig. 19-4, which has a turbo-type rotor.

When the ac generator is to be driven by a slow speed diesel engine or by a water turbine (up to 720 rpm), the salient-pole or projecting-pole rotor is used. The field poles are formed by fastening a number of steel laminations to a spoked-frame or spider. The heavy pole pieces produce a flywheel effect on the slow-speed rotor. This helps to keep the angular speed constant. It also reduces variation in the voltage and frequency of the generator output. In high-speed alternators (up to 3600 rpm), the smooth-surface turbo type rotor is used for two major reasons: (1) it has less air-friction (heating) loss, and (2) the windings can be placed so that they can withstand the centrifugal forces developed at high speeds. Turbo-type rotors are a solid-steel forging, a number of steel disks fastened together with the field coils locked in slots. These field coils are usually placed so they distribute the field flux evenly around the rotor, as shown in Fig. 19-4.

### Stators

In a rotating-field ac generator, the armature windings are stationary and are therefore the stator. The armature iron, being in a moving magnetic field, is laminated to reduce eddy current losses. A typical ac generator stator is shown in Fig. 19-5. In high-speed turbo-type generators, the stator laminations are ribbed to provide sufficient ventilation. This is necessary because the high temperature developed in the windings cannot be dissipated in the small air gap between the rotor and the stator. Figure 19-6 illustrates the close tolerance. In some large installations, the alternators are totally enclosed and cooled by hydrogen gas under pressure, which has greater heat-dissipating properties than air. Stator coils in high-speed alternators must be well braced. Bracing prevents coils from being pulled out of place when the alternator is operating with heavy loads.

**Fig. 19-5** *Typical alternator stator.*

**Fig. 19-6** *Note how close together the windings are.*

**Fig. 19-7** *Exciter armature and generator field mounted on the same shaft.*

friction-producing parts, such as brushes, brush holders, commutator, or slip rings. It needs very little maintenance (Fig. 19-8).

**Fig. 19-8** *Brushless rotor.*

## Exciters

Like many dc generators, ac generators need a separate dc source for their fields. This dc field current must be obtained from an external source called an exciter. The exciter used to supply this current is usually a flat, compound-wound dc generator designed to furnish from 125 to 250 V. The exciter armature may be mounted directly on the rotor shaft of the ac generator, or it may be belt driven. Figure 19-7 shows an exciter armature and generator field mounted on the same shaft.

Brushless exciters are also used to provide the dc fields. The brushless exciter is an ac generator that converts the ac power to dc. It does so by means of a diode rectifier assembly which is attached to, but insulated from, the generator shaft. The brushless exciter has no

**Static Exciters** Another method of field excitation commonly used is the static exciter. It is called a static exciter because it contains no moving parts. A portion of the ac current from each phase of the generator output is fed back through a system of transformers, rectifiers, and reactors to the field windings as dc excitation current. With this system an external source of dc current is necessary for initial excitation of the field windings. On engine-driven generators, the initial "field flash" may be obtained from storage batteries, which are also used to start the engine.

**Frame and Shaft** The frame and shaft of an alternator serve the same purpose as in the dc generator. The frame completes the magnetic circuit of the field. It also supports the parts and windings. The shaft, on which the rotor turns is supported by the end bells or end frames.

# TYPES OF ALTERNATORS
## Single-Phase

This type of alternator is seldom used except for special purposes. It is used as an emergency generator and for construction crews. As a rule, this type of alternator is low powered and self-excited. Figure 19-9 shows a typical alternator. Its construction is similar to a dc generator with an auxiliary ac winding on the dc armature. The dc winding on the rotating member is of the usual lap or wave-type construction. The winding is connected to the commutator bars in the usual way.

*Fig. 19-9* *Single-phase, portable generator.*

The dc winding output provides the current for the dc field excitation and other dc power applications. A second open wave winding is laid in the slots of the rotating member on top of the dc winding. This second winding, which is connected to slip rings, supplies the ac output.

## Two-Phase

This type of alternator is called a multiphase or polyphase ac generator. It has two or more single-phase windings symmetrically spaced around the stator. In a two-phase alternator, these are two single-phase windings physically spaced so that the ac voltage induced in one is 90° out of phase with the voltage induced in the other. When one winding is being cut by maximum flux, the other is cut by no flux.

Figure 19-10 shows a schematic diagram of a two-phase, four-pole alternator. This stator consists of two single-phase windings (phases) separated from each other. Each phase consists of four windings. These windings are connected in series so that their voltages add. The rotor is identical to the rotor used in the single-phase alternator. Note in Fig. 19-10B the waveforms produced by this type of generator.

*Fig. 19-10* *(A) Two-phase, four-pole alternator and its waveform; (B) generation of two-phase voltage.*

The two phases of a two-phase alternator can be connected together as shown in Fig. 19-11. Only three leads are brought out from the generator for connection

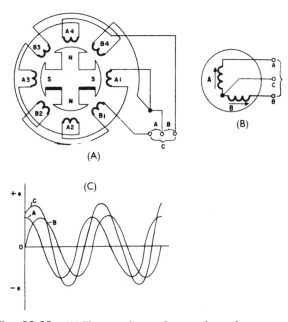

*Fig. 19-11* *(A) The two phases of a two-phase alternator connected together to produce single-phase power; (B) two-phase power brought out with all three connections available for connection to consuming devices; (C) waveforms produced by a two-phase alternator with all three connections brought out separately.*

Types of Alternators 311

to a load. This type of power is seldom used. However, in some parts of Europe this type of power is available for home and commercial use. It has an advantage when starting motors: ac motors that use two-phase power do not need a start winding or a switch to remove the windings from the circuit once the motor has reached operating speed, as is the case with single-phase motors.

## Three-Phase

As the name indicates, this type of alternator has three single-phase windings. These windings are spaced so that the voltage induced in each winding is 120° out of phase with the voltage in the other two windings. A schematic diagram of a three-phase generator showing all three coils is complex. Figure 19-12A shows how some of the various load options can be connected on three-phase power.

Figure 19-12B shows the output waveform or the alternator. Note that there are 120° of separation between each phase of the output.

Electrical power generated by power companies for use in homes and business is all produced as three-phase. The three phases are then divided by three separate transformers into single-phase power for three different subdivisions or three different customers. Some three-phase power is used by businesses to drive large motors. Three-phase motors do not require as much maintenance as single-phase motors.

## WYE AND DELTA CONFIGURATIONS
### Wye Connection

Instead of six leads coming out of the three-phase alternator, one of the leads from each phase, may be connected to form a common junction. The stator is then wye- or star-connected. Figure 19-13 shows a wye connection. The common lead may or may not be brought out of the machine. If it is brought out, it is called the neutral. One advantage of the neutral is the balancing of the load between or among all coils. The neutral serves as a common return circuit from all three phases. It maintains a voltage balance across the loads. No current flows in the neutral when the loads are balanced. The three-phase, four-wire system is widely used in industry and for aircraft ac power systems.

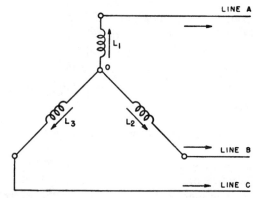

*Fig. 19-13* Current flow in three-phase windings, wye-connected.

*Fig. 19-12* (A) Various load options for three-phase alternator; (B) generation of three-phase voltage and the resultant wave forms.

### Delta Connection

A three-phase stator may also be connected in a delta configuration. In a delta-connected alternator, one phase winding and the start end of another are connected to the finish end of the third. The start end of the third is connected to the finish end of the first. The three junction points are connected to the line wires leading to the load. Figure 19-14 shows a delta connection. When the generator phases are properly

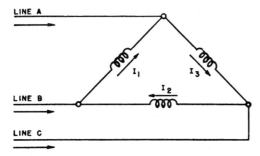

**Fig. 19-14** *Current flow in a delta-connected, three-phase alternator winding.*

connected in delta, no appreciable current flows within the delta loop when there is no external load connected to the alternator. If any one of the phases is reversed with respect to its correct connection, a short circuit current flows within the windings on no-load. This causes damage to the windings.

To avoid connecting a phase in reverse, it is necessary to test the circuit before closing the delta. This is done by connecting a voltmeter or fuse wire between the two ends of the delta loop before closing the delta. The two ends of the delta loop are never connected if there is an indication of any appreciable current or voltage between them when no load is connected to the alternator.

## Power in a Balanced Wye

The power delivered by a balanced three-phase wye-connected system is equal to three times the power delivered by each phase. The total true power is:

$$P_t = \sqrt{3} \times E_{phase} \times I_{phase} \times \cos \angle \theta$$

Since

$$E_{phase} = \frac{E_{line}}{\sqrt{3}} \quad \text{and} \quad I_{phase} = I_{line}$$

The total true power is:

$$P_t = \frac{E_{line}}{\sqrt{3}} \times I_{line} \times \cos \angle \theta$$

## Power in a Balanced Delta

The power delivered by a balanced three-phase delta connected system is also three times the power delivered by each phase (Fig. 19-15):

$$E_{phase} = E_{line} \quad \text{and} \quad I_{phase} = \frac{I_{line}}{\sqrt{3}}$$

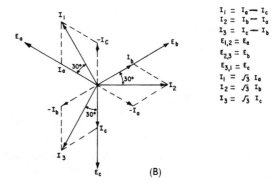

**Fig. 19-15** *(A) Power delivered by a balanced three-phase, delta-connected system; (B) current and voltage relationships in a three-phase alternator.*

The total true power is:

$$P_t = 3E_{line} \times \frac{I_{line}}{\sqrt{3}} \times \cos \angle \theta$$

Thus the expression for three-phase power delivered by a balanced delta-connected system is the same as the expression for three-phase power delivered by a balanced wye-connected system.

## FREQUENCY

The frequency of the alternator voltage depends on the speed of rotation of the rotor and the number of poles. The higher the frequency needed, the faster the alternator must turn. The lower the speed: the lower the frequency. The more poles on the rotor: the higher the frequency for a given speed. When a rotor has rotated through an angle such that two adjacent poles (a north and a south pole) have passed one winding, the voltage induced in that winding has varied through one complete cycle or hertz.

$$f = \frac{P}{2} \times \frac{N}{60} = \frac{PN}{120}$$

$P$ = number of poles,
$N$ = speed in rpm, and
$f$ = frequency.

# LOAD CHANGES

When the load of an alternator is changed, the terminal voltage carries the load. The amount of variation depends on the design of the generator and the power factor of the load. With a load having a lagging power factor (one with inductance dominating), the drop in terminal voltage with increased load is greater than for unity (1.00) power factor (that is, a totally resistive load). With a load, a power factor that is leading the terminal voltage tends to rise. The causes of terminal voltage changes with load changes are the armature resistance, armature reactance, and armature reaction.

## Armature Resistance

When current flows through a generator armature winding, there is an *IR* drop (voltage drop) due to the resistance of the winding. This drop increases with load. Thus the terminal voltage is reduced. The armature resistance drop is small because the resistance is low.

## Armature Reactance

The armature current in an alternator varies approximately as a sine wave. The continuously varying current in the generator armature is accompanied by an $IR_L$ voltage drop in addition to the *IR* drop. Armature reactance in an alternator may be from 30 to 50 times the value of armature resistance. This is because of the relatively large inductance of the coils in comparison with its resistance.

## Armature Reaction

When an alternator supplies no load, the dc field flux is distributed uniformly across the air gap. When an alternator supplies a reactive load, however, the current flowing through the armature conductors produces an armature magnetomotive force (MMF). That force influences the terminal voltage by changing the magnitude of the field flux across the air gap. When the load is inductive, the armature MMF opposes the dc field and weakens it. Thus the terminal voltage decreases. When a leading current flows in the armature, the dc field is aided by the armature MMF. The flux across the air gap is increased. Thus the terminal voltage increases.

# VOLTAGE REGULATION

Voltage regulation of an alternator is the change from full-load to no-load voltage. This is expressed in percentage of full-load volts with a constant speed and dc field current. For example, the no-load voltage of an alternator is 250 V. Its full-load voltage is 220 V. What is the percent of regulation?

$$\frac{250 - 220}{220} = 13.6\%$$

# STANDBY OR EMERGENCY POWER SOURCES

The loss of power can be very dangerous—catastrophic in some cases. When the power goes off it can interrupt ventilation fans, water pumps, milking machines, mechanical feeders, fallout shelters, refrigeration, furnace controls, and other vital modern production equipment that requires continuous electric service.

Storms, accidents, and equipment breakdown can all cause an interruption of electrical service. If a power outage lasts for any length of time, serious problems, such as animal suffocation in windowless animal shelters, food spoilage, frozen water pipes, or loss of production, can result. An ever-increasing dependence on a constant supply of electrical power causes increased interest in standby equipment for the generation of electricity.

Farms, hospitals, schools and businesses, and industry are all interested in a constant supply of electrical energy. With the widespread use of computers in industry, commerce, and education, the need for a dependable source of electricity is even more important.

## Transfer Switch

One of the most important parts of an emergency power source is the transfer switch. The *National Electrical Codes* requires that a standby generator be connected so as to prevent the inadvertent interconnection of two power sources. A double-pole; double-throw switch is usually installed between the power supplier's meter and the service entrance. If current transformers are used for metering, a pole-top transfer switch may be used. Figure 19-16 shows the typical transfer switch wiring.

The use of a double-pole, double-throw switch prevents power from feeding back into the power supplier's line and endangering the lives of linemen who may be working to restore power. It also prevents accidental-energizing of the farm, business, hospital, or other system and consequent burnout of the generator when regular power service is restored. Most standby equipment guarantees are voided if the transfer switch is not used.

The transfer switch shown in Fig. 19-16 is used at a home or farm. It shows the basis: method used to transfer power from a generator to a home or farm system and then back to the power company's line. Figure 19-17 shows the automatic transfer and bypass isolation switch.

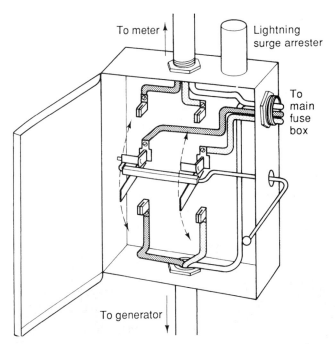

**Fig. 19-16** *Wiring of a manually operated transfer switch.*

**Fig. 19-18** *Portable engine generator with manual start.*

mounted unit (Fig. 19-19). Engine driven units can also be stationary or portable and can be either manual start or automatic start.

Standby generators are available to operate either as single- or three-phase. Some units are wired with four lead wires so that they can be operated either single- or three-phase. Generators must be matched to the power, voltage, and frequency used from the power company's lines. This is usually 120/240-V, single-phase, 60-Hz ac.

**Fig. 19-17** *Automatic transfer and bypass isolation switch.* (Automatic Switch)

## Types of Standby Generators

Standby generators can be divided into two types: engine driven and tractor driven (Fig. 19-18). Tractor driven units can be stationary or portable, as a trailer

**Fig. 19-19** *Tractor driven generator.*

**Standby or Emergency Power Sources** 315

**Fig. 19-20** *Typical indoor installation for a large diesel-engine system.*

There are no standard ratings for standby generators. That means overload or maximum capacity, limited to short intermittent periods, must be considered in selecting the size of generator. Manufacturers' ratings can vary from zero overload capacity to 100%. A large overload capacity permits a smaller unit, particularly where large electric motors are involved. Figure 19-20 shows a generator design for indoor installation. Larger units usually rely on a diesel engine for power.

## Automatic Transfer Switches

An automatic transfer switch can monitor the normal power source (Fig. 19-17). If there is a power outage, the switch signals the engine generator to start. When the generator reaches proper voltage and frequency, it transfers selected loads to it. When the normal source is then restored, the switch retransfers the load to it. It then shuts down the engine after a cool-down period.

At times a transfer switch handles more than its normal or continuous current rating. A reliable transfer switch must be able to handle all situations with no harmful effects on the switch. Motor starting current is one of these situations. When a motor starts up, it can draw as much as six times its running current. It can draw that much if it stalls while running. If the transfer switch operates at either of these times, it will have to interrupt that six times current. Thus the switch must be able to handle it; if not, the switch could be permanently damaged (Fig. 19-21).

All current goes through the transfer switch. Thus a short circuit on the load side will cause the maximum available current to go through the transfer switch until

**Fig. 19-21** *Transfer switch, 150 A.* (Automatic Switch)

**316** Power Generation

the circuit breaker opens the line. Therefore, the transfer switch must also withstand short-circuit loads.

Magnetic forces are so great during a short circuit that they can cause the contacts of an inadequately designed switch to open. Thus the switch must be able to lock its contacts closed until the circuit breaker operates.

Tungsten lamps can draw up to 16 times more current when they are cold. This means that a transfer switch feeding power to tungsten lamps must be able to handle this above-normal inrush current.

Motors can draw up to 15 times as much inrush current on starting as when they are running and is reconnected to a new normal or emergency source that is not synchronized with the motor. This can happen if the motor is as much as $180^0$ out of phase with the new source (Fig. 19-22).

**Fig. 19-22** *The automatic transfer switch maintains power to selected loads from either the normal or emergency power source.* (Automatic Switch)

## Electronic Loads

Almost every power system has some electronic load. Increased automation and the economies of solid state control suggest that even larger percentages of power system loads will be electronic in the future. Typical examples of electronic loads include data processing equipment, intensive-care-unit monitoring equipment, numerical control machinery, and security equipment. There is virtually no commercial or industrial facility in which electronic loads are not becoming a critical part of the overall load profile.

## Uninterruptible Power Systems

Electronic loads are very susceptible to voltage and frequency variations. As a typical example, real-time access computers require an ac power system that does not deviate from nominal voltage by more than $+8\%$ or less than $-10\%$. Allowable frequency deviations are typically $\pm 0.5$ Hz. In addition to the deviation stated, some data processing loads are affected by rate change of deviation as well. To protect computer systems, uninterruptible power systems (UPS) have been developed.

Essentially, there are two types of UPS systems: the motor-generator flywheel set and the static inverter with battery backup. Due to advances in solid state equipment, the more commonly used UPS is now the static inverter with battery backup. This is due primarily to the lower cost per kVA and the elimination of a high-starting kVA requirement.

A solid-state UPS is made up of three major sections. The first section is the rectifier that converts the ac input into dc. The second section is a dc bus on which floats a battery system. The third section is the static inverter that converts the dc back to a clean ac sine wave. The principal function of the UPS is to isolate the protected electronic load from power deviations that would affect the operation of the connected electronic load. Any short-term outage or transient that occurs on the ac power bus is filtered out or overridden by the battery in the dc bus section providing constant power into the inverter.

Sometimes the battery supply of the UPS is required to carry the output of the inverter for extended periods of time (in excess of 3 to 5 minutes). Upon reconnection of the ac power source to the rectifier, the UPS draws power not only to carry the output load, but also to recharge the batteries.

## Multiple-Engine Generator Sets

Where there are two critical loads with one more critical than the other, the three source priority load system can be used. This system supplies the priority load first, then the non-priority load (Fig. 19-23).

**Fig. 19-23** *Three-source priority load system.* (Automatic Switch)

The priority load system operates the same as the three-source system, with the following exceptions. The first engine generator set to reach acceptable output is connected to the emergency terminals of the automatic transfer switch 1 (ATS 1) by ATS 2, which is a six-pole, double-throw switch. The double-throw action of ATS 2 simultaneously connects the emergency terminals of ATS 3 to the other engine generator set. When that set reaches acceptable output, ATS 3 transfers the non-priority load to it. If the set carrying the priority load malfunction, ATS 2 transfers the other set to the priority load. Other features built into the system provide optimum protection for the non-priority load.

## Systems for Paralleling Multiple Power Sources

When it is necessary to start and parallel two or more engine generator sets to supply emergency power on a common bus, a system is needed that includes reverse power monitors, synchronizers, load sequencers, and other components to handle parallel-related operations.

A two-engine system is shown in Fig. 19-24. If the normal source fails or drops below acceptable levels, the controls start both engine generator sets. The first to reach adequate output is put onto the emergency bus. Assuming that ATS 1 is feeding the more critical load, it will transfer its load to the emergency bus as soon as the first engine is connected to the bus. When the other set reaches adequate output, the controls bring into synchronism with the first set, parallel it onto the same bus, and ATS 2 transfers the secondary load. Furthermore, the controls will operate to maintain power to the more critical load as long as power is available from any source.

Figure 19-25 shows a multiple-engine system. Multiple-engine generator sets are used for various reasons, such as economy, reliability, or to minimize downtime. A multiple-engine system operates the same way as a two-engine system with additional controls to handle required number of engine generator sets. Figure 19-26 shows a power control system.

**Fig. 19-25** *Multiple-engine Synchro-power system.* (Automatic Switch)

**Fig. 19-24** *Two-engine Synchro-power system.* (Automatic Switch)

## Selective Load Transfer Systems

The selective load transfer system is dependable and economical. It is also a limited supply of emergency power to be channeled to selected loads one at a time. Some of the applications for this system are: elevators, production processes, multiple-pump systems in sewage treatment plants, boiler feed-water pumps, HVAC chillers, chilled-water circulating pumps, equipment bays, and work-stations.

**Fig. 19-26** *Power control system.* (Automatic Switch)

The system is most frequently applied to elevator systems in both new construction and retrofit. In these cases, one elevator at a time can be operated when normal power fails. It uses the minimum amount of auxiliary power for operation because the standby generator can be sized for the necessary emergency load plus only one elevator. Once the first elevator reaches the main or selected floor, another elevator can be connected to the emergency source. This sequence continues until all elevators have safely been brought down—one at a time. Once the normal power is restored, all elevators automatically resume normal operation. Figure 19-27 shows a selective load emergency power transfer system.

*Fig. 19-28* Two-engine prime mover system. (Automatic Switch)

automatically synchronized, paralleled, and run together for a warm-up and stabilization period. Then the other set is disconnected, cooled down, and turned off. There is no interruption in power continuity during changeover.

If a set malfunctions while in service, the other is started and put on line automatically. There will be continuity of power for as long as it takes to start and connect the idle engine—usually less than 10 seconds.

## Multiple-Source Systems

Figure 19-29 shows a multiple-engine generator prime power system. More than two engine generator sets may be required, for reasons mentioned previously. However, rarely do they require continuous operation of all sets on line at the same time. Load requirements vary throughout the day and with the season. Furthermore, fuel is saved and engine wear is reduced by operating only enough sets to carry the load and have some on-line reserve capacity.

*Fig. 19-27* Selective-load emergency power transfer system. (Automatic Switch)

## PRIME MOVER SYSTEMS

To meet increased demands for clean and continuous reliable electric power, many installations are being supplied with on-site power generation. In some a cogeneration concept is employed where the heat from the power sources is used for heating and air conditioning.

Prime mover systems can consist of any number of engine generator sets. The number depends on the size and number of loads, and other factors, depending on individual applications. Basically, however, prime power applications fall into two categories: two-engine systems and multiple-engine systems.

## Two-Source Systems

Figure 19-28 shows a two-engine prime mover system. The prime mover system is similar to the two-engine generator emergency system except that typically in the prime power system either engine generator set can supply the total load. As a rule the sets are alternated on a weekly basis. That allows the idle set to be serviced easily. Changeover from one set to another can be initiated manually or automatically. The idle set is started,

*Fig. 19-29* Multiple-engine generator set prime mover system. (Automatic Switch)

Systems are available to measure the kilowatt load being drawn from the bus. They operate only the minimum number of sets necessary to carry the load. If the load increases, the controls start, synchronize, and parallel additional sets to meet it. If the load decreases, the controls remove sets from the line, run them for a cool-down period then turn them off.

When a running set malfunctions, the controls disconnect it, shed an appropriate amount of load so that the remaining sets are not overloaded, and start the next idle set in turn. When its output is adequate and synchronized, the set is paralleled onto the bus and the shed load is reconnected. As a rule the total from malfunctioning to reconnecting the shed load will not exceed 10 seconds.

## POWER MANAGEMENT SYSTEMS

A cogeneration system is defined as any system where a single source of thermal energy (fuel) drives two processes, the second process being driven by waste heat from the first process. Most power management systems for cogeneration usually involve the recovery and use of rejected heat (second process) produced in the generation of electricity (first process).

The electrical generation portion of the cogeneration system can operate in either of two ways: isolated from the utility or in parallel with the utility. When operating in parallel with the utility, the design must incorporate protective relaying, as required in guidelines issued by the local utility, in addition to the standard controls. Optimum cost-effectiveness, energy conversion efficiency, and system performance are achieved when the power management system operates in parallel with the utility system. These systems incorporate control functions that cause the on-line engines to produce only that amount of electricity which would provide recoverable heat to satisfy the heat load demand. Thus fuel consumption is minimized and efficiency is maximized.

Figure 19-30 shows a typical cogeneration system. Note that this diagram illustrates a cogeneration system operating in parallel with the utility. The recoverable heat may be applied to absorption chilling, as is usually the case in commercial buildings. In industrial buildings the recovered heat is more likely used for process heat.

As indicated, co-generating systems require control functions in addition to those controls supplied as standard in a prime power system. To ensure that these system controls meet all the specialized requirements of cogeneration, including operating in parallel with the utility, contact the manufacturer of such systems during the preliminary design stages for suggestions in selecting appropriate controls and to assure proper coordination.

### Peak Load-Sharing Systems

Peak sharing systems (peaking systems) allow designated building loads to be powered by on-site power generation whenever the building's total load demand exceeds a predetermined level. While an on-site generation system usually supplies emergency or standby power as its major function, such a system can be modified with additional controls to provide peaking service. These systems can be grouped into two categories: (1) isolated from the utility service, or (2) paralleled with the utility service.

### Utility-Isolated Peaking Transfer System

When on-site generation is isolated from the utility service, it is necessary to include transfer switches for the peaking loads (shown in PL in Fig. 19-31) when they are not the emergency loads. The transfer switches allow the designated peaking loads to draw power from either the utility or the engine generators as a function of total building demand. Only after the present level of demand is reached are the designated loads transferred to the generator sets. Thus the higher demand changes are avoided. When building demand falls below the retransfer level, the peak load is retransferred back to the utility service and the engine generators are signaled to shut down.

*Fig. 19-30* A typical cogeneration system. (Automatic Switch)

*Fig. 19-31* A typical dedicated load-peaking transfer system. (Automatic Switch)

When two or more generator sets provide power in a power management system, generators are brought on-line only as required. When the demand limit from the utility is exceeded, the first generator is started and one of the peaking loads is transferred to it. If the limit is again exceeded, the second generator is started, synchronized, and paralleled with the first, and the second peak load is transferred. This procedure can be repeated until all on-site generation is on-line, provided that sufficient peak loads are fed by transfer switches.

Priority interrupt logic is necessary in peak-shaving transfer systems because peak loads are sometimes not the same loads as emergency loads, as indicated in Fig. 19-31. This interrupt logic automatically suspends peak sharing upon sensing loads of adequate power to emergency loads. The logic initiates the retransfer or disconnect of the peak shaving loads from the standby source to enable the immediate transfer of the emergency loads to the standby source. The emergency loads are supplied with power without the usually startup delay of the generator sets since these sets are already up and on-line.

## Utility-Paralleled Peaking Systems

An optional form of peak sharing is shown in Fig. 19-32. The on-site generation is paralleled with the utility. In this case it is not necessary to designate particular loads as peak loads and add peak-shaving transfer switches since the utility and the standby generators share the building load.

**Fig. 19-32** *A typical utility paralleled peak.*

LEGEND
25 — Automatic Synchronizer
27 — Undervoltage Relay
50/51 — Overcurrent Relay
52 — Circuit Breaker
67 — Directional Overcurrent Relay
81 — Under Frequency Relay

As in the peaking transfer system, interrupt logic automatically suspends peak shaving upon loss of power to the emergency loads. The operation of the generator controls, however, is very different. The governors and voltage regulators are controlled to cause the engine generators to assume only that portion of the load in excess of the demand limit, not a fixed block of load as with peak transfer systems. The engine generators continue to assume all load in excess of the limit point and up to the capacity of the engine generator sets. Any further excess after all engine generators are carrying their rated load is assumed by the utility. In this manner, the on-site-generated kilowatt-hour value is minimized while kilowatt peak demand limiting is effectively accomplished. This results in the most economical mode of peaking operation.

## REVIEW QUESTIONS

1. How many revolutions of the alternator does it take to produce 1 Hz?
2. What type of rotor is used for a slow-speed alternator?
3. What is the purpose of brushless exciters?
4. What is the angular separation between phases in a three-phase power source?
5. What is the advantage of the neutral common in a three-phase power source?
6. What determines the frequency of the alternator voltage?
7. What effect on output voltage does armature resistance have?
8. How is percentage voltage regulation of an alternator found?
9. What does a standby generator do? When?
10. What is a transfer switch?
11. What is a static inverter?
12. What does UPS stand for?
13. What is a cogeneration system?
14. Why is a peak-load shaving system used?
15. What is an interrupt logic?
16. Why is a selective-load transfer system used?
17. What is a prime mover?
18. What determines how many generator sets are needed?
19. What are the two categories of prime power?
20. How can changeover from one set to another be initiated?

## REVIEW PROBLEMS

Power factor plays an important role wherever electric motors are connected to an ac line. The windings have inductance and resistance. A quick review of the way in which power factor is found mathematically will aid in the understanding of why this unit keeps showing up in the power calculations for motors. When reactance and resistance both exist in an ac circuit, the phase difference between the current and the voltage is no longer 90° as it is with the reactance alone. It is some quantity less than 90°. The phase angle (denoted by Greek letter "theta," $\Theta$), which is the angular difference between the voltage and the current, can quickly be converted to a power factor.

1. What is the phase angle of a circuit when the resistance is 100 $\Omega$ and the impedance is 100 $\Omega$? What does this condition represent?
2. What is the phase angle of a circuit if the resistance is 100 $\Omega$ and the impedance is 150 $\Omega$?
3. If a resistor of 500 $\Omega$ is placed in series with an inductor of 1 H on a 120-V line with 60 Hz, what is the phase angle?
4. What is the phase angle if the circuit is a series combination of resistance (1000 $\Omega$) and inductance with an inductive reactance of 628 $\Omega$ in a circuit where 1.660 H and a frequency of 60 Hz are present?

# 20
CHAPTER

# Power Distribution Systems

## PERFORMANCE OBJECTIVES

After studying this chapter, you will be able to:

1. Identify components of a power generation/distribution system.
2. List present-day commercial generating systems.
3. Describe the following radial power distribution systems: looped-primary, banked-secondary, primary-selective, secondary-selective.
4. Explain how a simple network distribution system operates.
5. Explain how fault currents are calculated.
6. Draw wattmeter schematic hookups for three-phase power circuits.
7. Explain the role of the transformer in power distribution systems.
8. Draw schematic drawings of transformer combinations used for power distribution.
9. Explain how switching transfers loads.
10. Describe a cable tray system of power distribution.

## SOURCES OF COMMERCIAL AC

The use of magnetism is the most common method of generating electricity in large quantities for businesses, homes, industries, hospitals, and other institutions. A simple ac generator consists of a single loop of wire wrapped around an iron core. As the loop rotates and cuts lines of force, it has an EMF induced in it according to the strength of the magnet and the polarity of the lines being cut. This has been discussed in chapter 19. What we are interested in here is the way the ac is generated by large devices called alternators. The alternators may be mounted on poles and operated by wind. They may also be used in automobiles and driven by the engine powering the auto. In some instances the power is generated in ac form and converted to dc to charge batteries and then converted to ac by electronics to produce the correct frequency for operating commercial and residential devices or excess energy may be placed on a grid that serves a larger area than the local source where generated.

### Falling-Water Generators

The principle of electricity generation is the same whether it uses mechanical energy produced by gasoline or diesel engines, or falling water or nuclear energy. The Niagara Power Project is an example of falling water used to generate huge amounts of electrical energy. It is one of the world's largest facilities of this type. Total installed power at Niagara's two generating plants is 2190 MW. There are 340 miles of transmission lines to interconnect this project with another at the dam on the St. Lawrence.

The generating section of the Robert Moses plant in Niagara Falls consists of 13 unit blocks. Each of these unit blocks consists of a Francis-type hydraulic (water-driven) turbine rated at 200,000 hp at 300 ft net head. The 150,000-kW generators operate at 120 rpm, three-phase, 60 Hz. They are mounted below the deck under removable hatch covers, as can be seen in Fig. 20-1. The crane travels along the top of them and can be positioned over the generator pits for handling the rotating parts of the turbine and generator assemblies. Transformers are located on the deck behind the generators and under what appears to be black rectangles with puffed-out pointed sides.

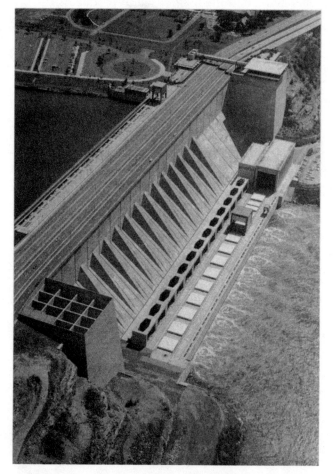

**Fig. 20-1** *Robert Moses plant, Niagara Falls, New York.*

## Switchyard

The Niagara switchyard is situated on a 35-acre site south of the power canal, halfway between the Robert Moses and Lewiston power plants. Purpose of the switchyard is to collect and meter power from the generators and send it out over transmission lines. The switchyard has three voltage sections: 115,000, 230,000, and 345,000 V. Power enters the switchyard from the Robert Moses plant through seven 115,000-V cable circuits and six 230,000-V cable circuits. The cables are installed in underground power tunnels. These cable circuits, 61,620 ft in length, are filled with a total of 138,000 gal of a special insulating oil (Fig. 20-2). Electrical power leaves the switchyard at 115,000, 230,000, and 345,000 V for use in many locations in New York State and as far north as Montpelier, Vermont. The power is generated at 13,800 V and stepped up by the generators located nearby to 138,000 and 230,000 V before it is sent to the switchyard.

## Nuclear Power Generators

Nuclear plants are similar to the fossil-fuel burning plants. The chief difference is in the way the heat is generated, controlled, and used to produce steam to run turbine generators. In a nuclear power plant, the furnace for burning coal, oil, or gas is replaced by a reactor that contains a core of nuclear fuel. Energy is produced in the reactor by a process called nuclear fission. This fission process splits the center, or nucleus, of certain atoms when they are struck by a subatomic particle called neutron. The resulting fragments, or fission products, then fly apart at great speed, generating heat as they collide with surrounding matter.

Figure 20-3A shows the gas-cooled, fast-breeder reactor used in some of today's older nuclear plants. Figure 20-3B shows the liquid-metal fast breeder reactor used in some other types of generating plants.

## Fossil-Fuel Power Generators

Steam needed for driving turbines, which in turn drive electric generators, must be produced by heat. The method of heat production often becomes a rather difficult engineering problem. With the development of some dependable sources of heat from fossil fuel, former design problems have been simplified. Almost any substance may be used as a fuel. If it can be pulverized and fed into a furnace with extremely high temperatures it will burn (Fig. 20-4).

**Fig. 20-2**  *Switchyard.*

**Fig. 20-3** *(A) Gas-cooled fast breeder reactor. (B) Liquid-metal fast breeder reactor.*

**Fig. 20-4** *Conventional fossil-fuel power plant.*

Figure 20-5 shows a steam generator (side elevation) with tangential firing system and natural circulation. Of great concern today are power plants that use inexpensive fuels that pollute the atmosphere. Sulfur contained in coal combines with the moisture in clouds to produce acid rain, which does a great deal of damage to both trees and streams.

**Fig. 20-5** *Steam generator, side elevation. Tangential firing system.*

Electricity may be generated by fossil fuel-powered generators, by atomic-powered generators, or by the use of falling water. It is also possible to generate power by engine-driven generators. In Alaska, for example, where long distribution lines are impractical due to ice and wind conditions that result in line damage, engine-driven generators are common.

## DISTRIBUTION

Once electricity is generated in sufficient amounts for consumption in large quantities, the second necessary step is to get the energy to the consumer. Herein lies a distribution problem: that of stringing and maintaining long lines (Fig. 20-6).

The best distribution system is one that will most economically and safely supply adequate electric service to both present and future probable loads. The best system for a given building or industrial plant is one that takes into consideration the needs and requirements of the electrical equipment. Various types of systems have been designed to furnish the proper power when-needed to different types of loads.

## Types of Distribution Systems

The simplest type of system is the radial. However, there are other modifications that have specific advantages when special problems or situations are present. In the great majority of cases, power is supplied to a building at the utilization voltage. In practically all of these cases, the distribution of power within the building is achieved through the use of a simple radial system. In some cases where service is available at the building at a voltage higher than the utilization voltage, there are a number of systems that may be used.

**Simple Radial System** The conventional simple radial system receives power at the utility supply voltage at a single substation and steps the voltage down to the utilization level. In some cases the utility supply is at the utilization voltage, and the substation then becomes a main distribution switchgear or switchboard.

*Fig. 20-6* Power generation system. Power plant to consumer.

Low-voltage feeder circuits run from the substation bus to switchgear or switchboard assemblies and panel boards that are located with respect to the loads as shown in Fig. 20-7. Each feeder is connected to the substation bus through a circuit breaker or other over current protective device. Relatively small circuits are used to distribute power to the loads from the switchgear or switchboard assemblies and panel boards.

**Fig. 20-7**  *Conventional simple radial system.* (Westinghouse)

The modern simple radial system is an improvement over the simple radial system. It distributes power at a primary voltage. The voltage is stepped down to utilization level in the several load areas within the building through power center transformers. The transformers are usually connected to their associated load bus through a circuit breaker (Fig. 20-8). Each power center is a factory-assembled unit substation consisting of a three-phase liquid-filled or air-cooled transformer, an integrally mounted primary switch, and low-voltage circuit breakers. Circuits are run to the load from these circuit breakers.

**Fig. 20-8**  *Modern simple-radial system.* (Westinghouse)

An improved version of the simple radial system is the modified modern simple radial system (Fig. 20-9). In this arrangement one primary feeder is used per transformer. The system is comparable in service continuity to the single substation form of the simple radial system. This system arrangement is usually very expensive. The cost can be materially reduced, however, by replacing automatic feeder circuit breakers with load-break switches, fused or unfused, and backing up a number of these switches with one automatic circuit breaker.

**Fig. 20-9**  *Modified modern simple-radial system.* (Westinghouse)

In this system a primary feeder fault interrupts service to all loads. Service can be restored to all loads except those associated with the fault system element by opening the load-break switch on the faulted circuit and closing the circuit breaker. Service can be restored to the remaining load when the necessary repairs are made.

**Loop Primary Radial System**  This system is similar to the modern form of a simple radial system. It provides for quick restoration of service when a primary feeder or transformer fault occurs, as does the modified modern simple radial system shown in Fig. 20-9, but at a lower cost. A sectionalized primary loop controlled by a single primary feeder breaker, rather than a radial primary feeder, is shown in Fig. 20-10.

**Banked-Secondary Radial System**  This system permits quick restoration of service to all loads following a primary feeder or transformer fault. It uses a secondary loop to provide an emergency supply when a fault occurs in a transformer or a section of the primary loop (Fig. 20-11).

The secondary loop gives a number of important advantages other than providing an emergency supply to

Distribution  **329**

**Fig. 20-10** (A) Loop primary radial system. (B) Load-break loop switches. (Westinghouse)

starting of large motors is involved, the use of this system may result in a savings in the cost of motor starting equipment. It is also the most satisfactory type of radial system for combined light and power secondary circuits.

**Primary-Selective Radial System** This type of system uses at least two primary feeder circuits in each load area (Fig. 20-12). It is designed so that when one primary circuit is out of service, the remaining feeder or feeders have sufficient capacity to carry the total load. While three or more primary feeders may be used, usually only two feeders are employed. Half of the transformers are normally connected to each of two feeders. When a fault occurs on one of the primary feeders, only half of the load in the building is dropped. In the systems discussed previously a primary feeder fault causes an outage to all loads in the building. This system is not as good as the banked-secondary radial system from the standpoint of restoring service to the loads after a primary feeder or transformer fault. It is only a little better than the loop-primary radial system in this respect. In some cases, particularly in fairly large buildings, with medium- or light-load density, about the same quality of service can be rendered at less cost by using the loop primary radial system employing two separate loops instead of one. In this case, half of the transformers are connected to each loop of the system.

**Fig. 20-11** Banked-secondary radial system. (Westinghouse)

restore service. It helps equalize the loads on all transformers and thus makes it unnecessary to match the transformer capacity at each load center to the load connected to each load bus. This system is better adapted for across-the-line starting of relatively large motors than is any other type of radial system because the starting current is supplied through several transformers in parallel rather than through a single transformer. Thus if the

**Fig. 20-12** Primary-selective radial system. (Westinghouse)

330　Power Distribution Systems

**Secondary-Selective Radial System** This system uses the same principle of duplication feed from the power supply point as the primary-selective radial system. In this system, however, the duplication is carried all the way to each load bus on the secondary side of the transformers instead of just to the primary terminals of the transformers. This arrangement permits quick restoration of service to all loads when a primary feeder or transformer fault occurs, as does the banked-secondary radial system.

The usual form of secondary-selective radial system is shown in Fig. 20-13. Each load area in the building is supplied over two primary feeders and through two transformers.

instead of all the loads in the area. Both systems permit quick restoration of service to all loads when a primary feeder or transformer fault occurs.

**Modified Secondary-Selective Radial System** A modified form of secondary-selective radial system will often be less costly than the common form described previously (Fig. 20-14). In this system there is only one transformer at each load center, instead of two. Pairs of adjacent load buses connected by secondary cables or bus permit picking up the load of any bus when a primary feeder or transformer fault occurs.

*Fig. 20-13* Secondary-selective radial system. *(Westinghouse)*

*Fig. 20-14* Modified secondary-selective radial system. *(Westinghouse)*

A bus-tie breaker is provided for connecting the two secondary or load bus sections at each load center. A primary feeder fault causes half of the load in the building to be dropped.

A fault in a transformer causes the associated primary feeder breaker to trip and interrupts service to half the building load. Service can be restored by opening the disconnecting switch on the primary and the circuit breaker on the secondary of the fault transformer, closing the associated bus-tie breaker, and re-closing the primary feeder breaker. This manual switching restores the system to normal operating conditions except that the faulted transformer is de-energized and its load is being supplied through the adjacent transformer.

From the standpoint of voltage fluctuation when a load such as a relatively large motor is thrown on the system, this system is not as good as the banked-secondary radial system. The advantage that the secondary selective radial system offers over the banked-secondary radial system is that a primary feeder or transformer fault causes only one-half the load to be dropped

Each tie circuit is connected to each of its two load buses through a secondary-tie circuit breaker. Each secondary-tie breaker is interlocked so that it cannot be closed unless one of the two transformer breakers is open.

Basically, the single-transformer-substation form of secondary-selective radial system, regardless of the number and size of the transformers used, functions as described in connection with the more usual form of this system.

**Simple Network System** The ac secondary network system is the system that has been used for many years to distribute electrical power in the high-density downtown areas of cities. Modifications of this type of system make it applicable to serve loads within buildings (Fig. 20-15).

The best known advantage of the secondary network system is continuity of service. No single fault anywhere on the system will interrupt service to more than a small part of the system load. Most faults will

Distribution  331

**Fig. 20-15** *Simple network system.* (Westinghouse)

be cleared automatically without interrupting service to any load. Another outstanding advantage that the network system offers is its flexibility to meet changing and growing load conditions at minimum cost and with minimum interruption of service to the other loads. In addition to flexibility and service reliability, the secondary network system provides exceptionally uniform and good voltage regulation, and its high efficiency materially reduces the cost of losses.

The chief purpose of the secondary loop is to provide an alternative supply to any load bus when the primary feeder that normally supplies it is de-energized. This prevents any service interruption when a transformer or feeder fault occurs. The secondary loop makes it unnecessary to match the load supplied from any load bus with the transformer capacity at that point. Loads can be served from the loop at load buses located between network transformers and the load circuits can be run directly to the loads from the nearest load bus. This permits the use of very short radial circuits and results in considerable saving in secondary cable and conduit compared with the load circuits of a simple radial system. The voltage regulation on a network system is such that both lights and power can be fed from the same load bus. Much larger motors can be started across the line than on a simple radial system. This often results in greatly simplified motor control and permits the use of relatively large low-voltage motors with their less expensive control.

To obtain satisfactory selectivity between limiters under all fault conditions, a minimum of two similar single-conductor cables per phase should be used in the secondary loop. When a fault occurs on a primary feeder or in a transformer, the fault is isolated from the system through automatic tripping of the primary feeder circuit breaker and all of the network protectors associated with that feeder circuit. This operation does not interrupt service to any load. When the necessary repairs have been made, the system is restored to normal operating condition by closing the feeder circuit breaker. All network protectors associated with that feeder will close automatically.

The simple spot-network system resembles the secondary-selective radial system in that each load area is supplied over two or more primary feeders through two or more transformers. In this system, however, the transformers are connected through network protectors to a single load bus, as shown in Fig. 20-16.

**Fig. 20-16** *Simple spot-network system.* (Westinghouse)

Simple spot-network systems are more economical than other forms of network systems for buildings where there are heavy concentrations of load covering small areas, with considerable distances between these concentrations and very little load in areas between them. They are commonly used in high-rise office buildings. The chief disadvantage of these systems is that, they are not nearly as flexible for growing and shifting loads as are the other forms of network systems. Simple spot-network systems are used where a high degree of service continuity is required, flexibility is of minor importance, and their cost is less than the network systems using a secondary loop. The simple spot-network system is most economical where three or more primary feeders are required. This is because supplying each load bus through three or more transformers greatly reduces the spare cable and transformer capacity that is required.

**Primary-Selective Network** This is the most generally applicable and widely used form of industrial secondary network system. Each transformer in the primary-selective network system is equipped with a primary selector switch arrangement (Figs. 20-17 and 20-18).

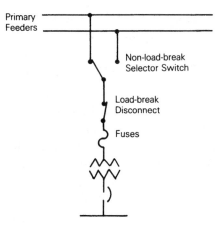

*Fig. 20-17* Non-load break selector switch location. *(Westinghouse)*

*Fig. 20-18* Load-break switch location. *(Westinghouse)*

Two primary feeders are run to each transformer as shown in Fig. 20-19. In a building requiring two primary feeders, each feeder must be capable of supplying the entire load in the building. Each transformer is connected to a load bus through a network protector. The radial circuits serving the loads are connected to the load bus through circuit breakers or fused switches. A secondary loop, such as is used in the banked-secondary radial and the simple network systems, connects each load bus to the two adjacent load buses. One half of the network transformers are normally connected to each primary feeder.

A primary feeder fault causes the faulty feeder and its transformers to be disconnected from the system by automatic operation of the primary feeder breaker and associated network protectors. The entire building load is then carried over the remaining primary feeder and one-half of the network transformers. The transformers associated with the faulty feeder can be connected to the good feeder by manually operating their selective selector switches. This relieves the overload on the transformers normally associated with the good feeder.

A transformer fault is isolated from the system by operation of the primary feeder breaker and the network protectors associated with the primary feeder. The defective transformer can be disconnected from the primary feeder by opening its selector switch. The feeder and its good transformer can be returned to service immediately. A primary feeder or transformer fault will not cause any interruption of service to the load.

As with the simple network system, the primary selective network may also take the form of a spot-network system. Each transformer in the primary-selective network system is equipped with a primary selector switch and arranged for connection to either of two primary feeders (Fig. 20-20). This largely eliminates the necessity for spare transformer capacity and the system is ordinarily designed without providing for such capacity.

*Fig. 20-20* Primary-selective spot-network system. *(Westinghouse)*

## SYSTEMS ANALYSIS

A major consideration in the design of a distribution system is to ensure that it provides the required quality of service to the various loads. This includes serving each load under normal conditions and under abnormal conditions providing the desired protection to service the system apparatus so that the interruptions of service are minimized consistent with good economic and mechanical design. Short-circuit currents and fault currents are to be considered in any system.

*Fig. 20-19* Primary-selective network system. *(Westinghouse)*

## Short-Circuit Currents

The amount of current available in a short-circuit fault is determined by the capacity of the system voltage sources and impedances of the system, including the fault. Constituting voltage sources are the power supply (utility or on-site generation) plus all rotating machines connected to the system at the time of the fault. A fault may be either an arcing or bolted fault. In all arcing faults, part of the circuit voltage is consumed across the fault and the total fault current is somewhat smaller than for a bolted fault, so the latter condition is the value sought in the fault calculations.

Basically, the short-circuit current is determined by Ohm's law except that the impedance is not constant since some reactance is included in the system. The effect of reactance in an ac system is to cause the initial current to be high and then decay toward a steady-state or Ohm's law value. The fault current consists of an exponentially decreasing direct-current component superimposed on a decaying alternating current. The rate of decay of both the dc and ac components depends on the ratio of reactance to resistance, X/R, of the circuit. The greater this ratio, the longer the current remains higher than the steady state that it would eventually reach.

## Fault Currents

The calculation of asymmetrical currents is a laborious procedure since the degree of asymmetry is not the same on all three phases. It is common practice to calculate the rms symmetrical fault current, with the assumption being made that the dc component has decayed to zero, and then apply a multiplying factor to obtain the first half-cycle rms asymmetrical current, which is called the momentary current.

To determine motor contribution to the first half-cycle fault current when the system motor load is known, the following assumptions generally are made:

1. 208Y/120-V systems:
   a. Assume 50% lighting and 50% motor load.
   b. Assume motor feedback contribution of twice full-load current of transformer.
2. 240/480/600-V three-phase, three wire systems:
   a. Assume 100% motor load.
   b. Assume motor feedback contribution of four times full-load current of transformer.
3. 480Y/277-V systems in commercial buildings:
   a. Assume 50% motor load.
   b. Assume 100% induction motors.
   c. Assume motor feedback contribution of two times full-load current of transformer source.
4. For industrial plants, make same assumptions as for three-phase, three-wire systems.

## Three-Phase Power

Three-phase power can be generated by a wye-connected generator or a delta-connected generator. Instead of having six leads come out of the three-phase ac generator, one of the leads from each phase can be connected to form a common junction. The stator is then called a wye. Sometimes the wye connection is also called a star connection. Tire common lead may not be brought out of the generator. If it is brought out, it is called the neutral (Fig. 20-21 ). If the wye connection has the neutral brought out of the generator, it is called a four-wire, three-phase system.

***Fig. 20-21*** *Single-phase voltage from a wye connection.*

The wye connection provides 1.73 times the phase voltage for any two of the three wires connected. The line currents are equal to the current in any phase winding. The advantage of a wye connection is its ability to produce more voltage. Note in Fig. 20-21 that windings 1 and 2 are in series with each other. Windings 2 and 3 are in series with each other also. If windings 1 and 3 are used for connections, they, too, are in series. Thus no matter which connections are used, there are two coils in series to produce the single-phase power needed. The current provided, however, is as in any series connection. It is that which can be provided by only one winding.

The true power delivered by a wye-connected generator is the same as that of a delta-connected generator. The total true power is

$$P_t = 1.73 E_{line} \times I_{line} \times \cos \angle \theta$$

The delta connection provides 1.73 times the phase current for any two connections (Fig. 20-22). The line voltages are equal to the voltage in any phase winding. The advantage to the delta connection is its ability to produce more current. Note that winding 1 is in parallel with

**Fig. 20-22** *Delta-connected with single-phase voltages identified.*

windings 2 and 3, and windings 2 and 3 are in series with each other. This produces the output labeled C. Winding 2 is in parallel with windings 1 and 3, and windings 1 and 3 are in series with each other. They produce output A. Winding 3 is in parallel with windings 1 and 2, and windings 1 and 2 are in series with each other. This combination produces output B. With delta connection, the current is the advantage. With the wye connection, voltage is the advantage. In summary, the delta connection produces line voltage equal to the phase voltage and the line current equal to the square root of 3 or 1.73 times the phase current. True power is the same in the delta connection as in the wye connection. In both cases it is merely a matter of multiplying the line current times the line voltage times the 1.73 factor and the cosine of the angle $\theta$.

## MEASUREMENT OF POWER

The wattmeter connections for measuring the true power in a three-phase system are shown in Fig. 20-23. The method shown in Fig. 20-23A uses three wattmeters with their current coils inserted in series. The line wires and their potential coils are connected between line and

**Fig. 20-23** *(A) Three-wattmeter and (B) two-wattmeter measuring methods.*

neutral wires. The total true power is equal to the arithmetic sum of the three wattmeter readings.

In Fig. 20-23B two wattmeters are used with their potential coils connected between these line wires and the common or third wire that does not contain the current coils. The total true power is equal to the algebraic sum of the two wattmeter readings. If one meter reads backward, its potential coil connections are reversed. Note the polarity indications placed there to prevent incorrect connections. The incorrect connection causes the meter to read up-scale. The total true power is then equal to the difference in the two wattmeter readings. If the load power factor is less than 0.5 and the loads are balanced, the true power will equal the difference between the two wattmeter readings. If the load power factor is 0.5, one meter will indicate zero. If the load factor is above 0.5, the total true power is the sum of the two wattmeter readings.

## TRANSFORMERS

Transformers are important in any power distribution system. A quick review of their basic principles will aid in understanding their use in various parts of a distribution system.

A transformer is a device that has no moving parts. It transfers energy from one circuit to another by electromagnetic induction. The energy is always transferred without a change in frequency.

Usually, changes in voltage and current are evident. A step-up transformer receives electrical energy at one voltage and delivers it at a higher voltage. Conversely, a step-down transformer receives energy at one voltage and delivers it at a lower voltage. Transformers require little care and maintenance because they are simple, rugged, and durable in construction. Since there are no moving parts, there is little to wear out. The efficiency can be more than 99%.

Transformers are responsible for the use of ac today in every phase of life.

## Transformer Construction

A transformer in its simplest form has two windings insulated electrically from each other. These windings are wound on a common magnetic circuit built of laminated steel sheet. The principal parts are the following:

1. *Core*: provides a circuit of low reluctance for the magnetic flux
2. *Primary winding*: receives the energy from the ac source
3. *Secondary winding*: receives the energy by mutual induction from the primary and delivers it to the load
4. *Enclosure*: prevents damage when the transformer is overloaded

When a transformer is used to step up voltage, the low-voltage winding is the primary. When a transformer is used to step down the voltage, the high-voltage winding is the primary. The primary is always connected to the source of the power. The secondary, is always connected to the load. It is common practice to refer to the windings as the primary and secondary rather than the high-voltage and low-voltage windings.

The principal types of transformer construction are the core type and the shell type. Figure 20-24 shows the two types. The cores are built of thin stampings of silicon steel called laminations. Eddy currents are generated in the core by the alternating flux as it cuts through the iron. These currents are minimized by using the thin laminations with insulating varnish.

Hysteresis losses are caused by the friction developed between magnetic particles as they rotate through each cycle of magnetization. These losses are minimized by using a special grade of heat-treated grain oriented silicon-steel laminations.

In the core-type transformer, the copper windings surround the laminated iron core. In the shell-type transformer the iron core surrounds the copper windings.

***Fig. 20-24*** *(A) Core-type and (B) shell-type transformers.*

Distribution transformers are generally of the core type. Some of the largest power transformers are of the shell type.

If the windings of a core-type transformer were placed on separate legs of the core, a relatively large amount of flux produced by the primary windings would fail to link the secondary winding. This would cause a large leakage flux. The effect of the leakage flux would be to increase the leakage reactance drop, XI, in both windings. To reduce the leakage flux and reactance drop, the windings are subdivided. Half of each winding is placed on each leg of the core. The windings may be cylindrical in form and placed one inside the other with the necessary insulation. The low-voltage winding is placed with a large part of its surface area next to the core. The high-voltage winding is placed outside the low-voltage winding to reduce the insulation requirements of the two windings. If the high-voltage winding were placed next to the core, two layers of high-voltage insulation would be required. One would be needed next to the core; the other would be needed between the two windings. In another method, the windings are built up in thin flat sections called pancake coils. The pancake coils are sandwiched together with the required insulation between them.

The complete core and coil assembly is placed inside a set of steel covers or a steel tank (Fig. 20-25). In commercial transformers, the complete assembly is usually immersed in a special mineral oil. This oil provides a means of insulation and cooling. No oil is used in the transformer enclosures shown in Fig. 20-25. This type requires air circulation to keep it cool.

Transformers are built in both single-phase and three-phase units. A three-phase transformer consists of separate insulated windings for the different phases.

***Fig. 20-25***  *75 VA for 12 V output from 120-V line.*

The windings are wound on a three-legged core capable of establishing three magnetic fluxes displaced 120° in time phase. Three-phase transformers are not used in locations where the operation of a single unit may affect the others.

## Three-Phase Transformer Connections

Power may be supplied through three-phase circuits containing transformers in which the primaries and secondaries are connected in various wye and delta combinations. For example, three single-phase transformers may supply three-phase power with four possible combinations of their primaries and secondaries. Possible combinations are:

1. Primaries in delta and secondaries in delta
2. Primaries in wye and secondaries in wye
3. Primaries in wye and secondaries in delta
4. Primaries in delta and secondaries in wye

If the primaries of three single-phase transformers are properly connected (either in wye or delta) to a three-phase source, the secondaries may be connected in delta (Fig. 20-26). A topographic vector diagram of the three-phase voltages is shown in Fig. 20-26A. The vector sum of these three voltages is zero. This can be seen by combining any two vectors. For example, check $E_A$ and $E_B$. Note that their sum is equal and opposite to the third vector, $E_C$. A voltmeter inserted within the delta will indicate zero voltage (Fig. 20-26B).

Assuming that all three transformers have the same polarity, the delta connection consists of the $X_2$ lead of winding A to the $X_1$ lead of B, the $X_2$ lead of B to $X_1$ of C, and the $X_2$ lead of C to $X_1$ of A. If any one of the three windings is reversed with respect to the other two windings, the total voltage within the delta will equal twice the value of one phase. If the delta is closed on itself, the resulting current will be of short circuit magnitude. The result is damage to the transformer windings and cores. The delta should never be closed until a test is made to determine that the voltage within the delta is zero or nearly zero. This may be done with a voltmeter, fuse wire, or a test lamp. In Fig. 20-26B, when the voltmeter is inserted between the $X_2$ lead of A and the $X_1$ lead of B, the delta circuit is completed through the voltmeter. The indication should be approximately zero. Then the delta is completed by connecting the $X_2$ lead of A to the $X_1$ lead of B.

If the three secondaries of an energized transformer bank are properly connected in delta and are supplying a balanced three-phase load, the line current

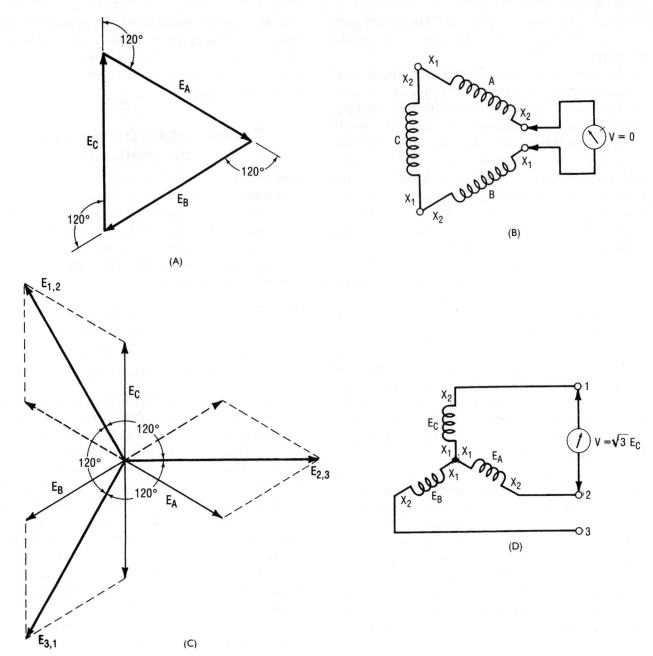

**Fig. 20-26** Delta and wye vectors and test for voltages.

will be equal to 1.73 times the phase current. If the rate current of a phase (winding) is 100 A, the rated line current will be 173 A. If the rated voltage of a phase is 120 V, the voltage between any two line wires will be 120 V.

The three secondaries of the transformer bank may be reconnected in wye to increase the output voltage. The voltage vectors are shown in Fig. 20-26C. If the phase voltage is 120 V, the line voltage will be 1.73 times 120 = 208 V. The line voltages are represented by vectors, E1,2, E2,3, and E3,1. A voltmeter test for the line voltage is represented in Fig. 20-26D.

If the three transformers have the same polarity, the proper connections for a wye-connected secondary bank are indicated in the figure. The $X_1$ leads are connected to form a common or neutral connection. The $X_2$ leads of the three secondaries are brought out to the line leads. If the connections of any one winding are reversed, the voltages between the three line wires will become unbalanced. Thus the loads will not receive their proper magnitude of load current. Also, the phase angle between the line currents will be changed. They will no longer be 120° out of phase with each other.

**338** Power Distribution Systems

Therefore, it is important to connect the transformer secondaries correctly to preserve the symmetry of the line voltages and currents.

Transformer installations with both primary and secondary windings delta connected are shown in Fig. 20-27. The $H_1$ lead of one phase is always connected to the $H_2$ lead of an adjacent phase. The $X_1$ lead is connected to the $X_2$ terminal of the corresponding adjacent phase, and so on. The line connections are made at these junctions. This arrangement assumes that the three transformers have the same polarity.

load. The line current is equal to the transformer phase current in the open-delta connection. In the closed-delta connection, the transformer phase current, $I_{phase}$ = Line/$\sqrt{3}$. Thus when one transformer is removed from a delta-connected bank of three transformers, the remaining two transformers will carry a current to $\sqrt{3}\ I_{phase}$. This value amounts to an overload current on each transformer of 1.73 times the rated current, or an overload of 73.2%.

Thus to prevent overloading transformers in an open-delta connection, the line current must be reduced so that the rated current of the individual transformers is not exceeded. The open-delta connection results in a

**Fig. 20-27** *Connections for a three-transformer bank.*

An open-delta connection results when any one of the three transformers is removed from the delta connected transformer bank without distributing the three-wire, three-phase connections to the remaining two transformers. These transformers will maintain the correct voltage and phase relations on the secondary to supply a balanced three-phase load. An open delta connection is shown in Fig. 20-28. The three-phase source supplies the primaries of the two transformers. The secondaries supply a three-phase voltage to the

**Fig. 20-28** *Open-delta feeding a resistive load.*

reduction in system capacity. The full-load capacity in a delta connection at unit power factor is:

$$P_\Delta = \sqrt{3}\, E_{phase}\, I_{phase} = \sqrt{3}\, E_{line}\, I_{line}$$

In an open-delta connection, the line current is limited to the rate phase current of $I_{line}/\sqrt{3}$. The full load capacity of the open-delta or V-connected system is

$$P_V = \sqrt{3}\, E_{line}\, \frac{I_{line}}{\sqrt{3}} = E_{line}\, I_{line}$$

The ratio of the load that can be carried by two transformers connected in open-delta to the load that can be carried by three transformers in closed-delta is

$$\frac{P_V}{P} = \frac{E_{line}\, I_{line}}{\sqrt{3}\, E_{line}\, I_{line}} = \frac{1}{\sqrt{3}} = 0.577$$

or 57.7% of the closed-delta rating.

For example, a 150-kW, three-phase balanced load operating at unity power factor is supplied at 250 V. The rating of each of three transformers in closed delta is 150/3 = 50 kW. The phase current is 50,000/250 = 200 A. The line current is 200 × √3 = 346 A. The removal of one transformer from the bank leaves two transformers that would be overloaded 346 − 200 = 146 A, or (146/200) × 100 = 73%. To prevent overload on the remaining two transformers, the line current must be reduced from 346 A to 200 A. Also, the total load must be reduced to

$$\frac{\sqrt{3} \times 250 \times 200}{1000} = 86.5 \text{ kW}$$

$$\frac{86.5}{150} \times 100 = 57.7\% \text{ of the original load}$$

The rating of each transformer in an open-delta necessary to supply the original 150-kW load is

$$\frac{E_{phase}\, I_{phase}}{1000} \quad \text{or} \quad \frac{250 \times 346}{1000} = 86.5 \text{ kW}$$

Two require a total rating of 2 × 86.5 = 173.0 kW. This compares with 150 kW for three transformers in closed-delta. Assume that two transformers are used in open-delta to supply the same load as three 50-kW transformers in closed-delta. Then the required increase in transformer capacity is:

$$173.0 - 150 = \frac{23.0}{150} \text{ kW}$$

or

$$\frac{23.0}{150} \times 100 = 15.3\%$$

Three single-phase transformers with both primary and secondary windings wye-connected are shown in Fig. 20-29. Only 57.7% of the line voltage ($E_{line}/\sqrt{3}$) is impressed across each winding. But full line current flows in each transformer winding. Three-phase transformers delta-connected to the primary circuit and

**Fig. 20-29** *Connections for a three-transformer bank.*

wye-connected to the secondary circuit are shown in Fig. 20-30. This connection provides four-wire, three-phase service with 208 V between lines wires A'B'C'. There is 208/√3 or 120 V between each line wire and neutral, N.

The wye-connected secondary is desirable when a large number of single-phase loads are to be supplied from a three-phase transformer bank. The neutral, or grounded wire is brought out from the midpoint of the wye connection. This permits the single-phase loads to be distributed evenly across the three phases; and three-phase loads can be connected directly across the line wires. The single-phase loads have a voltage rating of 120 V. The three-phase loads are rated at 208 V. This connection is often used in high-voltage power supply transformers for radar installations.

**Fig. 20-30** *Transformer bank connected delta to wye.*

**Fig. 20-31** *Transformer bank connected wye to delta.*

The phase voltage is 1/1.73, or 0.577 (57.7%) of the line voltage.

Three single-phase transformers with wye-connected primaries and delta-connected secondaries are shown in Fig. 20-31. This arrangement is used for stepping down the voltage from approximately 4000 V between line wires on the primary side to either 115 or 230 V, depending on whether the secondary windings of each transformer are connected in parallel, and the secondary output voltage is 115 V. There is an economy in transmission with the primary in wye. The line voltage is 73% higher than the phase voltage, and the line current is accordingly less. Thus line losses are reduced and the efficiency of transmission is improved.

## SWITCHING

Electrical distribution systems are often quite complicated. They cannot be made absolutely fail safe. Circuits are subject to destructive over currents. Harsh environments, general deterioration, accidental damage from natural causes, excessive expansion, and overloading of the electrical distribution system are factors that contribute to the occurrence of such over currents. Reliable protective devices prevent or minimize costly damage to transformers, conductors, motors, and many other components and loads that make up the complete distribution system. Reliable circuit protection is essential to avoid the severe momentary losses that can result from power blackouts and prolonged downtime of facilities. Switching is one of the means utilized to be able

Switching 341

to disable a circuit while the fuses or circuit breakers are replaced or transformers or other devices are repaired or maintained.

## Automatic Emergency Power Switching

When the utility power or the normal source fails, controls for automatic handling of the emergency power and the restoration of utility power are needed (Fig. 20-32). Controls are needed to start the engine and transfer the loads to the generator when it reaches proper voltage and frequency. Then, when the utility power is restored, the controls are needed to sense that power is restored and a need to retransfer the loads to the utility is evident. Allowing the engine to cool off before shutting down also needs to be handled automatically. The basic controls needed to accomplish this job are shown in Fig. 20-32.

All the abbreviations are defined below for ease in understanding some of the complexity of a system with an automatic emergency power source.

*ATS (Automatic Transfer Switch)*: switches the loads from the utility power (normal source) when it fails, to the engine generator (emergency source). Then retransfers the loads back to the utility when its power is restored. Switching is done automatically at the proper times under the supervision of the appropriate sensors, time delays and relays.

*ECC (Engine Control Contact)*: closes when the normal source fails and thereby initiates engine operation through the Automatic Engine-Starting Controls (AESC).

*ECO (Engine Control Contact)*: opens when the normal source fails. This may be used to trigger the AESCs, depending on the type of controls being used.

*FSE (Frequency Sensor, Emergency)*: senses the frequency of the power from the engine generator.

*PAP (Pre-alarm Panel)*: alerts personnel that engine oil pressure is dropping or that water temperature is rising before they reach a critical point. This panel also indicates low fuel level, low water temperature. In addition, it shows the system status and sounds an alarm when the engine is shut down from over crank, over speed, low oil pressure, and high water temperature.

*RAP (Remote Alarm Panel)*: located in an occupied area such as the maintenance engineer's, or building security office, this panel alerts personnel who are not in the vicinity of the power system and engine of the status of the complete system.

*SLG (Signal Light, Green)*: when lighted, indicates that the automatic transfer switch is connected to the utility (normal source).

*SLR (Signal Light, Red)*: when lighted, indicates that the automatic transfer switch is connected to the engine-generator set (emergency source).

*TDC (Time Delay, Cool-down)*: delays the engine from shutting down immediately when the load is retransferred to allow the engine to cool down from operating temperature.

*TDE (Time Delay, Emergency)*: delays the transfer switch from switching loads from the failed utility to the engine generator. This delay is usually set at zero unless there is more than one transfer switch handling loads for one engine generator; in which case, you may want to set the time delays so that all switches do not transfer their loads at the same time.

**Fig. 20-32** *Basic controls needed to handle a power failure and restoration.* (Automatic Switch)

*TDN (Tune Delay, Normal):* delays starting the engine when the utility power dips momentarily, then comes back up. This time delay avoids nuisance starts on the engine.

*TDR (Time Delay, Retransfer):* delays the transfer switch from retransferring the loads from the generator (emergency source) back to the utility (normal source) as soon as the utility power has been restored. This delay gives the utility time to establish itself.

*T'NS (Test Normal Switch):* simulates a failure of the utility power. When this switch is operated, the engine generator starts and runs. Then the transfer switch transfers the loads to the engine generator at the proper time. After a prescribed amount of engine-running time determined by the TDR setting, the transfer switch retransfers the loads back to the utility power and shuts down the engine. If it is important that loads not be interrupted when testing, a combination automatic transfer and bypass-isolation switch should be used.

*VSE (Voltage Sensor, Emergency):* senses when the engine generator is producing acceptable voltage, and if the frequency (see FSE) is also acceptable, signals the transfer switch to transfer the loads to the generator.

*VSN (Voltage Sensor, Normal):* senses when the utility power (normal source) drops below an acceptable value on any phase and, through the ECC, triggers the engine-starting controls. Also senses when the utility power is again acceptable and signals the transfer switch to return the loads back to the restored utility. Then signals the engine controls to begin cooldown and shutdown.

In addition to the items mentioned above, which are basic for a reliable automatic emergency power control system, other lights, meters, and specialized accessories may be added, depending on the installation. Accessories may include additional time delays, manual controls, engine-generator control accessories, and indicators. However, the items listed, if properly designed and coordinated, will provide reliable control for automatic transfer of two sources of power: one normal, the other emergency. Figure 20-33 has all the controls necessary to handle a power failure and restoration in a facility having one normal and one emergency source of power. One enclosure houses all the controls for switching purposes.

It is important to test and inspect the automatic, transfer switch because it is the heart and brains of the emergency power system. It senses the power failure, signals the engine to start, transfers the loads and then retransfers the load back to normal, and shuts down the engine. If the automatic transfer switch fails, nothing else in the system will respond to the need.

However, in hospitals and other facilities where an uninterrupted supply of power is vital to human life—places where the automatic transfer switches must be periodically tested—there is reluctance to do so because maintenance personnel do not want to interrupt the power even momentarily. For facilities such as these where power interruption and downtime cannot be tolerated, an automatic transfer and bypass isolation switch (Fig. 20-34) solves the problem. The automatic transfer and isolation switch unit goes into the same location in the electrical circuit as the automatic transfer switch (Fig. 20-35).

**Fig. 20-33** *All controls needed to handle a power failure.* (Automatic Switch)

Switching 343

**Fig. 20-34** *Automatic transfer and bypass isolation switch.*
(Automatic Switch)

**Fig. 20-35** *Automatic transfer switch, basic system.* (Automatic Switch)

## Switchboards

There are a number of manufacturers of switchboards. Each does the same thing but in a slightly different way. Square D switchboards with 200,000-A short-circuit rating are referred to as QMB. This type of switchboard uses fusible switches. The switches are of modular design (Fig. 20-36). Red and black operating handles and large on-off nameplates clearly identify the position of the switchblades. When the switch door

30 Ampere
2 or 3-Pole
240 V. or 600 V. ac, 250 V. dc
Twin branch switch units

60 Ampere
2 or 3-Pole
240 V. or 600 V. ac, 250 V. dc
Twin branch switch units

400, 600 or 800 Ampere
2 or 3-Pole
480 Y/277 V. ac
Single branch switch units

100 Ampere
2 or 3-Pole
240 V. or 600 V. ac, 250 V. dc
Twin branch switch units

200 Ampere
2 or 3-Pole
240 V. or 600 V. ac, 250 V. dc
Single branch switch units

400, 600 or 800 Ampere
2 or 3-Pole
240 V. or 600 V. ac, 250 V. dc
Single branch switch units

**Fig. 20-36** *Fusible switches QMB.* (Square D)

is open, the switchblades are completely visible and the entire length of the fuse is exposed for circuit testing or installation and replacement of fuses. They also have built-in fuse pullers (through 100 A) with dual cover interlocks and positive lock-off means. These switches are rated from 30 to 800 A. Switchboards are shown in Fig. 20-37. Circuit breaker units for motor starter applications are available with either thermal-magnetic circuit breakers or magnetic trip circuit breakers (Fig. 20-38).

**Fig. 20-37** *(A) Power style switchboard. (B) Speed-D switchboard.* (Square D)

**Fig. 20-38** *Circuit breaker units, 15 to 400 A.* (Square D)

Motor starter centers are available for 120 to 600 V with reversing or non-reversing types. They have push-buttons, pilot lights, control voltage transformers, and fuse blocks (Fig. 20-39).

**Fig. 20-39** *Motor starter center.* (Square D)

## Panel Boards

Panel boards are available from many different manufacturers. A wide range of panel types is available for applications to 600 V ac with 10,000 through 200,000 A maximum short-circuit current rating. Plug-in or bolt-on branch circuit breakers are available. Plug-in circuits are locked in position with dead front-panel cover, assuring positive contact.

Power panel boards have a wide range of mains capacities and circuit flexibility (Fig. 20-40). Main lugs are available from 225 to 1600 A. Branch circuit breakers

**Fig. 20-40** *Power panel board.* (Westinghouse)

or fusible switches are rated 15 to 1200 A, main breakers or main fusible switches to 1200 A, 600 V dc maximum. Figure 20-41 shows the typical panel board wiring diagram for single-phase and three-phase power.

## Raceways

In the distribution of electrical power to various locations within a building, a number of methods have been devised by various manufacturers; it would require an entire book just to give examples; however, the main

**Fig. 20-41** *Typical panel wiring diagrams.* (Westinghouse)

purpose is to distribute the power to the motor or device that needs it and do the job without exposing people or equipment to the dangers of high voltage and high-current-carrying cables.

## Aluminum Lay-In Wall Duct and Floor Trench Duct

Evolving technology in medical care facilities has identified the need for a nonferrous raceway system. Magnetic resonance imaging (MRI) is a diagnostic procedure that utilizes a magnetic field rather than x-rays. Because the equipment uses magnets, it is important that all or as much ferrous material as possible be eliminated from the room (Fig. 20-42). These rooms will normally utilize a raised floor in areas with the magnetic and electronic equipment. Along with the raised floor, cables must be routed back to control rooms. This can be accomplished with aluminum conduit, but aluminum lay-in wall duct and in some cases aluminum floor trench are preferred.

## Cable Tray

An economical raceway system designed to support and protect electrical wire and cable is available in aluminum and two types of galvanized steel for outdoor or indoor applications (Fig. 20-43).

Cable trays are not raceways. They are covered by Article 318 of the National Electrical Code. Cable

**Fig. 20-43** *Cable tray, ladder, trough, solid bottom channel.* (Square D)

**Fig. 20-42** *Aluminum lay-in wall duct and floor trench duct.* (Square D)

346  Power Distribution Systems

trays are open raceway-like assemblies made of steel, aluminum, or a suitable nonmetallic material. They are used in buildings to route cables and support them out of the way of normal building activities. A strong, sturdy support for cables is provided through troughs and ladders to route the cables to their destination or termination.

Figure 20-44 provides examples of how the trays can be routed and used to support heavy cables. The trays are made in straight sections, with matching fittings

**KEY TO DRAWING**

1. Solid Bottom Tray
2. Tray to Box Splice
3. 90° Vertical Outside Bend
4. 90° Vertical Inside Bend
5. Horizontal Cross
6. Solid Cover
7. 60° Horizontal Bend
8. Trough Tray
9. 90° Vertical Outside Bend
10. 45° Vertical Outside Bend
11. 45° Vertical Inside Bend
12. 90° Horizontal Bend
13. 90° Vertical Inside Bend
14. Ladder Tray
15. 90° Vertical Outside Bend
16. Horizontal Tee
17. Barrier Strip
18. Horizontal Barrier
19. Reducing Splice
20. 90° Horizontal Bend
21. Box Connector
22. 90° Vertical Barrier
23. Channel Tray
24. 90° Vertical Outside Bend

*Fig. 20-44* Various cable tray configurations.

Switching 347

*Fig. 20-45* Accessories needed to complete a cable tray installation.

to accommodate all changes of direction or quantity of cables. They are usually made of aluminum or zinc-coated steel.

Trough-type trays protect cables from damage and give good support and ample ventilation. Solid bottom fittings generally create no ventilation problems since they are a small part of the system. Cables are adequately ventilated through straight sections.

Ladder trays provide maximum ventilation to power cables and other heat-producing cables. However, cables are vulnerable to damage and covers are available. Various parts are needed to support the trays and covers (Fig. 20-45). Cables suitable for use in cable trays are marked CT (cable tray) on the outside of the jacket.

The cable system must be complete. It must be used as a complete system of straight sections, angles, offsets, saddles, and other associated parts to form a cable support system that is continuous and grounded as required by the NEC in Section 318-6(a). The system must be grounded as any raceway system must also be grounded. The Code treats the cable tray as a raceway and a wiring method. Limitations are placed on the number, size, and placement of conductors inside the tray. These limitations can be obtained by checking the Code.

## REVIEW QUESTIONS

1. What is the simple radial system?
2. What is the loop primary radial system?
3. What is the banked-secondary radial system?
4. What is the primary-selective radial system?
5. What is the secondary-selective radial system?
6. What is the modified secondary-selective radial system?
7. What is a simple network system?
8. What is a simple spot-network system?
9. What is a primary selective network?
10. What are fault currents?
11. What is the most generally applicable and widely used form of industrial secondary network system?
12. Differentiate between a core and a shell transformer.
13. Describe an automatic emergency power switching system.
14. Why is periodic testing of an emergency power system necessary?
15. What is a raceway?
16. What is a panel board?
17. Are cable trays classified as raceways?
18. How are cables for cable trays marked?
19. Why is reliable circuit protection needed?
20. What does VSF stand for in emergency power service?

## REVIEW PROBLEMS

Delta and wye circuits are commonplace in three-phase current sources such as most commercially generated electrical power. Delta and wye connections are used in connecting transformers and in connecting the loads to these power sources. Formulas for converting from delta to wye and wye to delta are useful in figuring three-phase resistances. The formulas shown here can be used for purposes of review.

Delta-to-Wye (Fig. P-1):

Wye-to-Delta (Fig. P-2):

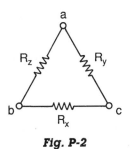

**Fig. P-2**

$$R_x = \frac{(R_a \times R_b) + (R_b \times R_c) + (R_c \times R_a)}{R_a}$$

$$R_y = \frac{(R_a \times R_b) + (R_b \times R_c) + (R_c \times R_a)}{R_b}$$

$$R_z = \frac{(R_a \times R_b) + (R_b \times R_c) + (R_c \times R_a)}{R_c}$$

**Fig. P-1**

$$R_a = \frac{R_y \times R_z}{R_x + R_y + R_z}$$

$$R_b = \frac{R_x \times R_z}{R_x + R_y + R_z}$$

$$R_c = \frac{R_x \times R_y}{R_x + R_y + R_z}$$

1. In the circuit shown in Fig. P-1, $R_a$ is 6000 Ω, $R_b$ is 2000 Ω, and $R_c$ is 3000 Ω. What is the value of $R_y$?
2. Find the value of $R_x$ in problem 1.
3. Find the value of $R_z$ in problem 1.
4. The delta circuit, shown in Fig. P-2, has all resistors of the same size (12,000 Ω). Find the value of the resistors in the equivalent wye circuit.
5. In the circuit shown in Fig. P-2, $R_x$ is 20,000 Ω, $R_y$ is 10,000 Ω and $R_z$ is 30,000 Ω. What is the value of $R_c$ in the equivalent wye circuit?

# CHAPTER 21

# Programmable Controllers

## PERFORMANCE OBJECTIVES

After studying this chapter, you will be able to:

1. Describe solid-state controllers that use PWM for operation.
2. List standard electronic features in controls.
3. Identify parts of a PLC system.
4. Define interfacing and input-output.
5. Differentiate between parallel and serial ports on a PLC.
6. Explain the RS-232C standard.
7. Identify PLC problems with electrical noise.
8. Explain solid-state reliability in PLCs.
9. List at least five processors available with Square D PLC systems.
10. List three types of display systems used with PLCs.
11. Explain the operation of a cell controller.
12. Explain the difference between a micro-cell and a mini-cell.
13. Discuss the future of PLCs.

## NEMA PLC Definition

The National Electrical Manufacturers Association gives the following definition for a programmable-logic controller (PLC): a digitally operating electronic apparatus which uses a programmable memory for the internal storage of instructions for implementing specific functions such as logic, sequencing, timing, counting and arithmetic to control, through digital or analog input/output modules, various types of machines or processes.

## SOLID-STATE ELECTRONICS

Solid-state electronics have gradually been making the control of motors easier and predictable in terms of load requirements and changes in loads. Electronics devices have been used to detect phase failure, phase reversals, open circuits, and short circuits. Electronic devices have also been used to provide reduced starting current and high starting torque. They also reduce the voltage to lightly loaded motors. Reducing the voltage in these applications can save energy.

Typical applications for controllers are in motors used in food-processing facilities, beverage bottling facilities, textile machinery, cranes, belt-driven equipment, conveyors, general machinery, materials handling facilities, compressors, machine tools, woodworking

**Fig. 21-1** Solid-state ac motor control circuitry. (Westinghouse)

equipment, and water treatment facilities. Figure 21-1 shows a solid-state ac motor control circuitry. Note the control logic and overload protection module. This type of electronics is utilized in the soft start and voltage reduction during light-loads situations.

The conversion of the fixed-frequency input to an adjustable-frequency output is also possible using electronics for adjustable-frequency drives. The rectifier converts incoming ac supply voltage to a fixed-potential dc bus level. The dc voltage is in turn inverted by a three-phase, pulse-width modulated (PWM) inverter section to an adjustable frequency output whose voltage is also adjusted proportionately to the frequency to provide constant voltage per hertz excitation to the motor terminals up to 60 Hz. Above 60 Hz, the voltage remains constant at nominal motor full voltage rating. In this way energy-efficient speed control is obtained in the range from 6 to 120 Hz.

## Standard Electronic Features in Controls

Some of the standard protective features provided by electronics circuitry are instantaneous power failure protection. The controller trips if power outage exceeds 15 milliseconds. Electronic instantaneous over current protection, inverse time overload protection, and under voltage protection are provided by electronics. DC bus overload protection and controller over temperature protection are provided by electronics. Torque (current) limit protect, if necessary, is automatically extended to accelerate or decelerate. When running under stated state conditions, current limit will reduce output frequency when the adjustable value is exceeded. Electronic ground-fault protection is also available in the package. Surge protection from input ac line transients is provided by a *snubber* network and electrical isolation is provided between the power and the logic circuits. All this and more can be accomplished by a programmable controller.

## THE PROGRAMMABLE CONTROLLER

The programmable controller (PLC)* is made up of a power supply, a processor unit, and I/O modules. Figure 21-2 shows a block diagram of a programmable controller system. The inputs can be a keyboard or limit switches, pressure switches, thermostats, or any number of devices that can provide an on-off status.

The controller has a memory so that it can be programmed and can retain the information presented to it and compare it to what has been programmed and then make a decision and send the required signal by way of the output modules to the motor it is controlling.

### Input-Output

The connections that interface with the outside world of the programmable controller are called input-output (I/O) ports. The input port allows data from a keyboard or other input device to be taken into the controller. The output port is used to send data to an output device such as a motor. Bus lines carry the signals to and from the major parts of the system.

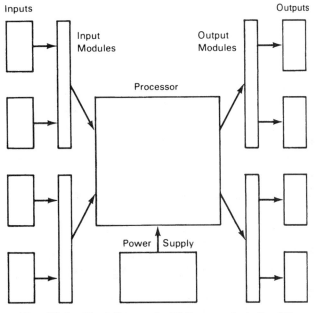

**Fig. 21-2** *Block diagram for PLC system, including I/Os.*

*Inasmuch as you see the abbreviation PC used to mean personal computer it is now suggested that **PLC** (for programmable logic controller) be used for programmable controller to differentiate it from the PC.

## Interfacing

Interfacing is used to describe what happens when the controller communicates with the motor it is controlling and the switches or sensing devices that sense require information needed for the process to be completed. The programmable controller is able to communicate with the rest of the equipment around it by being connected through ports. It is necessary to input information to the controller and it is necessary to receive information from the controller after it has processed it according to its program.

In some instances the programmable controller can be programmed using a computer keyboard and the information can be displayed on a cathode ray tube (CRT) and checked for accuracy before being stored in the controller's memory. LCD displays are also available.

## ASCII Code

Digital information is processed and communicated by using the ASCII code (Table 21-1). To communicate with the controller, it is necessary to be able to input instructions to the microprocessor or chip that makes up the processing unit. A special code has been designed so that the regular typewriter type of keyboard can be used to type in instructions. However, the keys of the keyboard are switches and they send a pulse to a decoder which generates a special binary code. That code is most often the American Standard Code for Information Interchange (ASCII).

**TABLE 21-1** *Seven-Bit ASCII Code for Digital Use*

| A | 1 000 001 | a | 1 100 001 | 0 | 0 110 000 |
|---|---|---|---|---|---|
| B | 1 000 010 | b | 1 100 010 | 1 | 0 110 001 |
| C | 1 000 011 | c | 1 100 011 | 2 | 0 110 010 |
| D | 1 000 100 | d | 1 100 100 | 3 | 0 110 011 |
| E | 1 000 101 | e | 1 100 101 | 4 | 0 110 100 |
| F | 1 000 110 | f | 1 100 110 | 5 | 0 110 101 |
| G | 1 000 111 | g | 1 100 111 | 6 | 0 110 110 |
| H | 1 001 000 | h | 1 101 000 | 7 | 0 110 111 |
| I | 1 001 001 | i | 1 101 001 | 8 | 0 111 000 |
| J | 1 001 010 | j | 1 101 010 | 9 | 0 111 001 |
| K | 1 001 011 | k | 1 101 011 | SP | 0 100 000 |
| L | 1 001 100 | l | 1 101 100 | ! | 0 100 001 |
| M | 1 001 101 | m | 1 101 101 | .. | 0 100 010 |
| N | 1 001 110 | n | 1 101 110 | # | 0 100 011 |
| O | 1 001 111 | o | 1 101 111 | $ | 0 100 100 |
| P | 1 010 000 | p | 1 110 000 | % | 0 100 101 |
| Q | 1 010 001 | q | 1 110 001 | & | 0 100 110 |
| R | 1 010 010 | r | 1 110 010 | • | 0 100 111 |
| S | 1 010 011 | s | 1 110 011 | ( | 0 101 000 |
| T | 1 010 100 | t | 1 110 100 | ) | 0 101 001 |
| U | 1 010 101 | u | 1 110 101 | * | 0 101 010 |
| V | 1 010 110 | v | 1 110 110 | + | 0 101 011 |
| W | 1 010 111 | w | 1 110 111 | ' | 0 101 100 |
| X | 1 011 000 | x | 1 111 000 | - | 0 101 101 |
| Y | 1 011 001 | y | 1 111 001 | . | 0 101 110 |
| Z | 1 011 010 | z | 1 111 010 | / | 0 101 111 |

The ASCII code is made up of seven binary bits, so there are 128 possible combinations. This is obtained when you take 2 to the seventh power ($2^7$). The 128 ($2^7$) possible combinations of 1's and 0's can represent all the letters of the alphabet, both upper and lower case, as well as the numbers 0 through 9 and several special codes that include punctuation and machine control information.

## Parallel Ports

Parallel ports are the outputs of the microprocessor or computer. The parallel port has a flat cable connected to it with eight conductors. Seven of these wires or conductors carry the information just mentioned. The eighth conductor carries the strobe line, the line that prevents the problem of switch bounce. When a switch is closed it also bounces or allows more than the on and off information to be given. It is necessary to eliminate this noise or incorrect signal information from being transmitted from the keyboard to the controller and then to the motor. It may cause the motor to make some incorrect on-off moves.

## Serial Ports

The serial format may also be used to transmit data in the ASCII code. The serial format allows the information to be transmitted along two wires. This is very convenient when transmitting over long distances. The parallel format is very good for short distances or connection between machines with the same work cell, but if the information has to be sent for a greater distance, it is better to send it by serial formatting.

The information may be transmitted as changes in voltage or changes in current. There are standards for both. In fact, there are two standards for each. The two voltage standards are known as RS-232C and TTL. The two current standards are the 60-mA current loop and the 20-mA current loop. As you can see, the raw data or the 20-mA is not enough to operate a motor, so it must be used to control a circuit that will turn the motor on and off.

The RS-232C standard says that the voltage of the signal will be between –3 and –25 V to represent the logic 1 or "on" condition. A voltage between +3 and +25 V is used to represent the logic 0 or "off" position. This standard was developed by the Electronic Industry Association (EIA). There is an advantage to this standard inasmuch as the line noise will have to be very high to make any false signals and the voltage losses along the line will not affect the signal level as much as lower voltages do. It does have a disadvantage, inasmuch as it has to be converted to transistor-transistor logic (TTL) at the port of the computer or microprocessor.

The TTL standard specifies that 5 V presents a logic 1 and 0 represents a logic 0. TTL standard is compatible with TTL logic and interfaces directly. There are problems with any transmission of data over a distance. If there is a line voltage of at least $1/2$ V, there is a possibility of receiving incorrect data. Since the peak is only 5 V there is always the possibility of picking up a noise signal when a wire is spread over a distance and in an electrical noise generating environment.

The 60-mA standard says that a current of 60-mA is logic 1 while zero represents logic 0. The main advantage is that the noise usually encountered over long-distance transmission lines does not affect the quality of the data being transmitted. However, the main disadvantage of this standard is that it has to be converted to voltage variations if used as inputs to a computer port.

The 20-mA standard is basically the same as the 60-mA standard except that it is 20 mA. The 20-mA level represents logic 1 and zero current represents logic 0. The same advantage is experienced with this standard as with the 60-mA standard. It is also necessary to convert the current variations to voltage variations if used as inputs to a computer port.

## THE ENVIRONMENT AFFECTS PERFORMANCE

Many solid-state controls are sensitive to various environment factors but not the same ones as those that generally affect electromechanical devices. Solid-state controls are generally less sensitive to shock and vibration since they contain no moving parts. At relatively high levels of shock and vibration, circuit boards may loosen, and crack or component leads may fail. Some electromechanical components are more susceptible to activation under shock and vibration. In this area solid-state controls may prove superior. Mounting position is also usually of little significance with the solid-state control, except in instances where airflow is required for cooling.

## ELECTRICAL NOISE

This noise is defined as any unwanted electrical signal that enters the equipment through various means. Electrical noise is capable of causing various types of malfunctions of solid-state equipment. Electrical noise may cover the entire spectrum of frequencies and exhibit any wave shape. Solid-state devices are especially sensitive to noise since they operate at low signal

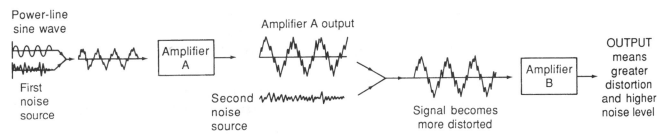

**Fig. 21-3** *Electrical noise.*

levels. Also, the speed of solid-state components allows them to respond to relatively high frequencies.

Electrical noise entering a solid-state control is often incapable of damaging components directly unless extremely high energy and/or voltage levels are encountered. Most of the malfunctions due to noise are temporary nuisance-type occurrences or operating errors, but could result in hazardous machine operations under certain conditions (Fig. 21-3).

Input lines, output lines, and power supply lines are the most common sources of noise entering solid state controls. Noise may be coupled into the lines electrostatically through capacitance between these lines and lines carrying other signals. A high potential is usually required or long closely spaced conductors are necessary. Magnetic coupling is also quite common when control lines are closely spaced to lines carrying large currents. The signals in this case are coupled through the mutual inductances as in a transformer. Electrostatic and magnetic noise may also be coupled directly into the control logic circuitry. This generally appears in unenclosed, unshielded control electronics and requires a strong noise field. Noise can occur in the form of electromagnetic radiation from remote sources.

Close coupling is not required and the various lines entering the control act as receiving antennas. Occasionally, the control circuitry itself is sufficiently sensitive to detect a radiated signal and respond to it. This type of noise is troublesome when it is encountered since it is usually of high frequency and occasionally difficult to filter and shield against on a generalized basis. A particular installation may require special treatment, since the various coupling elements, such as control lines, exhibit unpredictable characteristics, some of which could render built-in filters ineffective. Metal enclosures are effective shields where adequate electrical bonding around doors and bolted surfaces is provided.

Many designs have been tested and retested. Much effort has gone into filters, shielding, and the design of generally insensitive circuitry. It is, however, impossible to design a control that can handle every form of noise coupled into it, especially through the external wiring.

## Installation Practices

Noise must be minimized as much as possible from entering the control by using the appropriate installation practices, especially when the anticipated noise signal has characteristics very similar to the desired control input signal.

## Grounding

Grounding practices that are used have a significant effect on noise immunity. Each ground should be connected to its respective reference point by no more than one wire, called a single-point ground. Under no circumstance should two or more systems share a common single ground wire, either equipment ground or control common.

## SOLID-STATE RELIABILITY

Solid-state components exhibit a high degree of reliability when operated within their ratings. For example, a triac might have an average life of 450,000 hours or 50 years under typical operating conditions, but it fails at random even when operated within its ratings. The lifetime above is an average. The time of failure of an individual device cannot be predicted by observation as in the case of a relay, where patterns of wear might be watched. It is thus advisable to provide some sort of independent check on the operation of individual devices when they are controlling a critical or potentially hazardous operation. In addition, the predominant mode of failure of a solid-state output is in the "on" or short-circuit mode. This must be considered in certain critical applications.

## Backup Operations

For any size of solid-state control capable of performing potentially hazardous machine operations, emergency circuits for stopping operation must be routed outside the controller. For example, devices such as end-of-travel limit switches or emergency-stop pull cord switches should operate motor starters directly without being processed through controller logic. This forms a reliable means of control and should be implemented using a minimum number of simple, highly dependable components of an electromechanical nature, if possible. Thus, in the event of a complete controller failure, an independent rapid shutdown means is available. A convenient means of disconnecting the critical or potentially hazardous portions of a machine from the controller should be provided for use during troubleshooting or setup following maintenance.

## CONTROLLERS

A number of manufacturers make programmable controllers. Each has slightly different electronics packages. However, to make sense of the process and the equipment, one has been chosen for this discussion. It is the Sy/Max, made by Square D (Fig. 21-4). There are a number of processors for this particular manufacturer to suit as many of the requirements of customers as possible.

Programmable controllers are used in a variety of applications to replace conventional control devices, such as relays and solid-state logic. When compared with conventional control means, programmable controllers (**PLC**s) allow ease of installation, quick and efficient system modifications, more functional capability, troubleshooting diagnostics, and a high degree of reliability. Typical applications include automated material handling, machine tool, and assembly machine control, wood and paper processing control, injection molding machine control and process control applications such as film, chemical, food, and petroleum.

## System Hardware and Programming Equipment

The controller family consists of two groups of equipment: system hardware and programming equipment. System hardware is used to control the actual operation, while the programming equipment is used to enter the user control program into the system hardware. Once the program is entered, the programming equipment can be used for monitoring, program alteration, or message displays, but is not required for system operation (Fig. 21-5).

System hardware consists of a processor, one or more rack assemblies, power supplies, I/O modules, and various other modules that provide additional capabilities. The rack assemblies and associated I/O modules communicate with external I/O control devices such as limit switches, motor starters, and other devices.

## Processors

There are five processors available with the Sy/Max system. The smallest or least expensive, Model 50, can be programmed with an IBM or compatible personal computer using Sy/Mate software in addition to its own handheld programmer, or the process parameters can be fine tuned with the controller by using the control station shown in Fig. 21-6. Memory is available up

***Fig. 21-4*** *Processor family.* (Square D)

to 4K of EPROM with a battery-backed RAM or E²PROM. The I/O capacity is 256. To make a comparison of what is available, take a look at Model 300 processor. The Model 300 has an I/O capacity of 256. The processor has 128 internal relay equivalents and 96 four-digit storage registers for timers, counters, synchronous shift registers, and data storage. In addition to relay logic, counting, timing, and data manipulation, this processor offers modular construction for ease of installation and troubleshooting, fourfunction math, ASCII output to generate alarm messages and reports, plus peer-to-peer communication with other processors through either of two communication ports.

By upgrading to the Model 500 level, other features may be acquired. The 500 can perform square-root math functions, scan control, subroutines, timed interrupts, and matrix operations. It also has several levels of security that can prevent unauthorized access to data and program information through either of its two communications ports. It has up to 8K of battery-backed RAM or combination RAM/EPROM memory and has an I/O capacity of over 2000.

Another upgrade to Model 400 (See Fig. 21-7) produces a processor that can handle almost any application. In addition to Model 500 processor capabilities, the instruction set of the Model 400 includes trig-o-metric, transcendental, and statistical functions and the ability to perform these functions in integer or floating-point (result to $10^{\pm 38}$) format. Another unique feature is

**Fig. 21-5** *(A) Model 300 programmable controller. (B) CRT programmer. (Square D)*

**Fig. 21-6** *PID loop control station. (Square D)*

**Fig. 21-7** *Model 400 processor. (Square D)*

Controllers 357

the capability of an on-board battery to support the RAM memory and real-time clock upon removal of the processor from the rack. This model can also read ASCII data through either of its two independent communication ports, allowing it to interface directly with ASCII weight scales, bar code readers, and other such inputs for applications such as materials handling. It has up to 16K words of RAM or RAM/EPROM combination and has an I/O capacity of 4000. Keep in mind that these are some of the first electronic controls and many are still in use. For the latest information check with the internet web sites for the manufacturers of this equipment.

## Input-Output Modules

The I/O modules provides the interface between the processor and the field device that is being switched or controlled. The I/O modules shown in Fig. 21-8 are available in five versions. The standard four-function covers the range of operating voltages from TTL to 240-V ac/250-V dc. Standard I/Os have a single diagnostic LED for each point. Output fuses are accessible from the side of the module. The deluxe four-function module is interchangeable and compatible with the standard four-function modules but has a high-power (5 A per output) dual-point module and individual power circuit isolation when used with the isolated I/O rack. The 4-, 8-, 16-, and 32-function modules are capable of handling most processes.

The fiber-optic input module has field devices (switches) designed for use with its circuitry (Fig. 21-9). The fiber-optic interface module converts all programmable controller differential communication (remote I/O, programming and so forth.) to optical communication. Fiber-optic communications provides immunity from electromagnetic interference (EMI) and radio-frequency interference (RFI) and complete electrical isolation. It is intrinsically safe and uses lightweight, easily installed cable. Maximum length for the cable is 5 miles or 8 km. This type of communication system is useful in process control systems, petrochemical plants, utility power substations, and outdoor/underground installations.

## Intelligent I/O Modules

Intelligent (register) I/O modules such as multiplexed binary-coded decimal (BCD) and analog I/O and high-speed counter input, stepping motor output, and speech output modules provide special functions to the PLC. On-board microprocessors in each of these devices allow information to be directly transferred and stored in processor data registers. A plant floor microcomputer allows for a production report. Graphic generation is also available.

The speech module is a synthesized speech/message annunciator. It can provide audible alarm annunciation,

**Fig. 21-8**  *I/O systems. (Square D)*

**Fig. 21-9** *Fiber-optic interface and input module with field devices.* (Square D)

operator instructions, or directions to supplement or replace a visual display. The speech module is compatible with the processors and data controllers or any device that can generate ASCII output.

## Display Systems

There are a number of display systems that can utilize the features of a programmable controller. Several different types of color graphic displays are available. Color graphics are available with the hardware and software systems that interface with a computer such as that in Fig. 21-10.

The CRT programmer is a portable device that can monitor, program, and document the control logic of any processor mentioned previously (Fig. 21-11). In most industries the LCD flat screen monitor has been utilized for its ability to take up less space and produce less heat as well as consume less energy.

This programmer makes programming the processor easy. This is due to its set of multifunction soft keys. In addition, high-level functions such as math, shift registers, timers, and counters are programmed using simple fill-in-the-blank function boxes in many cases. Programming is fairly easy once you have learned the procedures.

**Fig. 21-10** *Programmers for Sy/Max.* (Square D)

Controllers 359

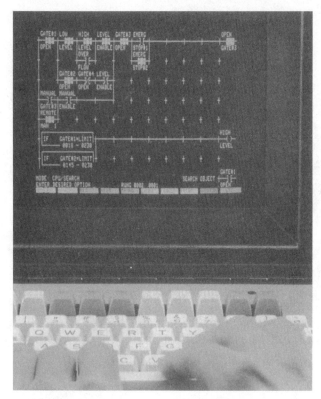

**Fig. 21-11** *CRT display with function names. (Square D)*

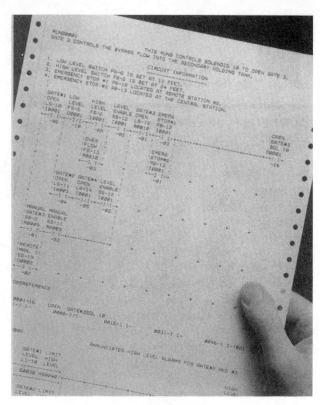

**Fig. 21-12** *Program documentation. (Square D)*

Off-line programming allows the processor control program to be developed, edited, and documented in a non-factory environment without a processor. The control program can then be stored on tape and downloaded into the processor on the plant floor, reducing overall system development time.

I/O function names improve the ability for maintenance and other plant personnel to understand a detailed control program by displaying the name (up to 12 alphanumeric characters) of the I/O device. This enhances using the CRT to diagnose control system faults. Each I/O element can have up to an 18-character alphanumeric name along with its address. In addition, each logic run can have up to a full-page description of its operation. See the program in Fig. 21-12.

## Cell Controllers

A cell controller is typically used to coordinate and manage the operation of a manufacturing cell, consisting of a group of automated programmable machine controls (programmable controllers, robots, and so forth) designed to work together and perform a complete manufacturing or process-related task.

## Micro-Cell Controller

The micro-cell controller is a programmable multifunction, data and program storage device, designed for small cell control applications (Fig. 21-13).

The mini-cell controller is the midrange member of the cell controller family. It is designed to perform basic control functions in addition to high-level functions such as data analysis, trending, statistical process control, statistical quality control, color graphic generation and serve as a communications gateway (Fig. 21-14).

**Fig. 21-13** *Micro-cell controller. (Square D)*

*Fig. 21-14* Mini-cell controller. *(Square D)*

## Local Area Network

One of the advantages of keeping the same type of programmable controller within a plant is its ability to become part of a local area communication network (Fig. 21-15). The network can have up to 200 controllers and other devices communicate with each other. The network consists of twin-axial cable up to 10,000 ft long and up to 100 network interface modules. Two devices (PLCs, computers, CRTs, printers, etc.) can be connected to each network interface module. The network allows any programmer or programmable package of the acceptable type to be used on the network. Several versions of network interface modules are available.

## Future of PLCs

The programmable controller (PLC) has a bright future inasmuch as it will probably be used in all new production facilities. In some instances it has found applications that were once thought to be robot jobs. The cost of robots and their limited reprogrammable nature makes the inexpensive PLC the device for the future.

Each manufacturer has a different training program for its particular devices. As you may have witnessed here it is impossible to cover PLCs in one chapter of a book. It is more a subject for an entire book and training program.

What has been done here is an introduction to some of the concepts, ideas, and equipment involved in a PLC system that is used to form part of a larger unit such as a manufacturing cell. Of course, as with any electronics equipment there are constant upgrades and improvements. Keeping up with the latest is time consuming and a task that is lifetime in nature.

Much more time and effort will be needed by all those involved in electrical motor control to keep abreast of how electronics is doing the job and becoming less expensive and more reliable while doing so.

A good background in digital electronics is necessary to be able to understand how the programmable controller operates and functions in computer integrated manufacturing (CIM) as part of a computer controlled manufacturing facility (Fig. 21-16).

*Fig. 21-15* Local area network. *(Square D)*

Controllers 361

MINICELL CONTROLLER

**Fig. 21-16** *Mini-cell controller. (Square D)*

**362 Programmable Controllers**

# REVIEW QUESTIONS

1. What does PC denote?
2. What does PLC stand for?
3. What are the parts needed to make up a programmable controller?
4. What does I/O mean?
5. What is a parallel port?
6. What is a serial port?
7. What does the strobe line do?
8. What does TTL mean?
9. Who uses the ASCII code?
10. What is the RS232C code?
11. What is electrical noise?
12. How does electrical noise affect electronic devices?
13. What is EMI?
14. How can a PC be programmed?
15. What does CRT mean?
16. What is a cell controller?
17. What is the difference between micro-cell and mini-cell controllers?
18. What are electronic devices used for in terms of detecting problems?
19. What are some typical applications for motor controllers?
20. How are fixed-frequency power source inputs changed to adjustable-frequency outputs?

# CHAPTER 22

# Troubleshooting and Maintenance

## PERFORMANCE OBJECTIVES

After studying this chapter, you will be able to:

1. Explain how preventive maintenance can prolong trouble-free operation of a system.
2. Explain how to prevent accidental shock.
3. Troubleshoot ac and dc motors.
4. Troubleshoot power supply disturbances.
5. Troubleshoot circuits using a volt-ohm-milliammeter (VOM).
6. Troubleshoot circuits using an oscilloscope.
7. Troubleshoot relays.
8. Troubleshoot solid-state motor control equipment.

## TROUBLESHOOTING AND THE ELECTRICIAN

Troubleshooting is another of the tasks performed by the electrician. It tests your ability to observe everything around you and your ability to understand how things work. One of the best ways to prevent trouble is to check off certain items as a routine procedure to catch trouble before it becomes a major item and causes fire, damage, and/or death. Electrical problems are many, and every connection and every device is a potential problem. Each device and service as well as circuits should have been properly wired, but that is not always the case.

One of the biggest problems is troubleshooting electric motors. A troubleshooting chart will aid in this task as well as some of the more obvious observations made by the person on the scene. All textbook troubleshooting can do is identify the logical problems. On-scene facts are not always detailed in textbooks, so a good observer must also be able to uncover the facts needed to make a diagnosis. Once the problem is found, it is usually easily corrected.

## PREVENTIVE MAINTENANCE
### Damp and Wet Areas

One of the areas that can cause problems in any home or shop wiring system is dampness and wetness. Watertight equipment should be installed wherever there is a danger of water coming in contact with live wires. One of the largest problems is the condensation of moisture inside the panel board (Fig. 22-1). Moisture condenses where the warm, moist air in the basement moves up to come in contact with the cold air outside making the riser cold. In areas where this is a

**Fig. 22-1** *Moisture in the panel board can cause problems.*

problem, either an underground entrance can be made or, an outside riser can be mounted alongside the building, which will make its entrance only when it reaches the panel board. An entrance as low as possible is preferred so that any moisture that does condense will easily drain out the bottom of the panel board without contacting the hot side of the distribution panel. Another problem is rust and corrosion. Anywhere there is moisture there is the possibility of rust and corrosion. Both rust and corrosion can cause contact problems with metals and remove or place high resistance in the path of a ground system. Removing a ground produces a situation that can be very hazardous (Fig. 22-2). In a single-phase system the current on the neutral of a properly installed 120/240-V system carries the difference between the current flowing on the hot lines. If the ground is removed by corrosion or rust preventing contact with the proper grounding lugs, it is the same as having an open ground.

One of the indications of this condition is that some lights in the building will appear very bright and others very dim. Turn off the main switch and locate the open or corroded ground connection before allowing continued operation. A situation of this sort will make it very dangerous for anyone who touches any of the conductors. That person or animal (in the case of barnyards) will complete the ground circuit and fatal shock may occur.

*Fig. 22-2* *Removing a ground.*

## Prevention of Accidental Shock

The ground-fault circuit interrupter (GFCI) is one device used to prevent accidental shock. However, GFCI protection should never be a substitute for good grounding practice, but should support a well-maintained grounding system. Figure 22-3 shows a device that can be used in various locations for the prevention of shock by checking the grounding system. This device is used in homes, plants, and businesses where people are employed and use electrical equipment. It is often encountered by the electrician whether in the home or on the job.

The arrangement shown in Fig. 22-4A checks polarity and grounding. It also diagnoses five other incorrect wiring conditions with a plug-in tester. Figure 22-4B shows the same tester being used to check continuity of the ground path of a tool. This is very important since the hand drill has a metal handle.

In Fig. 22-4C the meter is a ground loop tester; it measures the ground loop impedance of live circuits. It can also be used to check for grounding of tools, piping systems, and other equipment. The meter in Fig. 22-4D is used to check the 500- and 1000-V dc insulation resistance of de-energized circuits and electrical equipment. It also checks for continuity in low resistance circuits.

The ground-fault circuit interrupter shown in Fig. 22-4E mainly provides insurance that a tool will not develop a fault on the job, causing a serious personal injury. It tests tools to assure that any current leakage is below a hazard level. Keep in mind that nuisance tripping of a GFCI can be caused by a few drops of moisture or flecks of dust. One way to avoid this problem is to use watertight plugs and connectors on extension cords.

## Ground-Fault Receptacles

There are two different ways to wire up ground-fault receptacles (GFRs) (Figs. 22-5 and 22-6). The devices shown are not only GFCIs but also receptacles. They can be used, as shown in Fig. 22-6, to protect other downstream receptacles. This brings about problems in some places, inasmuch as the protected outlets are not always known by the persons using them, and when the GFR trips it takes them off-line also.

One way to check for a terminal installation is to check the red and gray wires. If they are capped with a wire nut, you know that GFR does not service any other outlets.

## Wiring Devices

Using the proper wiring devices is a form of preventive maintenance inasmuch as it prevents problems later. Shock hazards are minimized by the dielectric strength of the material used for the molded interior walls and the individual wire pocket areas. Nylon seems to be best for this job. Nylon devices withstand high impact in heavy-duty industrial and commercial applications. Each molded piece has to support adjacent molded pieces to result in good resiliency and strength. Devices made of vinyl, neoprene, urea, or phenolic materials can crack or be damaged under pressure.

*Fig. 22-3* *Ground monitor.*

**Fig. 22-4** *(A) Ground monitor used to check polarity and grounding. (B) Checking grounding path for a tool. (C) Ground loop tester. (D) Checking impedance of live circuit. (E) Testing tools to ensure that leakage is below hazardous level.*

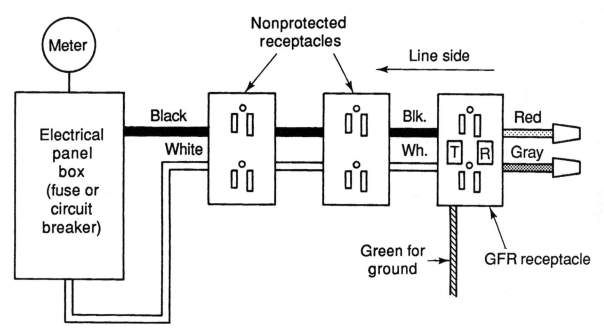

**Fig. 22-5**  *Ground-fault receptacle installed in a box.*

**Fig. 22-6**  *Ground-fault receptacle wired to protect downstream devices.*

Damage can be invisible and cause direct shorts and other hazards. Nylon also has the ability to withstand high voltages without breaking down.

## MAINTENANCE OF SMALL ELECTRIC MOTORS

Small motors usually operate with so little trouble that they are apt to be neglected. They should be thoroughly inspected twice yearly to detect wear and to remove any conditions that might lead to further wear. Special care must be taken to inspect motor bearings, cutouts, and other wearing parts. Make sure that dirt and dust are not interfering with ventilation or clogging moving parts.

### Adequate Wiring

When installing a new motor or transferring a motor from one installation to another, it is well to check the wiring. Be sure that adequate wire sizes are used to feed electrical power to the motor; in many cases, replacement of wires will prevent future breakdown. Adequate wiring assists in preventing overheating of motors and reduces electric power costs.

### Check Internal Switches

Start winding switches usually give little trouble; however, regular attention makes them last even longer. Use fine sandpaper to clean contacts. Make sure that the sliding member on the shaft that operates the start winding switch moves freely. Check for loose screws.

### Check Load Condition

Check the driven load regularly. Sometimes, additional friction develops gradually within the machine and imposes an overload on the motor, so watch the motor temperature. Protect motors with properly rated fuses or overload cutouts.

### Extra Care in Lubrication

A motor running three times as much as usual will need three times as much attention to lubrication. Motors should be lubricated according to the manufacturer's recommendation. Provide enough oil, but do not *overdo* it.

### Keep Commutators Clean

Do not allow a commutator on a dc motor to become covered with dust or oil. It should be wiped occasionally with a clean, dry cloth or one moistened with a solvent that does not leave a film. If necessary to use sandpaper, use No. 0000 paper or finer. Sandpaper or abrasive papers are available with ratings as high as 1500 grit.

### Motors Must Have Proper Service Rating

Sometimes it is necessary to move a motor from one job to another or to operate a machine continuously when it has previously been running for short periods of time. Whenever a motor is operated under different conditions or on a new application, make sure that it is rated properly. A motor is rated for intermittent duty because the temperature rise within the motor will not be excessive when it is operated for short periods. Putting such a motor on a continuous-duty application will result in excessive temperature rise, which will cause the insulation to deteriorate or may even cause burnout.

### Replace Worn Brushes

Brushes should be inspected at regular intervals so that replacements can be made if necessary. Whenever a brush is removed for inspection, be sure that it is replaced in the same axial position; that is, it must not be turned around in the brush holder when putting it back in the motor. If the contact surface, which has been "worn in" to fit the commutator, is not replaced in the same position, excessive sparking and loss of power will result. Brushes naturally wear down and should be replaced before they are less than $1/4$ in. in length. The commutator should also be inspected when brushes are removed. See the section "DC Motor Problems" later in the chapter.

## MOTOR PROBLEMS

Certain danger signals are presented before a motor overheats or burns out.

### Ball Bearing Motors

**Danger Signals**

- A sudden increase in the temperature differential between the motor and bearing temperatures is an indication of malfunction of the bearing lubricant.

- A temperature higher than that recommended for the lubricant warns of a reduction in bearing life. The rule of thumb is that grease life is halved for each 25°F increase in operating temperature.

- An increase in bearing noise, accompanied by a bearing temperature rise, is an indication of a serious malfunction of the bearing.

**Major Duties of Ball Bearing Lubricant**

- To dissipate heat caused by friction of bearing members under load.
- To protect bearing members from rust or corrosion.
- To offer maximum protection against entrance of foreign matter into the bearings.

**Causes of Bearing Failures**

- Foreign matter in bearing from dirty grease or ineffective seals.
- Deterioration of grease because of excessive temperature or contamination.
- Overheated bearings as a result of too much grease.

## Sleeve Bearing Motors

The lubricant used with sleeve bearings must actually provide an oil film that completely separates the bearing surface from the rotating shaft member, and ideally, eliminate metal-to-metal contact.

**Lubricant** Oil, because of its adhesive properties and because of its viscosity or resistance to flow, is dragged along by the rotating shaft of the motor and forms a wedge-shaped film between the shaft and the bearing. The oil film forms automatically when the shaft begins to turn and is maintained by the motion. The forward motion sets up a pressure in the oil film, which in turn supports the load. This wedge-shaped film of oil is an absolutely essential feature of effective hydrodynamic, sleeve bearing lubrication. Without it no great load can be carried, except with high friction loss and resultant destruction of the bearing. When lubrication is effective and an adequate oil film is maintained, the sleeve bearing serves chiefly as a guide to ensure alignment. In the event of failure of the oil film, the bearing functions as a safeguard to prevent actual damage to the motor shaft.

**Selection of Oil** The selection of the oil that will provide the most effective bearing lubrication and not require frequent renewal, merits careful consideration. Good lubricants are essential to low maintenance cost. Top-grade oils are recommended, as they are refined from pure petroleum, are substantially noncorrosive as far as metal surfaces to be lubricated are concerned, are free from sediment, dirt, or other foreign materials, and are stable with respect to heat and moisture encountered in the motor. In performance terms, the higher-priced oils prove to be cheaper in the long run.

An oil film is built up of many layers or laminations that slide upon one another as the shaft rotates. The internal friction of the oil, which is due to the sliding action of the many oil layers, is measured as the viscosity. The viscosity of the oil chosen for a particular application should provide ample oiliness to prevent wear and seizure at ambient temperature, low speeds, and heavy loads, before the oil film is established and operating temperature is reached. Low-viscosity oils are recommended for use with fractional-horsepower motors, as they offer low internal friction, permit fuller realization of the motor's efficiency, and minimize the operating temperature of the bearing.

**Standard Oils** High ambient temperatures and high motor operating temperature will have a destructive effect on sleeve bearings lubricated with standard temperature range oil by increasing the bearing operating temperature beyond the oil's capabilities. Such destructive effects include reduction in oil viscosity, an increase in corrosive oxidation products in the lubricant, and usually a reduction in the quantity of the lubricant in contact with the bearing. Special oils are available, however, for motor applications at high temperatures and also for motor applications at low temperatures. The care exercised in selecting proper lubricant for expected extremes in bearing operating temperatures will have a decided influence on motor performance and bearing life.

**Wear** Although sleeve bearings are less sensitive to a limited amount of abrasive or foreign materials than are ball bearings, owing to the ability of the relatively soft surface of the sleeve bearing to absorb hard particles of foreign materials, good maintenance practice recommends that the oil and bearing be kept clean. Frequency of oil changing will depend on local conditions such as severity and continuity of service and operating temperature. A conservative lubrication maintenance program should call for periodic inspections of the oil level and cleaning and refilling with new oil every 6 months.

**Warning:** Over lubrication should be avoided. Insulation damage by excess motor lubricant represents one of the most common causes of motor winding insulation failure in both sleeve and ball bearing motors.

## COMMON MOTOR PROBLEMS AND THEIR CAUSES

Easy-to-detect symptoms, in many cases, indicate exactly what is wrong with a fractional-horsepower motor. However, where general types of trouble have similar symptoms, it becomes necessary to check each possible cause separately. Table 22-1 lists some of the more common ailments of small motors, together with

**Table 22-1**  *Squirrel-Cage Motor Problems*

| Symptom and Possible Cause | Possible Remedy |
|---|---|
| **Motor Will not Start** | |
| (a) Overload control tripped | (a) Wait for overload to cool. Try starting again. If motor still does not start, check all the causes as outlined in the following. |
| (b) Power not connected | (b) Connect power to control and control to motor. Check clip contacts. |
| (c) Faulty (open) fuses | (c) Test fuses. |
| (d) Low voltage | (d) Check motor name-plate values with power supply. Also check voltage at motor terminals with motor under load to be sure wire size is adequate. |
| (e) Wrong control connections | (e) Check connections with control wiring diagram. |
| (f) Loose terminal lead connection | (f) Tighten connections. |
| (g) Driven machine locked | (g) Disconnect motor from load. If motor starts satisfactorily, check driven machine. |
| (h) Open circuit in stator or rotor winding | (h) Check for open circuits. |
| (i) Short circuit in stator winding | (i) Check for shorted coil. |
| (j) Winding grounded | (j) Test for grounded winding. |
| (k) Bearing stiff | (k) Free bearings or replace. |
| (l) Grease too stiff | (l) Use special lubricant for special conditions. |
| (m) Faulty control | (m) Check control wiring. |
| (n) Overload | (n) Reduce load. |
| **Motor Noisy** | |
| (a) Motor running single phase | (a) Stop motor, then try to start. (It will not start on single phase.) Check for "open" in one of the lines or circuits. |

suggestions as to possible causes. Most common motor problems can be checked by some test or inspection. The order of making these tests rests with the troubleshooter, but it is natural to make the simplest ones first. For instance, when a motor fails to start, you first inspect the motor connections, since this is an easy and simple thing to do.

## Problem Diagnosis

In diagnosing problems, a combination of symptoms will often give a definite clue to the source of the trouble and hence eliminate other possibilities. For instance, in the case just cited of a motor that will not start, if heating occurs, it offers the suggestion that a short or ground exists in one of the windings and eliminates the likelihood of an open circuit, poor line connection, or defective starter switch.

## Centrifugal Switches

Centrifugal starting switches, found in many types of single-phase fractional-horsepower motors, occasionally are a source of trouble. If the mechanism sticks in the running position, the motor will not start. On the other hand, if stuck in the closed position, the motor will not attain speed and the starting winding heats up quickly. The motor may also fail to start if the contact points of the switch are out of adjustment or coated with oxide. It is important to remember, however, that any adjustment of the switch or contacts should be made only at the factory or authorized service station.

## Commutator-Type Motors

More maintenance is required by motors with commutators. High-speed series-wound motors should not be used on long, continuous-duty cycle applications because the commutator and brushes are a potential source of trouble. Gummy commutators and oil-soaked brushes can cause sluggish action and severe sparking. The commutator can be cleaned with fine sandpaper. However, if pitted spots still appear, the commutator should be reground.

## TROUBLESHOOTING AIDS
## Connection Diagrams

Figure 22-7 shows motor connection diagrams as an aid in troubleshooting. Knowing the arrangement of coils aids in checking out the shorts and grounds as well as opens.

## Small Three-Phase Motor Rating Data

Knowing the current expected to be drawn in normal operation aids in troubleshooting. It is possible to use a clamp-on type meter and check the current drawn by the motor to see if it is excessive or incorrect. Table 22-2 shows the ampere rating of ac motors that operate on three-phase power.

Ampere ratings of motors vary somewhat depending on the type of motor. The values given below are for drip-proof class B-insulated (T frame) 1.15 service

**Fig. 22-7** *Motor connection diagrams.*

Troubleshooting AIDS 373

Split-phase – Reversible only from rest (by transposing leads).

Shunt-wound – Reversible by transposing leads.

Compound-wound – Reversible by transposing armature leads.

Permanent-split capacitor, 4-lead – Reversible by transposing leads.

Permanent-split capacitor 3-lead – Reversible by connecting either side of capacitor to line.

Capacitor-start – Reversible only from rest (by transposing leads).

Two-value capacitor – Reversible only from rest (by transposing leads).

Two-phase 4-lead – Reversible by transposing either phase leads with line.

Three-phase, single voltage – Reversible by transposing any two leads.

3-phase, star-delta 6-lead reversible – For 440 volts connect together white, yellow, and green: Connect to line black, red, and blue. To reverse rotation, transpose any two line leads. For 220 volts connect white to blue, black to green, and yellow to red: Then connect each junction point to line.

To reverse rotation, transpose any two junction points with line.

Reluctance synchronous split-phase – Reversible only from rest (by transposing leads).

Reluctance synchronous permanent-split capacitor, 3-lead – Reversible by connecting either side of capacitor to line.

Reluctance synchronous permanent split-capacitor, 4-lead – Reversible by transposing lead.

Reluctance synchronous capacitor-start – Reversible only from rest (by transposing leads).

***Fig. 22-7*** (Continued)

*Table 22-2* Ampere Rating of Three-Phase, 60-Hertz, AC Induction Motors[a]

| Hp | Syn. Speed (rpm) | 115V | 230V | 380V | 460V | 575V | 2200V | Hp | Syn. Speed (rpm) | 115V | 230V | 380V | 460V | 575V | 2200V |
|---|---|---|---|---|---|---|---|---|---|---|---|---|---|---|---|
| 1/4 | 1800 | 1.90 | 0.95 | 0.55 | 0.48 | 0.38 | | 25 | 3600 | 121.5 | 60.8 | 36.8 | 30.4 | 24.3 | |
| | 1200 | 2.80 | 1.40 | 0.81 | 0.70 | 0.56 | | | 1800 | 129.8 | 64.8 | 39.2 | 32.4 | 25.9 | |
| | 900 | 3.20 | 1.60 | 0.93 | 0.80 | 0.64 | | | 1200 | 131.2 | 65.6 | 39.6 | 32.8 | 26.2 | |
| 1/3 | 1800 | 2.38 | 1.19 | 0.69 | 0.60 | 0.48 | | | 900 | 134.5 | 67.3 | 40.7 | 33.7 | 27.0 | |
| | 1200 | 3.60 | 1.80 | 1.04 | 0.90 | 0.72 | | 30 | 3600 | 147. | 73.7 | 44.4 | 36.8 | 29.4 | |
| | 900 | 3.60 | 1.80 | 1.04 | 0.90 | 0.72 | | | 1800 | 151. | 75.6 | 45.7 | 37.8 | 30.2 | |
| 1/2 | 1800 | 3.44 | 1.72 | 0.99 | 0.86 | 0.69 | | | 1200 | 158. | 78.8 | 47.6 | 39.4 | 31.5 | |
| | 1200 | 4.30 | 2.15 | 1.24 | 1.08 | 0.86 | | | 900 | 164. | 81.8 | 49.5 | 40.9 | 32.7 | |
| | 900 | 4.76 | 2.38 | 1.38 | 1.19 | 0.95 | | 40 | 3600 | 193. | 96.4 | 58.2 | 48.2 | 38.5 | |
| 3/4 | 1800 | 4.92 | 2.46 | 1.42 | 1.23 | 0.98 | | | 1800 | 202. | 101. | 61.0 | 50.4 | 40.3 | |
| | 1200 | 5.84 | 2.92 | 1.69 | 1.46 | 1.17 | | | 1200 | 203. | 102. | 61.2 | 50.6 | 40.4 | |
| | 900 | 6.52 | 3.26 | 1.88 | 1.63 | 1.30 | | | 900 | 209. | 105. | 63.2 | 52.2 | 41.7 | |
| 1 | 3600 | 5.60 | 2.80 | 1.70 | 1.40 | 1.12 | | 50 | 3600 | 241. | 120. | 72.9 | 60.1 | 48.2 | |
| | 1800 | 7.12 | 3.56 | 2.06 | 1.78 | 1.42 | | | 1800 | 249. | 124. | 75.2 | 62.2 | 49.7 | |
| | 1200 | 7.52 | 3.76 | 2.28 | 1.88 | 1.50 | | | 1200 | 252. | 126. | 76.2 | 63.0 | 50.4 | |
| | 900 | 8.60 | 4.30 | 2.60 | 2.15 | 1.72 | | | 900 | 260. | 130. | 78.5 | 65.0 | 52.0 | |
| 1 1/2 | 3600 | 8.72 | 4.36 | 2.64 | 2.18 | 1.74 | | 60 | 3600 | 287. | 143. | 86.8 | 71.7 | 57.3 | |
| | 1800 | 9.71 | 4.86 | 2.94 | 2.43 | 1.94 | | | 1800 | 298. | 149. | 90.0 | 74.5 | 59.4 | |
| | 1200 | 10.5 | 5.28 | 3.20 | 2.64 | 2.11 | | | 1200 | 300. | 150. | 91.0 | 75.0 | 60.0 | |
| | 900 | 11.2 | 5.60 | 3.39 | 2.80 | 2.24 | | | 900 | 308. | 154. | 93.1 | 77.0 | 61.5 | |
| 2 | 3600 | 11.2 | 5.60 | 3.39 | 2.80 | 2.24 | | 75 | 3600 | 359. | 179. | 108. | 89.6 | 71.7 | |
| | 1800 | 12.8 | 6.40 | 3.87 | 3.20 | 2.56 | | | 1800 | 365. | 183. | 111. | 91.6 | 73.2 | |
| | 1200 | 13.7 | 6.84 | 4.14 | 3.42 | 2.74 | | | 1200 | 368. | 184. | 112. | 92.0 | 73.5 | |
| | 900 | 15.8 | 7.90 | 4.77 | 3.95 | 3.16 | | | 900 | 386. | 193. | 117. | 96.5 | 77.5 | |
| 3 | 3600 | 16.7 | 8.34 | 5.02 | 4.17 | 3.34 | | 100 | 3600 | 461. | 231. | 140. | 115. | 92.2 | |
| | 1800 | 18.8 | 9.40 | 5.70 | 4.70 | 3.76 | | | 1800 | 474. | 236. | 144. | 118. | 94.8 | 23.6 |
| | 1200 | 20.5 | 10.2 | 6.20 | 5.12 | 4.10 | | | 1200 | 478. | 239. | 145. | 120. | 95.6 | 24.2 |
| | 900 | 22.8 | 11.4 | 6.90 | 5.70 | 4.55 | | | 900 | 504. | 252. | 153. | 126. | 101. | 24.8 |
| 5 | 3600 | 27.1 | 13.5 | 8.20 | 6.76 | 5.41 | | 125 | 3600 | 583. | 292. | 176. | 146. | 116. | |
| | 1800 | 28.9 | 14.4 | 8.74 | 7.21 | 5.78 | | | 1800 | 584. | 293. | 177. | 147. | 117. | 29.2 |
| | 1200 | 31.7 | 15.8 | 9.59 | 7.91 | 6.32 | | | 1200 | 596. | 298. | 180. | 149. | 119. | 29.9 |
| | 900 | 31.0 | 15.5 | 9.38 | 7.75 | 6.20 | | | 900 | 610. | 305. | 186. | 153. | 122. | 30.9 |
| 7 1/2 | 3600 | 39.1 | 19.5 | 11.8 | 9.79 | 7.81 | | 150 | 3600 | 687. | 343. | 208. | 171. | 137. | |
| | 1800 | 43.0 | 21.5 | 13.0 | 10.7 | 8.55 | | | 1800 | 693. | 348. | 210. | 174. | 139. | 34.8 |
| | 1200 | 43.7 | 21.8 | 13.2 | 10.9 | 8.70 | | | 1200 | 700. | 350. | 210. | 174. | 139. | 35.8 |
| | 900 | 46.0 | 23.0 | 13.9 | 11.5 | 9.19 | | | 900 | 730. | 365. | 211. | 183. | 146. | 37.0 |
| 10 | 3600 | 50.8 | 25.4 | 15.4 | 12.7 | 10.1 | | 200 | 3600 | 904. | 452. | 274. | 226. | 181. | |
| | 1800 | 53.8 | 26.8 | 16.3 | 13.4 | 10.7 | | | 1800 | 915. | 458. | 277. | 229. | 184. | 46.7 |
| | 1200 | 56.0 | 28.0 | 16.9 | 14.0 | 11.2 | | | 1200 | 920. | 460. | 266. | 230. | 184. | 47.0 |
| | 900 | 61.0 | 30.5 | 18.5 | 15.2 | 12.2 | | | 900 | 964. | 482. | 279. | 241. | 193. | 49.4 |
| 15 | 3600 | 72.7 | 36.4 | 22.0 | 18.2 | 14.5 | | 250 | 3600 | 1118. | 559. | 338. | 279. | 223. | |
| | 1800 | 78.4 | 39.2 | 23.7 | 19.6 | 15.7 | | | 1800 | 1136. | 568. | 343. | 284. | 227. | 57.5 |
| | 1200 | 82.7 | 41.4 | 25.0 | 20.7 | 16.5 | | | 1200 | 1146. | 573. | 345. | 287. | 229. | 58.5 |
| | 900 | 89.0 | 44.5 | 26.9 | 22.2 | 17.8 | | | 900 | 1200 | 600. | 347. | 300. | 240. | 60.5 |
| 20 | 3600 | 101.1 | 50.4 | 30.5 | 25.2 | 20.1 | | 300 | 1800 | 1356. | 678. | 392. | 339. | 274. | 69.0 |
| | 1800 | 102.2 | 51.2 | 31.0 | 25.6 | 20.5 | | | 1200 | 1368. | 684. | 395. | 342. | 274. | 70.0 |
| | 1200 | 105.7 | 52.8 | 31.9 | 26.4 | 21.1 | | 400 | 1800 | 1792. | 896. | 518. | 448. | 358. | 91.8 |
| | 900 | 109.5 | 54.9 | 33.2 | 27.4 | 21.9 | | 500 | 1800 | 2220. | 1110. | 642. | 555. | 444. | 116. |

Source: Courtesy of Bodine Electric Company, Chicago.
[a] Ampere ratings of motors vary somewhat. The values given here are for drip-proof class B insulated (T frame) where available, 1.15 service factor, NEMA design B motors. The values represent an average full-load motor current that was calculated from the motor performance data published by several motor manufacturers. In the case of high-torque squirrel-cage motors, the ampere ratings will be at least 10% greater than the values shown.

factor NEMA design B motors. These values represent an average full-load motor current that was calculated from the motor performance data published by several motor manufacturers. In the case of high-torque squirrel-cage motors, the ampere ratings will be at least 10% greater than the values given below.

## POWER SUPPLY DISTURBANCES

Maintenance of equipment is affected by the quality of the power supplied to it. There are a number of problems associated with various line power disturbances. Three types of irregularity that affect power supply are voltage fluctuations, transients, and power outages.

## Voltage Fluctuations

In many states the appropriate public service commissions establish allowable voltage tolerances for utilities. These tolerances are continually monitored, and in most instances every reasonable precaution is taken to stay within these limits. However, some equipment is so sensitive that fluctuations within the tolerance limits can still cause problems (Fig. 22-8).

*Fig. 22-8* Normal power sine wave.

Voltage fluctuations can usually be detected by visible flickering of lights. High- or low-voltage conditions can result in damage to equipment, loss of data, and erroneous readings in monitoring systems (Fig. 22-9).

*Fig. 22-9* Over voltage condition.

Under voltage can result from overloaded power circuits. Intermittent low voltage is typically caused by starting a large heavily loaded motor such as an air conditioner. Over voltage conditions are less common but are more damaging and are seen frequently in facilities with rapidly varying loads (Fig. 22-10).

## Transients

**Voltage Spikes** Short-duration impulses in excess of the normal voltage are called spikes or surges. Although their duration is incredibly brief, a spike may exceed the normal voltage level five- or tenfold. Spikes

*Fig. 22-10* Under voltage condition.

can wipe out data stored in memory, produce output errors, or cause extensive equipment damage. Besides the immediate damage, there are also harder-to-detect effects, particularly reduced service life. Subsequent random failures can be particularly annoying and expensive (Fig. 22-11).

*Fig. 22-11* Voltage spike on a sine wave.

The leading day-to-day cause of small low-energy spikes is the switching on and off of an electrical motor (inductive load switching). Air conditioners, electrical power tools, furnace ignitions, electrostatic copy machines, arc welders, and elevators are particularly guilty of creating voltage spikes. The problems created by the inductive load switching are very common in industrial plants. Larger spikes are typically caused by lightning. A direct lightning hit, of course, is catastrophic but of very low probability. However, a distant lightning strike several miles away may be transmitted through utility power lines and show up as a voltage spike all along the line.

**Electrical Noise** As contrasted to outright equipment damage, computer "glitches" are caused by electrical noise (Fig. 22-12). The same causes of voltage spikes can (at a lower voltage magnitude) cause noise interference. Other electrical noise generators include radio transmitters, fluorescent lights, computers,

**Fig. 22-12** *Noise interference on a power line sine wave.*

business machines, and electrical devices such as light sockets, wall receptacles, plugs, and loose electrical connections. Interaction between system components may generate sufficient noise to cause errors. Although most electronic equipment has some internal noise filtering, equipment located in severe noise environments may encounter some interference. Transients are by far the most common sort of power disturbance and fortunately, often the easiest to correct. However, they may be difficult to detect since they last such a short time.

## Power Outages

Power outages are a total interruption of power supply. An interruption of a mere 15 milliseconds is considered a blackout to sensitive equipment. Power outages can cause problems for equipment users, most critical of which are loss of valuable data and expensive time-consuming reprogramming.

Outages tend to be caused by a larger-scale problem than a transient. Interruptions may be caused, for example, by utility or on-site load changes, on-site equipment malfunction, or by faults on the power system (Fig. 22-13).

**Fig. 22-13** *Power outage with sine wave diminishing to zero.*

## Looking for Shorts

Shorted turns in the winding of a motor behave like a shorted secondary of a transformer. A motor with a shorted winding will draw excessive current while running at no load. Measurement of the current can be made without disconnecting lines. This means that you engage one of the lines with the split-core transformer of the tester. If the ammeter reading is much higher than the full-load ampere rating on the nameplate, the motor is probably shorted.

In a two- or three-phase motor, a partially shorted winding produces a higher current reading in the shorted phase. This becomes evident when the current in each phase is measured.

## MOTORS WITH SQUIRREL-CAGE ROTORS

Loss in output torque at rated speed in an induction motor may be due to opens in the squirrel-cage rotor.

To test the rotor and determine which rotor bars are loose or open, place the rotor in a growler. Engage the split-core ammeter around the lines going to the growler, as shown in Fig. 22-14. Set the switch to the highest current range. Switch on the growler and then set the test unit to the approximate current range. Rotate the rotor in the growler and take note of the current indication whenever the growler is energized. The bars and end rings in the rotor behave similarly to a shorted secondary of a transformer. The growler windings act as the primary. A good rotor will produce approximately the same current indications for all positions of the rotor. A defective rotor will exhibit a drop in the current reading when the open bars move into the growler field.

**Fig. 22-14** *Using a growler to test a rotor.*

## TESTING THE CENTRIFUGAL SWITCH IN A SINGLE-PHASE MOTOR

A defective centrifugal switch may not disconnect the start winding at the proper time. To determine conclusively that the start winding remains in the circuit, place the split-core ammeter around one of the start-winding leads. Set the instrument to the highest current range. Turn on the motor switch. Select the appropriate current range. Observe if there is any current in the start-winding circuit. A current indication signifies that the centrifugal switch did not open when the motor came up to speed (Fig. 22-15).

*Fig. 22-15* Checking the centrifugal switch with a clamp-on meter. (Amprobe Instrument)

## TESTING FOR SHORT CIRCUITS BETWEEN RUN AND START WINDINGS

A short between run and start windings may be determined by using the ammeter and line voltage to check for continuity between the two separate circuits. Disconnect the run- and start-winding leads and connect the instrument as shown in Fig. 22-16. Set the meter on voltage. A full-line voltage reading will be obtained if the windings are shorted to one another.

*Fig. 22-16* Finding a shorted winding using a clamp-on meter. (Amprobe Instrument)

## CAPACITOR TESTING

Defective capacitors are very often the cause of trouble in capacitor-type motors. Shorts, opens, grounds, and insufficient capacity in microfarads are conditions for which capacitors should be tested to determine whether they are good. You can determine a grounded capacitor by setting the instrument on the proper voltage range and connecting it and the capacitor to the line as shown in Fig. 22-17. A

*Fig. 22-17* Finding a grounded capacitor with a clamp-on meter. (Amprobe Instrument)

full-line voltage indication on the meter signifies that the capacitor is grounded to the can. A high-resistance ground is evident by a voltage reading that is somewhat below the line voltage. A negligible reading or a reading of no voltage will indicate that the capacitor is not grounded.

## Measuring Capacity of a Capacitor

To measure the capacity of the capacitor, set the test unit's switch to the proper voltage range and read the line-voltage indication. Then set to the appropriate current range and read the capacitor current indication. During the test, keep the capacitor on the line for a very short period of time because motor starting electrolytic capacitors are rated for intermittent duty (Fig. 22-18). The capacity in microfarads is then computed by substituting the voltage and current readings in the following formula, assuming that a full 60-Hz line is used:

$$\text{microfarads} = \frac{2650 \times \text{amperes}}{\text{volts}}$$

*Fig. 22-18* Finding the size of a capacitor with a clamp-on meter. (Amprobe Instrument)

An open capacitor will be evident if there is no current indication in the test. A shorted capacitor is easily detected. It will blow the fuse when the line switch is turned on to measure the line voltage.

## USING METERS TO CHECK FOR PROBLEMS

The voltmeter and the ohmmeter can be used to isolate various problems. You should be able to read the schematic and make the proper voltage or resistance measurements. An incorrect reading will indicate the possibility of a problem. Troubleshooting charts will aid in isolating the problem to a given system. Once you have arrived at the proper system that may be causing the symptoms noticed, you will then need to use the ohmmeter with the power off to isolate a section of the system. Once you have zeroed in on the problem, you can locate it by knowing what the proper reading should be. Deviation from a stated reading of over 10% is usually indicative of a malfunction, and in most cases the component part must be replaced to assure proper operation and no call-backs.

### Using a Volt-Ammeter for Troubleshooting Electric Motors

Most electrical equipment will work satisfactorily if the line voltage differs ± 10% from the actual nameplate rating. In a few cases, however, a 10% voltage drop may result in a breakdown. Such may be the case with an induction motor that is being loaded to its fullest capacity on both start and run. A 10% loss in line voltage will result in a 20% loss in torque.

The full-load current rating on the nameplate is an approximate value based on the average unit coming off the manufacturer's production line. The actual current for any unit may vary as much as ±10% at rated output. However, a motor whose load current exceeds the rated value by 20% or more will reduce the life of the motor due to higher operating temperatures, and the reason for excessive current should be determined. In many cases it may simply be an overloaded motor. The percentage increase in load will not correspond with the percentage increase in load current. For example, in the case of a single-phase induction motor, a 35% increase in current may correspond to an 80% increase in torque output.

Operating conditions and behavior of electrical equipment can be analyzed only by actual measurement. A comparison of the measured terminal voltage and current will check whether the equipment is operating within electrical specifications.

A voltmeter and an ammeter are needed for the two basic measurements. To measure voltage, the test leads of the voltmeter are in contact with the terminals of the line under test. To measure current, the conventional ammeter must be connected in series with the line so that the current will flow through the ammeter.

To insert the ammeter means that you have to shut down the equipment, break open the line, and connect the ammeter and then start up the equipment to read the meter. You have to do the same to remove the meter once it has been used. Then there are other time-consuming tests that may have to be made to locate the problem. However, all this can be eliminated by the use of a clamp-on volt-ammeter (Fig. 22-19).

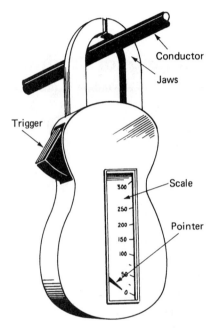

**Fig. 22-19** *Clamp-on volt-ammeter. (Amprobe Instrument)*

### Clamp-On Volt-Ammeter

The pocket-size volt-ammeter shown in Fig. 22-19 is the answer to most troubleshooting problems on the job. The line does not have to be disconnected to obtain a current reading. The meter works on the transformer principle that picks up the magnetic lines surrounding a current-carrying conductor and then presents this as a function of the entire amount flowing through the line. Remember that earlier we discussed how the magnetic field strength in the core of the transformer determines the amount of current in the secondary. Well, the same principle is used here to detect the flow of current and how much.

To get transformer action, the line to be tested is encircled with the split-type core simply by pressing the trigger button. Aside from measuring terminal voltages and load currents, the split-core ammeter-voltmeter can be used to track down electrical difficulties in electric motor repair.

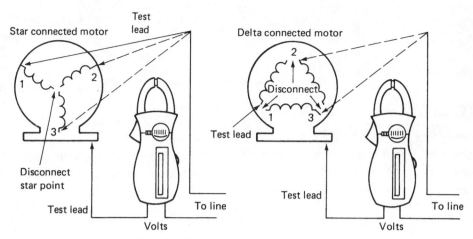

*Fig. 22-20*  Grounded phase of a motor.

## Looking for Grounds

To determine whether a winding is grounded or has a very low value of insulation resistance, connect the unit and test leads as shown in Fig. 22-20. Assuming that the available line voltage is approximately 120 V, use the unit's lowest voltage range. If the winding is grounded to the frame, the test will indicate full-line voltage.

A high-resistance ground is simply a case of low insulation resistance. The indicated reading for a high-resistance ground will be a little less than line voltage. A winding that is not grounded will be evidenced by a small or negligible reading. This is mainly due to the capacitive effect between the windings and the steel lamination.

To locate the grounded portion of the windings, disconnect the necessary connection jumpers and test. Grounded sections will be detected by a full-line voltage indication.

## Looking for Opens

To determine whether a winding is open, connect test leads as shown in Figs. 22-21 and 22-22. If the winding is open, there will be no voltage indication. If the circuit

*Fig. 22-21*  Isolating an open phase.

is not open, the voltmeter indication will read full-line voltage.

## TROUBLESHOOTING GUIDE

One of the quickest ways to troubleshoot is to check out symptoms and possible causes in a chart. This allows for quick isolation of the cause and suggests possible corrections. Both three-phase motors and their starters can be checked quickly this way. Tables 22-3 and 22-4 will aid in troubleshooting motors and starters.

*Fig. 22-22*  Finding an open phase.

**380**  Troubleshooting and Maintenance

*Table 22-3* Three-Phase Motor Troubleshooting Guide

| Symptom | Possible Causes | Correction |
|---|---|---|
| High input current (all three phases) | Accuracy of ammeter readings | First check accuracy of ammeter readings on all three phases. |
| Running idle (discounted from load) | High line voltage: 5 to 10% over nameplate | Consult power company—possibly decrease by using lower transformer tap. |
| Running loaded | Motor overloaded | Reduce load or use large motor. |
| | Motor voltage rating does not match power system voltage | Replace motor with one of correct voltage rating. Consult power company—possibly correct by using a different transformer tap. |
| Unbalanced input current (5% or more deviation from the average input current) | Unbalanced line voltage due to:<br>a. Power supply<br>b. Unbalanced system loading<br>c. High-resistance connection<br>d. Undersized supply lines | Carefully check voltage across each phase *at the motor terminals* with good, properly calibrated voltmeter. |
| *Note*: A small voltage imbalance will produce a large current imbalance. Depending on the magnitude of imbalance and the size of the load, the input current in one or more of the motor input lines may greatly exceed the current rating of the motor. | Defective motor | If there is doubt as to whether the trouble lies with the power supply or the motor, check as follows:<br>Rotate *all three input power lines* to the motor by one position (i.e., move line 1 to motor lead 2, line 2 to motor lead 3 and 3 to motor lead 1).<br>a. If the unbalanced current pattern follows the *input power lines*, the problem is the power supply<br>b. If the unbalanced current pattern follows the *motor leads*, the problem is in the motor.<br>Correct the voltage balance of the power supply or replace the motor, depending on the answer to a and b above. |
| Excessive voltage drop (more than 2 or 3% of nominal supply voltage) | Excessive starting or running load | Reduce load. |
| | Inadequate power supply | Consult power company. |
| | Undersized supply lines | Increase line sizes. |
| | High-resistance connections | Check motor leads and eliminate poor connections |
| | Each phase lead runs in separate conduits | All three-phase leads must be in a single conduit, according to the *National Electrical Code*\* (This applies only to metal conduit with magnetic properties.) |
| Overload relays tripping upon starting 9 see also "Slow Starting" | Slow starting (10–15 seconds or more) due to high inertia load | Reduce starting load. Increase motor size if necessary. |
| | Low voltage at motor terminals | Improve power supply and/or increase line size. |
| Running loaded | Overloaded | Reduce load or increase motor size. |
| | Unbalanced input current | Balance supply voltage |
| | Single phasing | Eliminate. |
| | Excessive voltage drop | Eliminate (see above). |
| | Too frequent starting or intermittent overloading | Reduce frequency of starts and overloading or increase motor size. |
| | High ambient starter temperatures | Reduce ambient temperature or provide outside source of cooler air. |
| | Wrong-size relays | Correct size per nameplate current of motor. Relays have built in allowances for service factor current. Refer to *National Electrical Code*® |
| Motor runs excessively hot | Overloaded | Reduce load or load peaks and number of starts in cycle or increase motor size |
| | Blocked ventilation<br>a. TEFCs<br>b. ODPs | Clean external ventilation system; check fan.<br>Blow out internal ventilation passages.<br>Eliminate external interference to motor ventilation. |
| | High ambient temperature over 40°C or 105°F | Reduce ambient temperature or provide outside source of cooler air. |
| | Unbalanced input current | Balance supply voltage. Check motor leads for tightness. |
| | Single phased | Eliminate. |
| Won't start (just hums and heats up) | Single phased | Shut power off. Eliminate single phasing<br>Check motor leads for tightness. |
| | Rotor or bearings locked | Shut power off. Check shaft for freeness of rotation.<br>Be sure proper sized overload relays are in each of the three phases of starter.<br>Refer to *National Electric Code*® |

(*Continued*)

**Table 22-3** Three-Phase Motor Troubleshooting Guide (Continued)

| Symptom | Possible Causes | Correction |
|---|---|---|
| Runs noisy under load | Single phased | Shut power off. If motor cannot be restarted, it is single phased. Eliminate single phasing. Be sure that proper-sized overload relays are in each of the three phases of the starter. Refer to *National Electrical Code®*. |
| Slow starting (10 or more seconds on small motors; 15 or more seconds on large motors) | | |
| Across the line start | Excessive voltage drop (5–10% voltage drop causes 10–20% or more drop in starting torque) | Consult power company; check system. Eliminate voltage drop. |
| | High-inertia load | Reduce starting load or increase motor size. |
| Reduced voltage start | Excessive voltage drop | Check and eliminate. |
| | Loss of starting torque | |
| Wye-delta | Starting torque reduced to 33% | Reduce starting load or increase motor size. |
| PWS | Starting torque reduced to 50% | Choose starting method with higher starting torque. |
| Autotransformer | Starting torque reduced 25 to 64 | Reduce time delays between first and second steps on starter; get motor across the line sooner. |
| Load speed appreciably below nameplate speed | Overload | Reduce load or increase voltage. |
| | Excessively low voltage | *Note*: A reasonable overload or voltage drop of 10–15% will reduce speed only 1–2%. A report of any greater drop would be questionable. |
| | Wrong nameplate | If speed is off appreciably (i.e., from 1800 to 1200 rpm, check Lincoln code stamp (on top of stator) with nameplate. If codes do not agree, replace with motor of proper speed. |
| | Inaccurate method of measuring rpm | Check meter using another device or method |
| Excessive vibration (mechanical) | Out of balance<br>a. Motor mounting<br>b. Load<br>c. Sheaves or coupling<br>d. Motor<br>e. Misalignment on close-coupled application | Be sure motor mounting is tight and solid<br>Disconnect belt or coupling; restart motor. If vibration stops, the unbalance was in load.<br>Remove sheave or coupling; securely tape 1/2 key in shaft keyway and restart motor. If vibration stops, the imbalance was in the sheave or coupling.<br>If the vibration does not stop after checking a, b, and c above, the imbalance is in the motor; replace the motor<br>Check and realign motor to the driven machine. |
| Noisy bearings (listen to bearings) | | |
| Smooth midrange hum | Normal fit | Bearing OK. |
| High whine | Internal fit of bearing too tight | Replace bearing; check fit. |
| Low rumble | Internal fit of bearing too loose | Replace bearing; check fit. |
| Rough clatter | Bearing destroyed | Replace bearing; avoid:<br>a. Mechanical damage<br>b. Excess greasing<br>c. Wrong grease<br>d. Solid contaminants<br>e. Water running into motor<br>f. Misalignment on close-coupled application.<br>g. Excessive belt tension |
| Mechanical noise | Driven machine or motor noise? | Isolate motor from driven machine; check difference in noise level. |
| | Motor noise amplified by resonant mounting | Cushion motor mounting or dampen source of resonance. |
| | Driven machine noise transmitted to motor through drive | Reduce noise of driven machine or dampen transmission to motor. |
| | Misalignment on close-coupled application | Improve alignment. |

Source: Courtesy of Lincoln Electric Co.

*Table 22-4* Troubleshooting Motor Starters

| Symptoms | Possible Causes | Correction |
|---|---|---|
| **Magnetic and Mechanical Parts** | | |
| Noisy magnet Humming | Misalignment or mismating of magnet pole faces | Realign or replace magnet assembly. |
| | Foreign matter on pole face (dirt, lint, rust, etc.) | Clean (but do not file) pole faces and realign if necessary. |
| | Low voltage applied to coil | Check system and coil voltage. Observe voltage variations during startup time. |
| Loud buzz | Broken shading coil | Replace shading coil and/or magnet assembly. |
| Failure to pick up and seal in | Low voltage | Check system voltage, coil voltage, and watch for voltage variations during start. |
| | Wrong magnet coil or wrong connection | Check wiring, coil nomenclature, etc. |
| | Coil open or shorted | Check with an ohmmeter and when in doubt, replace. |
| | Mechanical obstruction | Disconnect power and check for free movement of magnet and contact assembly. |
| Failure to drop out | "Gummy" substance on pole faces or magnet slides | Clean with nonvolatile solvent, degreasing fluid, possibly gasoline (with caution). |
| | Voltage or coil not removed | Shorted seal-in contact (exact cause found by checking coil circuit). |
| | Worn or rusted parts causing binding | Clean or replace worn parts. |
| | Residual magnetism due to lack of air gap in magnet path | Replace any worn magnet parts or accessories. |
| **Clean** | | |
| Contact clatter (source is probably from magnetic assembly | Broken shading coil | Replace assembly. |
| | Poor contact continuity in control circuit | Improve contact continuity or use holding-circuit interlock (three-wire control). |
| | Low voltage | Correct voltage condition. Check momentary voltage dip during start. |
| Welding | Abnormal inrush of Current | Use larger contactor or check for grounds, shorts, or excessive motor load current. |
| | Rapid jogging | Install larger device rated for jogging service or caution operator. |
| | Insufficient tip pressure | Replace contact springs; check contact carrier for deformation or damage. |
| | Low voltage preventing magnet from sealing | Correct voltage condition. Check momentary voltage dip during starting. |
| | Foreign matter preventing contacts from closing | Clean contacts with nonvolatile solvent. Contactors, starters, and control accessories used with very small current or low voltage should be cleaned with solvent and then with acetone to remove the solvent residue. |
| | Short circuit | Remove short fault and check to be sure that fuse or breaker size is correct. |
| Short contact life or overheating | Filing or dressing | Do not file silver contacts. Rough spots for discoloration will not harm them or impair their efficiency |
| | Interrupting excessively high currents. | Install larger device or check for grounds, shorts, or excessive motor currents. |
| **Coils** | | |
| Open circuit | Mechanical damage | Handle and store coils carefully. Replace coil. |
| Cooked coil (overheated) | Overvoltage or high ambient temperature | Check application and circuit. Coils will operate over a range of 85–110% rated voltage. |
| | Incorrect coil | Check rating and, if incorrect, replace with proper coil. |
| | Shorted turns caused by mechanical damage or corrosion | Replace coil. |
| | Undervoltage, failure of magnet to seal in | Correct system voltage. |
| | Dirt or rust on pole faces, increasing air gap | Clean pole faces. |
| | Sustained low voltage | Remedy according to local code requirements, low voltage system protection, etc. |
| **Overload Relays** | | |
| Nuisance tripping | Sustained overload | Check for motor or electrical equipment grounds and shorts, as well as excessive motor currents due to overload. Check motor winding resistance to ground. |
| | Loose connections | Clean connections and tighten. This includes load wires and heater element mounting screws. |
| | Incorrect heater | Check heater sizing and ambient temperature. |
| Failure to trip out (causing motor burn-out) | Mechanical binding, dirt, corrosion, etc. | Clean or replace. |

*(Continued)*

**Table 22-4** Troubleshooting Motor Starters (Continued)

| Symptoms | Possible Causes | Correction |
|---|---|---|
| | Incorrect heater or heaters omitted and jumper wires used instead | Recheck ratings and heater size. Correct if necessary. |
| | Wrong calibration adjustment | Consult factory. Calibration adjustment is not normally recommended unless factory supervised. It is customary to return units to factory for check and calibration. |
| **Manual Starters** | | |
| Failure to operate (mechanically) | Mechanical parts, including springs, worn or broken | Replace parts as needed. |
| | Welded contacts due to misapplication or other abnormal cause | Replace contacts and recheck operation. |
| Trips out prematurely | Motor overload, incorrect heaters, or misapplication | Check conditions and replace or adjust as needed. |
| **Pushbuttons** | | |
| Button inoperable Mechanical | Shaft binding due to dirt or residue | Check, clean, and clear. |
| | Contact board spring broken | Replace contact board. |
| Electrical | Contaminated contacts and corrosion | Clean. |
| | Excessive jogging | Install larger device or check rated for jogging or caution operator. |
| | Weak contact pressure | Replace contact springs; check contact carrier for deformation or damage. |
| | Dirt or foreign matter on contact surface | Clean contacts with nonvolatile solvent. |
| | Short circuits | Remove short fault and check to be sure fuse or breaker size is correct. |
| | Loose connection | Clean and tighten. |
| | Sustained overload | Install larger device or check for excessive motor load current. |
| | Excessive wear | Higher-than-normal voltage will cause unnecessary forces that may result in mechanical wear. |
| Contacts, supports, discoloring | Loose connections | Tighten hardware or replace. |

Source: Courtesy of Square D.
Note: Any contact replacement should include a complete set replacement, including support springs, screws, etc.

## MOTOR LIFE

The stator windings of integral horsepower ac motors are capable of full-power operation for many years. However, winding life can be shortened by any combination of the following:

- **Mechanical damage** produces weak spots in the insulation. It can occur during maintenance of the motor or result from such operating problems as severe vibration.

- **Excessive moisture** encountered in service causes deterioration of the insulation.

- **High dielectric stress** such as voltage surges or excess input current, can cause overheating and insulation deterioration.

- **High temperature** reduces the ability of the insulation system to withstand mechanical or electrical abuse. Over temperature is usually a result of poor installation or misapplication of the motor.

Regardless of the reason for failures, the obvious result is thermal degradation of the insulation-or burnouts. The rate of insulation degradation is increased by higher temperature. In fact, insulation life is reduced by about half for each 10°C increase of winding temperature. Therefore, long winding life requires normal operating temperatures.

## Ventilation

Forced ventilation is generally an inherent design feature of induction motors. Decreased cooling air volume caused by blocked air passages, blower failure, or low air density at higher altitudes leads to overheating and shortened winding life.

## Ambient Temperature

The insulation system of motors is usually designed to operate at a maximum 40°C (104°F) ambient temperature. Any increase of ambient over 40°C requires de-rating the motor or its expected life will be shortened. Factors that raise input air temperature include placement of motors in discharge airstreams from other equipment and high-temperature locations.

To calculate the de-rating required for high ambient temperatures, multiply the ambient factor obtained from Fig. 22-23 by the rated horsepower of the motor. The ambient factor from Fig. 22-23 can also be used to up-rate motors used in ambient temperatures under 40°C.

**Fig. 22-23** *De-rating a motor by ambient temperature.* (Lincoln Electric Co.)

Whenever a motor is de-rated or up-rated, the starting, pull-up, and breakdown torques remain the same as the nameplate rating, but the bearings, shaft, and other components may be subjected to life that is a function of the new rating.

# PERFORMANCE CHARACTERISTICS

Performance characteristics of motors change when they are operated on high or low voltages at rated frequency. Table 22-5 shows the characteristic and the change when operated 10% above rated voltage and 10% below rated voltage. Winding failures resulting from extreme voltage variation are identical to those of overloads because the input current is uniformly excessive.

## Voltage Unbalance

Voltage unbalance occurs when the phase voltages differ from one another. At one extreme this phenomenon occurs as single phasing. It can also occur in a more subtle form as voltage differences among the three phases.

**Single Phasing**  A single-phase motor at standstill will not start and the large inrush current overheats the windings rapidly. If the single phasing occurs when running, the motor will continue to supply torque, but the input current to the remaining two phases increases, causing overheating.

**Voltage Unbalance**  By definition

$$\text{voltage unbalance (\%)} = 100 \times \frac{\text{maximum voltage deviation from average voltage}}{\text{average voltage}}$$

When the voltage varies among the three phases, the unbalanced voltage raises the current in one or two phases, causing overheating.

*Table 22-5* Performance Characteristic Changes When Motors Are Operated on High or Low Voltages (at Rated Frequency)

| Performance Characteristic | Change, Relative to Performance at Rated Voltage | |
|---|---|---|
| | When Actual Voltage Is 10% above Rated Voltage | When Actual Voltage Is 10% below Rated Voltage |
| Starting-Pull-Up–Breakdown | | |
|   Current | Increases 10–12% | Decreases 10–12% |
|   Torque | Increases 21–25% | Decreases 19–23% |
| Idle | | |
| Current | | |
|   1800/1200/900-rpm motors | Increases 12–39% | Decreases 10–21% |
|   3600-rpm motors | Increases 28–60% | Decreases 21–34% |
| Rated Load | | |
| Current (1800- and 3600-rpm motors only) | | |
|   143T–182T | Varies −4 to +11% | Varies −11 to +4% |
|   184T–256T | Decreases 1–10% | Increases 1–10% |
|   284T-445T | Decrease 0–7% | Increases 0–7% |
| Current (1200- and 900-rpm motors only): These motors do not follow the pattern above. Most will respond to a 10% overvoltage with an increase (as much as 15%) in input current. A similar reduction in input current is characteristic of a 10% reduction in input voltage. | | |
| Power factor[a] | Decreases 5–8% | Increases 2% |
| Efficiency[a] | Little change | Decreases 2% |
| Speed | Increases 1% | Decreases 1.5% |
| Percent slip | Decreases 1.0 | Increases 1.5 |

[a] Note at $\frac{3}{4}$ and $\frac{1}{2}$ load, these changes are approximately the same.

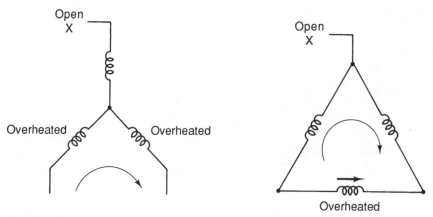

**Fig. 22-24** Heating effects on wye- or delta-connected windings when voltage is unbalanced or single phasing occurs.

Voltage unbalances are caused by the following conditions:

- Unequal loading of the three-phase system
- Unequal tap settings on transformers
- Poor connections in the power supply
- Open delta transformer systems
- Improper function of capacitor banks

The winding failure pattern from either single phasing or voltage unbalance will be similar. Figure 22-24 shows two phases that are overheated in a Y-connected motor. A delta-connected motor will have one overheated phase.

## MOTOR PROTECTION

In each of the abnormal operating conditions listed over loading, voltage regulation, and voltage unbalance—the risk of excessive winding temperature exists because input current is higher than nameplate current. Motor controls and protective devices must prevent input current from exceeding nameplate current for extended periods.

The *National Electrical Code* has established standards that apply to the control and protection of motors and associated circuits. For example, Fig. 22-25 is a general circuit for the installation of squirrel-cage inductor motors. Consult the *National Electrical Code* for complete requirements for all installations.

**Fig. 22-25** Diagram showing protective devices.

Motor controllers and their associated circuits will vary among different models as well as different controller manufacturers. Figure 22-26 is a functional schematic diagram of a motor controller that incorporates overload protection. Good engineering practice emphasizes using three overload relays, one in each phase line of the motor, to give protection from voltage unbalance conditions. The size of the controller and overload protection relays must be in accordance with the controller manufacturer's specifications as well as the *National Electrical Code*.

**Fig. 22-26** A motor control consists of a contactor and overload relays.

As a further safeguard against motor failure due to prolonged locked rotor conditions, the overload protection should trip out in 15 seconds or less at locked-rotor current.

## DC MOTOR PROBLEMS

One of the problems associated with dc motors is high maintenance costs in both time and materials. They do have a tendency to need attention at the commutator and brush level. Figure 22-27 shows some of the problems associated with worn or damaged commutators.

(A) PITCH BAR-MARKING produces low or burned spots on the commutator surface that equals half or all the number of poles on the motor.

(B) STREAKING on the commutator surface denotes the beginning of serious metal transfer to the carbon brush.

(C) HEAVY SLOT BAR-MARKING involves etching of the trailing edge of commutator bar in relation to the numbered conductors per slot.

(D) THREADING of commutator with fine lines is a result of excessive metal transfer leading to resurfacing and excessive brush wear.

(E) COPPER DRAG is an abnormal amount of excessive commutator material at the trailing edge of bar. Even though rare, flashover may occur if not corrected.

(F) GROOVING is caused by an abrasive material in the brush or atmosphere.

**Fig. 22-27** *Worn or damaged commutators.* (Reliance)

Figure 22-27A. The pitch bar-marking produces low or burned spots on the commutator surface that equals half or all the number of poles of the motor.

Figure 22-27B. Streaking on the commutator surface denotes the beginning of serious metal transfer to the carbon brush.

Figure 22-27C. Heavy slot bar-marking involves etching of the trailing edge of the commutator bar in relation to the numbered conductors per slot.

Figure 22-27D. Threading of the commutator with fine lines is a result of excessive metal transfer leading to resurfacing and excessive brush wear.

Figure 22-27E. Copper drag is an abnormal amount of excessive commutator material at the trailing edge of the bar. Although rare, flashover may occur if not corrected.

Figure 22-27F. Grooving is caused by an abrasive material in the brush or the atmosphere. Satisfactory commutator surfaces are shown in Fig. 22-28. Determine desirable and undesirable commutator surfaces by checking these with those shown in the previous figure. Figure 22-29 shows what can be done to correct the situation.

Figure 22-28A. Light tan film over the entire commutator surface is a normal condition.

(A) LIGHT TAN FILM over entire commutator surface is a normal condition.

(B) MOTTLED SURFACE with random film patterns is satisfactory.

(C) SLOT BAR MARKINGS appearing on bars in a definite pattern depicts normal wear.

(D) HEAVY FILM with uniform appearance over entire commutator surface is acceptable.

**Fig. 22-28** *Satisfactory commutator surfaces.* (Reliance)

| | Vibration | Brush Pressure (light) | Unbalanced Shunt Field | Armature Connection | Light Electrical Load | Electrical Overload | Electrical Adjustment | Contamination | | Type of Brush In Use | |
|---|---|---|---|---|---|---|---|---|---|---|---|
| | | | | | | | | Gas | Abrasive Dust | Abrasive Brush | Porous Brush |
| Pitch bar-marking | ■ | | ■ | ■ | | ■ | | | | ■ | |
| Slot bar-marking | | ■ | | | | ■ | ■ | | | ■ | |
| Copper Drag | ■ | | ■ | | ■ | ■ | | ■ | | ■ | ■ |
| Streaking | | ■ | | ■ | | ■ | | ■ | ■ | | ■ |
| Threading | | ■ | | ■ | | ■ | | ■ | ■ | ■ | ■ |
| Grooving | | | | | | | | | ■ | ■ | ■ |

**Fig. 22-29** *Troubleshooting causes of worn commutators.* (Reliance)

**Fig. 22-30** (A) Brush terminology. (B) Special style brushes. (Helwig Carbon)

DC Motor Problems 389

(B)

*Fig. 22-30* (Continued)

Figure 22-28B. Mottled surface with random film patterns is satisfactory.

Figure 22-28C. Slot bar markings appearing on bars in a definite pattern depicts normal wear.

Figure 22-28D. Heavy film with uniform appearance over entire commutator surface is acceptable.

Figure 22-30 shows 30 illustrations of brushes used in dc motors and generators.

## SOLID-STATE EQUIPMENT TROUBLESHOOTING

Troubleshooting is easier with solid-state equipment since it is usually made with printed circuit boards or modules that can be identified as the cause of the problem and removed to be exchanged by another board or circuit module of the same type. Many are removed by pulling the card from clips and others have plugs that have to be disconnected.

Most of the devices are self-diagnosing. LEDs are used to indicate possible problems not only in the circuitry or equipment as a whole but also in individual devices. In fact, many pieces of equipment are self-diagnosing to the point of telling you what to replace to repair the device.

In many of these devices a diode is either shorted or open. You can check the diode using a VOM and changing the probes across the diode. The diode, if good, will read high resistance in one direction and low resistance in the other. If the diode is shorted, there will be a low-resistance reading in both directions. If the diode is open, there will be an infinite reading in both directions. If a diode is found open or shorted, it must be replaced with one of the same size and rating.

## REVIEW QUESTIONS

1. What is an indication of losing a ground in an electrical system?
2. How does the GFCI work?
3. What is the main duty of ball bearing lubrication?
4. What causes bearing failure?
5. Why is oil viscosity important to a motor?
6. How often should motors be oiled?
7. Why are commutator motors in need of more maintenance?
8. What three irregularities in a power supply affect motor power supplies?
9. Define voltage spikes. What causes them?
10. What is electrical noise? What is its source?
11. What are transients? What is their source?
12. What does a growler do?
13. How do you test for a grounded capacitor?
14. What is the main advantage of a clamp-on meter?
15. What does high temperature do to a motor?
16. What does ambient temperature mean?
17. What causes motor winding failures?
18. Is a light tan film over an entire commutator surface normal?
19. How is a diode checked with an ohmmeter?

# 23 CHAPTER

# Careers in Electricity

## PERFORMANCE OBJECTIVES

After studying this chapter, you will be able to:

1. Identify the prominent career areas in electricity.
2. List the educational requirements necessary to become an electrical engineer.
3. Identify the job responsibilities of a maintenance electrician.
4. Define entrepreneur.
5. List the four similar problems faced by all entrepreneurs.
6. Identify the qualities of an effective leader.
7. Identify the responsibilities of an industrial electrician.
8. Identify the responsibilities of a motor control electrician.

## Industrial Electrician

The industrial electrician requires a 4 to 5-year apprenticeship with 28 credits required to graduate. During this time the apprentice learns to install, maintain, and repair machinery for an industrial employer. Gain knowledge about ac/dc fundamentals, motors, hydraulics/pneumatics, programmable logic controls, and transformers. In addition, study topics such as codes, safety issues, schematic drawing and schematic print reading are mastered.

## INDUSTRIES THAT EMPLOY ELECTRICIANS

The field of electricity offers a variety of career opportunities. Electricians are employed by many different industries, from steel plants to drug manufacturers, from coal mines to the merchant marine.

Many industries that rely heavily on electrical equipment hire full-time electricians to keep that equipment in good working order. For example, the aluminum industry needs electricians to install and repair electrical fixtures, apparatus, and control equipment. In the iron and steel industry, electricians install wiring and fixtures, and hook up electrically operated equipment. Electrical repairers (motor inspectors) keep the wiring, motors, switches, and other electrical equipment in good working condition. The paper industry employs electricians to repair wiring, motors, control panels, and switches.

Railroad companies employ electrical workers to install and maintain the wiring and electrical equipment in locomotives, cars, and railroad buildings. Some workers also lay and maintain power lines.

Ships at sea need their own crew of maintenance workers on board. In the merchant marine industry, the ship's electrician repairs and maintains electrical equipment, such as generators and motors. The electrician tests wiring for short circuits and removes and replaces fuses and defective lights.

In hospitals, the lives of patients often depend on equipment powered by electricity. Proper maintenance and quick repair of this equipment are both vital. Many hospitals employ full-time electricians and electronics technicians for this type of maintenance.

Factories that make medicines also rely on electricians to keep equipment working properly and preventing costly breakdowns. Electricians who work for these companies also install and repair various types of electrical equipment.

Electricians are involved in the coal mining industry. They check and install electrical wiring in and around the mines. In addition, they help, repair, and maintain the machinery used for mining. Today's cars have a number of electrically operated devices and systems as well as onboard computers. In the auto industry, electrical engineers design the car's electrical systems, such as on board computers, the ignition system, lights, and accessories.

In the construction industry, electricians assemble install wire systems for heat, light, power, air-conditioning, and refrigeration. Electrical inspectors then check these systems to make sure they work properly and comply with electrical codes and standards. The inspectors visit worksites to inspect new and existing wiring, lighting, sound and security systems, as well as the generating equipment. They also check the installation of all electrical wiring for heating and air-conditioning systems, kitchen appliances, motors, and other components.

## CAREERS IN ELECTRICITY

Although each industry has its own special requirements, careers in electricity can be divided into general categories:

- Electrical engineer
- Industrial electrician
- Construction electrician
- Maintenance electrician
- Electric motor control electrician

There are other types of electricans who specialize:

- Industrial electrician (motor controls)
- Residential electrician (homes and apartments)
- Maintenance electrician (all areas where needed)

- Inside electrician (usually a maintenance electrician for an institution)
- Limited energy electrician (electrician who works with low voltage installations—closed circuit TV, communications systems, and cable problems)

## Electrical Engineer

Electrical engineers design, develop, and supervise the manufacture of electrical and electronic equipment. The projects they work on include electric motors and generators, communications equipment, electronic equipment such as heart pacemakers, pollution-measuring devices, radar, computers, lasers, missile guidance systems, and electrical applicances of all kinds. They also design and operate facilities for the generation and distribution of electric power.

Usually, electrical engineers specialize in one of the major areas such as electronics, computers, electrical equipment, or power. Within these areas, there are still more specialized fields such as missile guidance and tracking systems.

About 400,000 people are employed as electrical engineers. Most work for makers of electrical and electronic equipment, aircraft and parts, business machines, and professional and scientific equipment. Many work in the communications industry for electric light and power companies. Electrical engineers also work for government agencies, for construction firms, for engineering consultants, or as independent consulting engineers. Others teach in colleges and universities.

Electricial engineers are responsible for design and development of a wide variety of equipment. A bachelor's degree in engineering is required for most beginning engineering jobs. In a typical 4-year curriculum, the first 2 years are spent studying basic sciences (mathematics, physics, introductory engineering) and the humanities, social sciences, and English. The last 2 years are devoted mostly to specialized engineering courses.

Graduate training is now required for many jobs, particularly for teaching and for jobs in specialized areas. A number of colleges and universities offer 5-year master's degree programs.

All 50 states and the District of Columbia require licensing for engineers whose work may affect life, health, or property, or who offer their services to the public. In order to obtain a license, an engineer usually must have a degree from an accredited engineering college, 4 years of related work experience, and must pass a state examination. They are called professional engineers (PEs).

Engineering graduates usually begin work under the supervision of experienced engineers. With experience and proven ability, they advance to positions of greater responsibility. Electrical engineers should be able to work as part of a team and have creativity, an analytical mind, and a capacity for detail. They should be able to express their ideas well, both orally and in writing.

The employment outlook for electrical engineers is good. Increased demand for computers, for electrical and electronic consumer goods, for military electronics, and for communications and power generating equipment will increase the need for engineers in these fields.

## Construction Electrician

As stated earlier, construction electricians assemble and install wire systems for heat, light, power, air-conditioning, and refrigeration. They also install electrical machinery electronic equipment, controls, and signal and communications systems.

Construction electricians must be able to read blueprints and specifications. For safety reasons, they must follow the regulations in the *National Electrical Code*, and their work must also conform to local electrical codes.

There are an estimated 400,000 construction electricians in the United States. Most work for electrical contractors or are self-employed as contractors. Some work for government agencies and businesses that do their own electrical work. Training for a career as a construction electrician usually involves a 4-year apprenticeship program. Apprenticeship programs are sponsored and supervised by local union-management committees. The programs include both classroom and on-the-job training. To qualify for an apprenticeship, an applicant usually must be a high school or vocational school graduate. Courses in electricity, electronics, mechanical drawing, science, and shop provide a good background.

Physical strength is not essential, but skill with one's hands, agility, and good health are important. Good color vision is essential because electrical wires are often identified by color.

Electricians who install residential wiring and equipment are usually referred to as construction electricians since they usually work with new or remodeled housing. This type of work also serves to introduce most entry level people to electrical work. By doing on the job tasks for a number of years, the electrician's helper is also an assistant who may not be an apprentice working with a union contractor, but with a small town-skilled electrician.

In most cities a license is required for employment. The electrician must pass a test which requires a thorough knowledge of the craft and of state and local building codes.

Employment outlook is good; the number of construction electricians is expected to grow faster than the average for all occupations. In any given year, the number of employed persons depends on the amount of construction activity. When jobs are not available, however, construction electricians may be able to transfer to related work. For example, they may work as maintenance electricians.

## Maintenance Electrician

Maintenance electricians keep lighting systems, transformers, generators, and other electrical equipment in good working order. They may also install new equipment.

The duties of maintenance electricians depend on where they are employed. Electricians working in large factories may repair particular items such as motors and welding machines. Those in office buildings and small plants usually fix all kinds of electrical equipment. Regardless of location, electricians spend much of their time doing preventive maintenance. They check equipment to locate and correct defects before breakdowns occur. An estimated 300,000 people work as maintenance electricians. More than half are employed by manufacturing industries. Many others work for utilities, mines, railroads, or the government.

Most maintenance electricians learn their trade on the job or through apprenticeship programs. The apprenticeship programs and the requirements for getting into the programs are the same as those for construction electricians. Hand skills, agility, and good health are important. Good color vision is required. Maintenance electricians check equipment to locate defects before machine failure results.

Because of increased use of electrical and electronic equipment by industry, the demand for maintenance electricians will increase. Growth in the number of job openings is expected to be steady since the demand for maintenance electricians is not very sensitive to ups and downs in the economy.

## Limited Energy Electrician

Limited energy electricians install, maintain, replace, and repair electrical systems and equipment of under 100-V amperes.

The work may include:

- Basic telephone cable installation and termination
- Telephone station and control equipment installation for telephone systems
- Fire alarms system installation and repair
- Intrusion/burglar alarm system installation and repair
- Audio/visual system installation and repair
- Closed-circuit television systems
- Central control systems
- Nurse call and emergency call systems
- Installation and repair of sound and public address systems
- Fiber optics
- Programmable and controlling instruments
- Local area networks (LANs) and structured wiring systems

This type of electrician works inside and outside. The job requires standing, bending, reaching in cold and wet conditions and working either in confined crawl spaces or at heights.

## GENERAL INFORMATION

The electrician may find a position in the physical plant departments of higher education institutions and other state institutions. Some positions in the class may require:

- A willingness to work in an environment containing dust, grime, odor, fumes, and high levels of noise
- You to walk, stand, and work while bending and stooping for extended periods
- You to lift and carry heavy objects; to work from ladders, scaffolds, and other above ground locations such as roofs
- You to work overhead while lying on your back, or to work outdoors in inclement weather

The *inside wireman's* job is to distribute and connect the customer's electrical equipment to the outside power source. The inside wireman installs and maintains all of the various types of electrical systems found in commercial and industrial facilities.

This equipment may include lighting, receptacles, motors, heating equipment, systems that control the operation of all of a facility's energy usage.

The inside wireman installs conduit systems that contain the wire from the motor control centers or panel boards to all of the equipment that use electricity. Those conduits may contain power cables or control cables. Many of the conduit systems are exposed and must be installed to exacting standards using neat and workmanlike craftsmanship.

The work of an inside wireman can vary. One day the inside wireman could be installing a fire alarm system or security system in a high rise building and the next day he may be installing conduit in a ditch on the outside of the building.

Inside wiremen also install electrical systems in

- Industrial facilities such as chemical plants
- Power plants
- Chip manufacturing facilities
- Distribution centers

Each type of installation has specific electrical needs and systems to support those needs. While there are many tasks associated with the inside wireman classification, the apprenticeship training provides all of the knowledge necessary for an individual to perform these tasks in a professional manner while helping the individual to sharpen natural abilities so as to become the best worker in the electrical construction and maintenance industry.

Most states require licensing of anyone performing electrical work.

1. In most states you must be registered in a state approved apprenticeship program to receive an apprentice license.
2. Upon completion of the apprenticeship you will be required to pass a state examination in order to receive a general journeyman electrician's license.

Electricians currently earn around $31.30 per hour and can expect some time off work due to construction market conditions in a typical year.

## Information Sources

For additional information about a career as an electrical engineer, contact:

Institute of Electrical and Electronic Engineers
http://www.ieee.org

For information about a career as a construction electrician, contact local electrical contractors, a local union of the International Brotherhood of Electrical Workers, a local union-management apprenticeship committee, or the nearest office of the state employment service or state apprenticeship agency. Some state employment service offices screen applicants and give aptitude tests.

Information about apprenticeships or other work opportunities for maintenance electricians is available from local firms that employ maintenance electricians and from local union-management apprenticeship committees. In addition, the local office of the state employment service may provide information about training opportunities. Some state employment service offices screen applicants and give aptitude tests.

For general information about the work of electricians, contact:

- International Brotherhood of Electrical Workers
- National Electrical Contractors Association
- National Joint Apprenticeship and Training Committee for the Electrical Industry

## STARTING YOUR OWN BUSINESS

In your town, there are many businesses—some large and some small. Some of the small businesses may be owned and managed by one person. Others may be medium-sized, employing several people. There may also be large industries in your town. For example, there may be a large manufacturing plant. Or, the headquarters of a large company may be located in your town. Either of these could employ hundreds or even thousands of people.

Most large businesses started as small businesses. This, also, is true of some large corporations. Building up a business requires skill and hard work. It can be difficult to imagine the energy needed to start a business. Anyone starting a small business faces a variety of tasks. The person, first, must have the money needed. (This money may have to be borrowed.) He, also, must have a product or service that can be sold at a profit. He must have a way of distributing the product or service to the public. The smaller the business, the more of these jobs the owner/operator may have to do personally. In some small businesses, the owner/operator has all of these responsibilities. In larger businesses, such duties are divided. For example, one person (or one department) may be in charge of making the product. Another may be in charge of advertising it. A third person may be in charge of delivering or shipping the item. Usually, the larger the business, the greater the number of people needed for each of these duties.

The person who starts a business is usually an entrepreneur. An entrepreneur is anyone who organizes and manages a business. This person also assumes the risks of the business. This means that the entrepreneur is responsible for paying the business expenses. All responsibility for the success of the business rests with the entrepreneur.

A person who is self-employed, or in business for himself, is an entrepreneur. All entrepreneurs face four similar problems. These problems are:

- Identifying a need
- Finding a product to satisfy the need

- Financing the business
- Selling the product

Now, let us look at each of these concerns:

- All successful businesses have one thing in common. They were started because someone noticed that people needed or wanted an item or service.
- Once an entrepreneur has noticed a need, he can then find a product to fill the need. This product might be an item (such as an electrical device). It might also be a service (such as an electrical product repair).
- Starting a business is not easy. Succeeding in business is even harder. Anyone who starts a business is an entrepreneur, but not all entrepreneurs are successful. For an entrepreneur to be successful, his business must be successful.

## THE DEVELOPMENT OF LEADERSHIP

Our society has always prized resourcefulness and initiative. These qualities were essential to the westward expansion of the American frontier and the early shaping of the values of the infant Republic. Some of the values of our American culture have been shaped by our history which, especially in its early days, required strong decisions and firm enterprise. The events which are now simply history, required fast commitments and unwavering resolution from those who took part in them.

Though the accidentals of historical influence may change, it's the basic forces which remain much the same. These resulting forces are the problems and challenges, our society thinks, as puzzling and difficult as those posed for our ancestors. Because ours is a democratic society, every man and woman is prompted to develop leadership qualities.

Good leaders are characterized by effective communication, firmness of will and singleness of purpose, and moral integrity. While it seems, these qualities are more readily apparent in some, all of us can develop, in some degree, the skills needed for effective leadership. Membership in student clubs—such as VICA or Technology Education Clubs—can help develop leadership ability. A club, by its nature (bringing together as it does individuals with a single common interest) has perhaps less diversity than society generally. Still, it can offer you an opportunity to practice effective communication and learn the skills needed to work within a group, such as a committee. The valuable qualities of leadership are often exercised in the most quiet situations. Each of you will be given opportunities to exert the force of leadership to bring about a decision, resolve a crisis, or prompt an action. Leadership is essential to active participation in a democratic society.

## *Employment of Electricians

Electricians held about **656,000** jobs in 2004 (latest year for statistics). Nearly two-thirds of wage and salary workers were employed in the construction industry; while the remainder worked as maintenance electricians in other industries. In addition, about one in ten electricians were so employed.

Because of the widespread need for electrical services, electrician jobs are found in all parts of the country.

## *Earnings

In May 2004, median hourly earnings of electricians were $20.33. The middle 50% earned between $15.43 and $26.90. The lowest 10% earned less than $12.18, and the highest 10% earned more than $33.63. Median hourly earnings in the industries employing the largest numbers of electricians in May 2004 were as follows:

| | |
|---|---|
| Motor vehicle parts manufacturing........ | $30.04 |
| Local government............................ | $22.24 |
| Nonresidential building construction..... | $19.99 |
| Building equipment contractors............. | $19.76 |
| Employment services............................ | $15.62 |

Apprentices usually start at between 40 and 50% of the rate paid to fully trained electricians, depending on experience. As apprentices become more skilled, they receive periodic pay increases throughout the course of their training.

Some electricians are members of the International Brotherhood of Electrical Workers. Among unions representing maintenance electricians are the International Brotherhood of Electrical Workers; the International Union of Electronic, Electrical, Salaried, Machine, and Furniture Workers; the International Association of Machinists and Aerospace Workers; the International Union, United Automobile, Aircraft and Agricultural Implement Workers of America; and the United Steelworkers of America.

## Related Occupations

To install and maintain electrical systems, electricians combine manual skill and knowledge of electrical materials and concepts. Workers in other occupations involving similar skills include heating, air-conditioning,

---

*Reference has been made to: U.S. Department of Labor, Bureau of Labor Statistics: Occupational Outlook Handbook. www.bls.gov

and refrigeration mechanics and installers; line installers and repairers; electrical and electronics installers and repairers; electronic home entertainment equipment installers and repairers; and elevator installers and repairers.

## *Sources of Additional Information

For details about apprenticeships or other work opportunities in this trade, contact the offices of the state employment service, the state apprenticeship agency, local electrical contractors or firms that employ maintenance electricians, or local union-management electrician apprenticeship committees. This information also may be available from local chapters of the Independent Electrical Contractors, Inc.; the National Electrical Contractors Association; the Home Builders Institute; the Associated Builders and Contractors; and the International Brotherhood of Electrical Workers.

For information about union apprenticeship and training programs, contact:

*Reference has been made to: U.S. Department of Labor, Bureau of Labor Statistics: Occupational Outlook Handbook. www.bls.gov

- National Joint Apprenticeship Training Committee (NJATC), 301 Prince George's Blvd., Upper Marlboro, MD 20774. Internet: http://www.njatc.org
- National Electrical Contractors Association (NECA), 3 Metro Center, Suite 1100, Bethesda, MD 20814. Internet: http://www.necanet.org
- International Brotherhood of Electrical Workers (IBEW), 1125 15th St. NW., Washington, DC 20005. Internet: http://www.ibew.org

For information about independent apprenticeship programs, contact:

- Associated Builders and Contractors, Workforce Development Department, 4250 North Fairfax Dr., 9th Floor, Arlington, VA 22203. Internet: http://www.trytools.org
- Independent Electrical Contractors, Inc., 4401 Ford Ave., Suite 1100, Alexandria, VA 22302. Internet: http://www.ieci.org
- National Association of Home Builders, Home Builders Institute, 1201 15th St. NW., Washington, DC 20005. Internet: http://www.hbi.org
- National Center for Construction Education and Research. P.O. Box 141104, Gainesville, FL 32614-1104. Internet: http://www.nceer.org

# Glossary

**AC generator** A generator that produces alternating current.

**Alternator** Another name for an AC generator.

**Ampere** The current produced when $6.25 \times 10^{18}$ electrons flow past a given point in 1 second.

**Ammeter** A meter capable of measuring amperes.

**Armature** Rotating part of an electric motor or generator. May also be the moving part of a relay, buzzer, or speaker.

**Atoms** The smallest particles of an element that retains all the properties of that element.

**Automatic load transfer** A method of starting and switching to emergency power plants when the utility power is interrupted.

**Ballast** A choke or inductor used in a gaseous discharge (fluorescent) lamp circuit.

**Battery** Two or more cells connected together in series or parallel.

**Bonding** Effective bonding means that the electrical continuity of the grounding circuit is assured by proper connections between service raceways, service cable armor, all service equipment enclosures containing service conductors, and any conduit or conductor that forms part of the grounding conductor to the service raceway.

**Bonding jumpers** Are used to assure continuity around concentric or eccentric knockouts that are punched or otherwise formed in such a manner that they impair electrical current flow.

**Branch circuit** A circuit that is used to feed small electrical loads such as lamps, small appliances, and kitchen equipments.

**Brushes** Devices that make contact with the rotating connections to the armature of a motor or generator. Usually made of carbon or metal.

**Canadian Standards Association (CSA)** A parallel organization to UL in the United States. It is set up to develop voluntary national standards, provide certification services for national standards, and to represent Canada in international standards activities.

**Capacitor** A device used to store an electrical charge. Consists of two plates and a dialectric.

**Cell** A device used to generate an electrical current or emf.

**Circuit** A path for electrons to flow or electricity to move.

**Circular mil** A measurement unit for cross-sectional area of a round wire; equal to one-thousandth (0.001) of an inch in diameter.

**Coil** A device made by turns of insulated wire wound around a core. A coil sometimes has a hollow center portion. Also called an inductor.

**Color code** The carbon composition resistors are marked with their resistance with color bands. The bands are colored according to an established code as follows:

| Black | 0 |
| Brown | 1 |
| Red | 2 |
| Orange | 3 |
| Yellow | 4 |
| Green | 5 |
| Blue | 6 |
| Violet | 7 |
| Gray | 8 |
| White | 9 |

**Commutator** A device made of segments of copper insulated by mica or some other material. Used to reverse direction of current flow from a generator or to a motor. Brushes are usually placed to make contact with the surface of the commutator. Segments of a commutator are connected to ends of the armature coils in a motor or generator.

**Complete circuit** Is made up of a source of electricity, a conductor, and a consuming device.

**Conductors** Materials through which electrons move.

**Configuration** An arrangement of wires or components in a circuit.

**Construction electrician** One who assembles, installs, and wires systems for heat, light, power, air conditioning, and refrigeration. They also install electrical machinery, electronic equipment, controls, and signal communications systems.

**Convenience outlet** Outlets placed for the convenience of the home owner in everyday use of electricity.

**Coulomb** The unit of measurement of $6.25 \times 10^{18}$ electrons.

**Current** The movement of electrons in a negative-to-positive direction along a conductor.

**Current flow** The movement of free electrons in a given direction.

**Dielectric** An insulating material used in a capacitor or other electrical device.

**Diode** A semiconductor or vacuum tube device that is used to allow current to flow easily in one direction but retards or stops its flow in the other direction. Diodes are used in rectifier circuits and in switching circuits.

**Direct current (DC)** Current that flows in one direction only.

**Draftsperson** (also referred as **Drafter**) One who draws plans and electrical schematics with the aid of mechanical devices.

**Electrical engineer** One who designs, develops, and supervises the manufacture of electrical and electronic equipment.

**Electrician** One who works with electrical equipment and wiring.

**Electricity** The flow of electrons along a conductor.

**Electrodynamometer** A type of meter that uses no permanent magnet, but two fixed coils to produce a magnetic field. The meter also uses two moving coils. This meter can be used as a voltmeter or an ammeter.

**Electrolyte** A solution capable of conducting electric current, the liquid part of a battery.

**Electrolytic** A capacitor that has parts separated by an electrolyte. Thin film formed on one plate provides the dielectric. The electrolytic capacitor has polarity ($-$ and $+$).

**Electromagnet** A magnet produced by current flow through a coil of wire. The core is usually used to concentrate the magnetic lines of force.

**Electromagnetism** Magnetism produced by current flowing through a coil of wire or other conductor.

**Electron** Smallest particle of an atom with a negative charge.

**Elements** The most basic materials in the universe. Ninety-four elements, such as iron, copper, and nitrogen, have been found in nature. Every known substance is composed of elements.

**Encapsulation** The embodiment of a component or assembly in a solid or semisolid medium such as tar, wax, or epoxy.

**Energy** Ability to do work.

**Entrance signals** Usually consist of a door bell, chime, or some other device used to alert the home occupant to someone outside wishing to see them.

**Equipment ground** The grounding of exposed conductive materials such as conduit, switch boxes, or meter frames that enclose conductors and equipment. This is to prevent the equipment from exceeding ground potential.

**Explosion proof** Term used to describe apparatuses that are enclosed in a case that is capable of withstanding an explosion within the case without igniting flammable materials outside it.

**Farad (F)** A unit of measurement for capacitance.

**Fatal current** The amount of current needed to kill. Currents between 100 and 200 mA are lethal.

**Feeder circuit** A circuit that is used to feed others or take electrical energy to where it can then be branched off to service other locations.

**Filaments** Small coils of resistance wire in light bulbs and vacuum tubes that heat up to glow either red or white hot. The filament of a vacuum tube boils off electrons from the cathode. The filament in a light bulb glows to incandescence to produce light.

**Fluorescence** A term that means to glow or give off light. Fluorescent lamps fluoresce, or produce light, by having ions of mercury collide and strike a fluorescent coating inside the tube.

**Fluorescent lamp** A lamp that produces light through action of ultraviolet rays striking a fluorescent material on the inside of a glass tube.

**Four-way switch** Used where it is necessary to turn a light or circuit on or off from three or more locations. It also needs a minimum of two three-way switches to be able to do its job.

**Fuse** A safety device designed to open if excessive current flows through a circuit.

**Generator** A device that turns mechanical energy into electrical energy.

**Ground fault circuit interrupter (GFI or GFCI)** A fast-operating circuit breaker that is sensitive to very low levels of current leakage to ground. It is designed to limit electric shock to a current and time duration value below that which can produce serious injury.

**Grounding** Effective grounding means that the path to ground is permanent and continous; and it has a low impedance to permit all current carrying devices on the circuit to work properly.

**Grounding conductor** A wire attached to the housing or other conductive ports of electrical equipment that are not normally energized. The conductor carries current for them to the ground.

**Henry (H)** Unit of measurement for inductance.

**Hertz (Hz)** Unit of measurement for frequency.

**Horsepower** The unit of measurement that equates work done electrically with the work done by a horse. One horsepower is the energy consumed to produce the equivalent work done by a horse lifting 33,000 pounds for 1 foot in 1 minute. It takes 746 watts to equal 1 horsepower.

**Impedance (Z)** Total circuit opposition to alternating current. Measured in ohms.

**Incandescent** A term that means to glow white hot. The filament in an incandescent lamp glows white hot to produce heat and light.

**Incandescent lamp** A lamp or bulb that produces light by heating a filament to incandescence (white hot).

**Induced current** Current produced by electromagnetic induction, usually from a coil or transformer.

**Inductance (L)** That property of a coil that opposes any change in circuit current. Measured in henrys.

**Inductive reactance** Opposition presented to alternating current by an inductor. The symbol is $X_L$. Measured in ohms.

**Inerting** Consists of mixing a chemical inert, nonflammable gas with a flammable substance, displacing the oxygen until the percentage of oxygen in the mixture is too low to allow combustion.

**Input** Term used to describe the energy applied to a circuit, device, or system.

**Insulator** Nonconducting material lacking a sufficient supply of free electrons to allow the movement of electrons without exceptional force or high voltage.

**Inverter** A device used to convert DC to AC.

**Joule** Metric unit of electrical work done by 1 coulomb flowing with a potential difference of 1 volt.

**Kilo** A prefix that means one thousand (1000).

**Kilowatt** One thousand watts (1000 W), kW.

**Kilowatthour meter** Used to measure the power used by a consumer for a month's period of time.

**Kirchoff's law of voltages** The sum of all voltages across resistors or loads is equal to the applied voltage.

**Load** Anything that may draw current from an electrical power source.

**Magnet** A device which possesses a magnetic field.

**Magnetism** A force produced by an electrical current in a wire or found in nature in certain materials.

**Magnetohydrodynamic generator** A generator of electricity that uses hot plasma to produce electricity. Electrons in the gas (plasma) are deflected by a magnetic field. Between collisions with the particles in the gas, they make their way to one of the electrodes to produce electricity.

**Magnet wire** Copper wire used to wind coils, solenoids, transformers, motors; usually coated with varnish or other insulation material.

**Maintenance electrician** One who keeps lighting systems, transformers, generators, and other electrical equipment in good working order. He or she may also install new equipment.

**Mega** A prefix that means one million (1,000,000).

**Meter** Means to measure. Or, a unit of measurement in the metric system.

**Mica** Insulating material that can withstand high voltages and elevated temperatures. No longer used in the manufacture of appliances and capacitors.

**Micro** A prefix that means one-millionth (0.000001). Symbol is Greek letter μ.

**Microammeter** A meter limited to measuring microamperes.

**Microampere** One-millionth (0.000001) of an ampere (μA).

**Microvolt** One-millionth (0.000001) of a volt (μV).

**Milli** A prefix that means one-thousandth (0.001).

**Milliammeter** A meter limited to measuring milliamperes.

**Milliampere** One-thousandth (0.001) of an ampere (mA).

**Millivolt** One-thousandth (0.001) of a volt (mV).

**Milliwatt** One-thousandth (0.001) of a watt (mW).

**Motor** A device used to change electrical to mechanical energy.

**Multimeter** A meter capable of measuring (in most instances) volts, ohms, and milliamperes.

**Multiplier** A resistor placed in series with a meter movement to handle the extra voltage applied to the movement. It serves to extend the range of the meter.

**National Electrical Code (NEC)** A book of standards produced by the *National Fire Protection Association* every 3 years. It describes the proper installation of various electrical machines and devices for safe operation.

**Neutrons** Tiny particles that have no electrical charge.

**Ohmmeter** A device used to measure resistance.

**Ohm's law** Georg Ohm's law states that the circuit current is equal to the voltage divided by the resistance.

**Omega** The Greek symbol (Ω) is used to indicate the unit of resistance, the ohm.

**OSHA** Occupational Safety and Health Act.

**Outlet** Socket or receptacle that accepts a plug to make electrical contact, usually to provide power.

**Parallel circuit** A circuit where each load (resistance) is connected directly across the voltage source.

**Piezoelectrical effect** The process of using crystals under pressure to produce electrical energy.

**Photosensitive** Sensitive to light.

**Potentiometer** A variable resistor that has three contact points.

**Power** Rate of doing work; abbreviated as **P** and measured in watts.

**Proton** Smallest particle of an atom with a positive charge.

**Pulsating direct current** Current produced when changed to DC and left unfiltered; abbreviated as **PDC**.

**Rectifier** A device that changes AC to DC; allows current to flow in only one direction.

**Relay** An electromatic device that is used for remote switching.

**Remote control**  The ability to start or control a device from a location other than that of the device being controlled.
**Residential wiring**  The electrical wiring used to supply a home with electrical outlets where needed.
**Resistance**  Opposition to the movement of electrons. Measured in ohms.
**Resistor**  A device that opposes current flow.
**Rheostat**  Variable resistor; usually has only two points connected into the circuit. Used to control voltage by increasing and decreasing resistance.
**Root-mean-square (rms)**  A type of reading obtained by using a standard voltmeter or ammeter.
**Rotor**  A moving part of a motor or generator.
**Rural grid**  Consists of radial feeders that leave the substation as three-phase circuits with each of the phases fanning out to serve the countryside as single-phase circuits.

**Series circuit**  A circuit with one resistor or consuming device located in a string or one after another.
**Series-parallel circuit**  A combination of series and parallel circuits where a minimum of three resistors or devices are connected with at least one in series and with at least two in parallel.
**Shock**  A process whereby an outside source of electricity is such that it overrides the body's normal electrical system and causes the muscles to react involuntarily.
**Short circuit**  A circuit that has extremely low resistance.
**Shunt**  A resistor placed in parallel with a meter movement to handle or shunt most of the current around the movement. It serves to extend the range of the meter.
**Slip ring**  A ring of copper mounted on the shaft of a motor or generator through which a brush makes permanent (or constant) contact with the end of the rotor windings. Always used in pairs.
**Solar cell**  A cell that turns light energy into electrical energy. It is usually made of silicon.
**Solenoid**  A coil of wire wrapped around a hollow form, usually with some type of core material sucked into the hollow. The movement of the core is usually to move a switch or to open valves.
**Splice**  A form of electical connection in which wires are joined directly to each other.
**Static electricity**  A form of energy present when there are two charges of opposite polarity in close proximity. Static electricity is generated by friction.

**System ground**  The grounding of the neutral conductor, or ground leg, of the circuit to prevent lightning or other high voltages from exceeding the design limits of the circuit.

**Terminal**  A connecting point for wires. It is usually present on batteries, cells, switches, relay, motors, and electrical panels.
**Thermocouple**  A device made of two different kinds of metals joined at one end. When the junction is heated, an emf is generated across the open ends.
**Thermostat**  A device that acts as a switch which is operated when heat causes two metals to expand at different rates.
**Three-way switch**  Used where it is necessary to turn a light or circuit on or off from more than one location.
**Toggle switches**  Devices used to turn various circuits on and off or to switch from one device to another. They are made in a number of configurations.
**Transformer**  A device that can induce electrical energy from one coil to the other by using magnetic lines of force, stepping voltage up or down.

**Underwriters' Laboratories**  An organization which tests electrical devices, systems, and materials to see if they meet certain safe operation standards. The trademark is UL. It is a non-profit organization.
**Universal motor**  A motor that operates on AC or DC.

**Volt (V)**  Unit of measurement of electromotive force.
**Voltage**  Electromotive force (emf) that causes electrons to move along a conductor.
**Voltmeter**  An instrument used to measure voltage.

**Watt (W)**  Unit of electrical power.
**Watthour meter**  A meter that measures electrical power consumed in an hour.

**X**  Symbol for reactance. Measured in ohms.

**Z**  Symbol for impedance. Measured in ohms.
**Zener**  A type of semiconductor diode that breaks down intentionally at a predetermined voltage. Usually used for voltage regulation circuits.

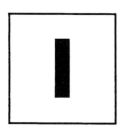

# DC Motor Trouble Chart

| Symptom and Possible Cause | Possible Remedy |
|---|---|
| **Motor will not start** | |
| (a) Open circuit in controller | (a) Check controller for open starting resistor, open switch, or open fuse. |
| (b) Low terminal voltage | (b) Check voltage with nameplate rating. |
| (c) Bearing frozen | (c) Recondition shaft and replace bearing. |
| (d) Overload | (d) Reduce load or use larger motor. |
| (e) Excessive friction | (e) Check bearing lubrication to make sure that the oil has been replaced after installing motor. Disconnect motor from driven machine and turn motor by hand to see if trouble is in motor. Strip and reassemble motor; then check part by part for proper location and fit. Straighten or replace bent or sprung shaft (machines under 5 hp [3.73 kW]). |
| **Motor stops after running short time** | |
| (a) Motor is not getting power | (a) Check voltage at the motor terminals; also fuses, coils, and overload relay. |
| (b) Motor is started with weak or no field | (b) If adjustable-speed motor, check rheostat for correct setting. If correct, check condition of rheostat.<br>Check field coils for open winding.<br>Check wiring for loose or broken connection. |
| (c) Motor torque insufficient to drive load | (c) Check line voltage with nameplate rating. Use larger motor or one with suitable characteristic to match load. |
| **Motor runs too slow under load** | |
| (a) Line voltage too low | (a) Check and remove any excess resistance in supply line, connections, or controller. |
| (b) Brushes ahead of neutral | (b) Set brushes on neutral. |
| (c) Overload | (c) Check to see that load does not exceed allowable load on motor. |
| **Motor runs too fast under load** | |
| (a) Weak field | (a) Check for resistance in shunt-field circuits. |
| (b) Line voltage too high | (b) Correct high-voltage condition. |
| (c) Brushes back of neutral | (c) Set brushes on neutral. |
| **Sparking at brushes** | |
| (a) Commutator in bad condition | (a) Clean and reset brushes. |
| (b) Commutator eccentric or rough | (b) Grind and true commutator.<br>Undercut mica. |
| (c) Excessive vibration | (c) Balance armature.<br>Check brushes to make sure they ride freely in the holders. |
| (d) Broken or sluggish brush-holder spring | (d) Replace spring and adjust pressure to manufacturer's recommendations. |
| (e) Brushes too short | (e) Replace brushes. |

# Wound-Rotor Motor Trouble Chart

| Symptom and Possible Cause | Possible Remedy |
|---|---|
| **Motor runs at low speed with external resistance cut out** | |
| (a) Wires to the control unit too small | (a) Use larger cable to the control unit. |
| (b) Control unit too far from motor | (b) Bring control unit nearer motor. |
| (c) Open circuit in rotor circuit (including cable to the control unit) | (c) Test by ringing out circuit and repair. |
| (d) Dirt between brush and ring | (d) Clean rings and insulation assembly. |
| (e) Brushes stuck in holders | (e) Use right-size brush. |
| (f) Incorrect brush tension | (f) Check brush tension and correct. |
| (g) Rough collector rings | (g) File, sand, and polish. |
| (h) Eccentric rings | (h) Turn down on lathe, or use portable tool to true-up rings without disassembling motor. |
| (i) Excessive vibration | (i) Balance motor. |
| (j) Current density of brushes too high (overload) | (j) Reduce load. If brushes have been replaced, make sure they are of the same grade as originally furnished. |

# Fractional-Horsepower Motor Trouble Chart

| Symptom and Possible Cause | Possible Remedy |
|---|---|

**Split-Phase Induction Motor**

**Failure to start**

(a) No voltage

(a) Check for voltage at motor terminals with test lamp or voltmeter. Check for blown fuses or meter. Check for blown fuses or open overload device in starter. If motor is equipped with a slow-blow fuse, see that the fuse plug is not open and that it is screwed down tight.

(b) Low voltage

(b) Measure the voltage at the motor terminals with the switch closed. Voltage should read within 10% of the voltage stamped on the motor nameplate. Overload transformers or circuits may cause low voltage. If the former, check with the power company. Overloaded circuits in the building can be found by comparing the voltage at the meter with the voltage at the motor terminals with the switch closed.

(c) Faulty cutout switch operation

(c) Cutout switch operation may be observed by removing the inspection plate in the front end bracket. The mechanism consists of a cutout switch mounted on the front end bracket and a rotating part called the governor weight assembly, which consists of a Bakelite disk so supported that it is moved back and forth along the shaft by the operation of the governor weights. At standstill, the disk holds the cutout switch closed. If the disk does not hold the switch closed, motor cannot start. This may call for adjustment of the end-play washers. Dirty contact points may also keep the motor from starting. See that the contacts are clean. After the motor has accelerated to a predetermined speed, the disk is withdrawn from the switch, allowing it to open. With the load disconnected from the motor, close the starting switch. If the motor does not start, start it by hand and observe the operation of the governor as the motor speeds up, and also when the switch has been opened and the motor slows down. If the governor fails to operate, the governor weights may have become clogged. If it operates too soon or too late, the spring is too weak or too strong. Remove motor to service shop for adjustment. Governor weights are set to operate at about 75% of synchronous speed. Place rotor in balancing machine and, with a tachometer, determine if the governor operates at the correct speed.

(d) Open overload device

(d) If the motor is equipped with a built-in micro switch or similar overload device, remove the cover plate in the end bracket on which the switch is mounted and see if the switch contacts are closed. Do not attempt to adjust this switch or to test its operation with a match. Doing so may destroy it. If the switch is permanently open, remove the motor to the service shop for repairs.

(e) Grounded field

(e) If the motor overheats, produces shock when touched, or if idle watts are excessive, test for a field ground with a test lamp across the field leads and frame. If grounded, remove the motor to the service shop for repairs.

| Symptom and Possible Cause | Possible Remedy |
|---|---|
| (f) Open-circuited field | (f) These motors have a main and a phase (starting) winding. Apply current to each winding separately with a test lamp. Do not leave the windings connected too long while rotor is stationary. If either winding is open, remove the motor to the repair shop for repairs. |
| (g) Short-circuited field | (g) If the motor draws excessive watts, and, at the same time lacks torque, overheats, or hums, a shorted field is indicated. Remove to the service shop for repairs. |
| (h) Incorrect end play | (h) Certain types of motors have steel-enclosed cork washers at each end to cushion the end thrust. Too great an end thrust, hammering on the shaft, or excessive heat may destroy the cork washers and interfere with the operation of the cutout switch mechanism. If necessary, install new end-thrust cushion bumper assemblies. End play should not exceed 0.01 in. (0.254 mm); if it does, install additional steel endplay washers. End play should be adjusted so that the cutout switch is closed at standstill and open when the motor is operating. |
| (i) Excessive load | (i) This may be approximately determined by checking the ampere input with the nameplate marking. Excessive load may prevent the motor from accelerating to the speed at which the governor acts and cause the phase winding to burn up. |
| (j) Tight bearings | (j) Test by turning armature by hand. If adding oil does not help, bearings must be replaced. |

**Motor overheats**

| | |
|---|---|
| (a) Grounded field | (a) Test for a field ground with a test lamp between the field and motor frame. If grounded, remove the motor to the service shop for repair. |
| (b) Short-circuited field | (b) Test for excessive current draw, lack of torque, and presence of hum. Any of these symptoms indicates a shorted field. Remove the motor to the service shop for repair. |
| (c) Tight bearings | (c) Test by turning armature by hand. If oiling does not help, new bearings must be installed. |
| (d) Low voltage | (d) Measure voltage at motor terminals with switch closed. Voltage should be within 10% of nameplate voltage. Overloaded transformers or power circuits may cause low voltage. Check with power company. Overloaded building circuits can be found by comparing the voltage at the meter with the voltage at the motor terminals with the switch closed. |
| (e) Faulty cutout switch | (e) See Paragraph (c) under *Failure to Start*. |
| (f) Excessive load | (f) See Paragraph (i) under *Failure to Start*. |

**Motor does not come up to speed**
Same possible causes and possible remedies as under *Motor Overheats*.

**Excessive bearing wear**

| | |
|---|---|
| (a) Belt too tight | (a) Adjust belt to tension recommended by manufacturer. |
| (b) Pulleys out of alignment | (b) Align pulleys correctly. |
| (c) Dirty, incorrect, or insufficient oil | (c) Use type of oil recommended by manufacturer. |
| (d) Dirty bearings | (d) Clean thoroughly. Replace worn berarings. |

**Excessive noise**

| | |
|---|---|
| (a) Worn bearings | (a) See Paragraphs (a), (b), (c), and (d) under *Excessive Bearing Wear*. |
| (b) Excessive end play | (b) If necessary, add additional end-play washers. |
| (c) Loose parts | (c) Check for loose hold-down bolts, loose pulleys, etc. |
| (d) Misalignment | (d) Align pulleys correctly. |
| (e) Worn belts | (e) Replace belts. |
| (f) Bent shaft | (f) Straighten shaft, or replace armature or motor. |
| (g) Unbalanced rotor | (g) Balance rotor. |
| (h) Burrs on shaft | (h) Remove burrs. |

**Motor produces shock**

| | |
|---|---|
| (a) Grounded field | (a) See Paragraph (e) under *Failure to Start*. |
| (b) Broken ground strap | (b) Replace ground strap. |
| (c) Poor ground connection | (c) Inspect and repair ground connection. |

**Rotor rubs stator**

| | |
|---|---|
| (a) Dirt in motor | (a) Thoroughly clean motor. |
| (b) Burrs on rotor or stator | (b) Remove burrs. |

| Symptom and Possible Cause | Possible Remedy |
|---|---|
| (c) Worn bearings | (c) Replace bearings and inspect shaft for scoring. |
| (d) Bent shaft | (d) Repair and replace shaft or rotor. |

**Radio interference**

| | |
|---|---|
| (a) Poor ground connection | (a) Check and repair any defective grounds. |
| (b) Loose contacts or connections | (b) Check and repair any loose contacts on switches or fuses, and loose connections on terminals. |

### Repulsion-Start Induction Brush-Lifting Motors

**Failure to start**

| | |
|---|---|
| (a) Fuses blown | (a) Check capacity of fuses. They should not be greater in ampere capacity than recommended by the manufacturer, and in no case smaller than the full-load ampere rating of the motor, and with a voltage capacity equal to or greater than the voltage of the supply circuit. |
| (b) No voltage or low voltage | (b) Measure voltage at motor terminals with switch closed. See that it is within 10% of the voltage stamped on the nameplate of the motor. |
| (c) Open-circuited field or armature | (c) Indicated by excessive sparking in starting, refusal to start at certain positions of the rotor, or by a humming sound when the switch is closed. Examine for broken wires, loose connections, or burned segments on the commutator at the point of loose or broken connections. Inspect the commutator for a foreign metallic substance that might cause a short between the commutator segments. |
| (d) Incorrect voltage or frequency | (d) Requires new motor built for operation on local power supply. Dc motors will not operate on ac circuit, or vice versa. |
| (e) Worn or sticking brushes | (e) When brushes are not making proper contact with the commutator, the motor will have a weak starting torque. This can be caused by worn brushes, brushes sticking in holders, weak brush springs, or a dirty commutator. The commutator should be polished with fine sandpaper (never use emery). The commutator should never be oiled or greased. |
| (f) Improper brush setting | (f) Unless a new armature has been installed, the brush holder or rocker arm should be opposite the index and locked in position. If a new armature has been installed, the position may be slightly off the original marking. |
| (g) Improper line connection | (g) See that the connections are made according to the connection diagram sent with the motor. The motor may, through error, be wired for a higher voltage. |
| (h) Excessive load | (h) If the motor starts with no load, and if all the foregoing conditions are satisfactory, then failure to start is most likely due to an excessive load. |
| (i) Shorted field | (i) Take separate current readings on each of the two halves of the stator winding. Unequal readings indicate a short. Shorted coil may also feel much hotter than the normal coil. An increase in hum may also be caused by a shorted winding. |
| (j) Shorted rotor | (j) Remove the brushes from the commutator and impress full voltage on the stator. If there is one or more points at which the rotor "hangs" (fails to revolve easily when turned), the rotor is shorted. Forcing the rotor to the position where it is most difficult to hold will cause the shorted coil to become hot. Do not hold in position too long or the coil will burn out. |

**Motor operates without lifting brushes**

| | |
|---|---|
| (a) Dirty commutator | (a) Clean with fine sandpaper. (Do not use emery.) |
| (b) Governor mechanism or brushes sticking, or brushes worn too short for good contact | (b) See that brushes move freely in slots and that governor mechanism operates freely by hand. Replace worn brushes. |
| (c) Frequency of supply circuit incorrect | (c) Run motor idle. After brushes throw off, speed should be slightly in excess of full-load speed shown on nameplate. An idle speed varying more than 10% from nameplate speed indicates that motor is being used on an incorrect supply frequency. A different motor will be required. |
| (d) Low voltage | (d) See that voltage is within 10% of nameplate voltage with the switch closed. |
| (e) Line connection improperly or poorly made | (e) See that contacts are good and that connections correspond with diagram sent with motor. |
| (f) Incorrect brush setting | (f) Check to see that rocker-arm setting corresponds with index mark. |
| (g) Incorrect adjustment of governor | (g) The governor should operate and lift brushes at approximately 75% of speed stamped on nameplate. Below 65% or over 85% indicates incorrect spring tension. |

| Symptom and Possible Cause | Possible Remedy |
|---|---|
| (h) Excessive load | (h) An excessive load may be started but not carried to and held at full-load speed, which is beyond where the brushes lift. Tight motor bearings may contribute to overload. This is sometimes indicated by brushes lifting and returning to the commutator. |
| (i) Shorted field | (i) See Paragraph (i) under *Failure to Start*. |
| **Excessive bearing wear** | |
| (a) Belt too tight, or an unbalanced line coupling | (a) Correct the mechanical condition. |
| (b) Improper, dirty, or insufficient oil | (b) The lubrication system of most small motors provides for supplying the right amount of filtered oil to the bearings. It is necessary only for the user to keep the wool yarn saturated with a good grade of machine oil. |
| (c) Dirty bearnings | (c) When bearings become clogged with dirt, the motor may need protection from excessive dust. The application may be such that a specially constructed motor should be used. |
| **Motor runs hot** | |
| (a) Bearing trouble | (a) See Paragraphs (a), (b), and (c) under *Excessive Bearing Wear*. |
| (b) Short-circuited coils in stator | (b) Make separate wattmeter reading on each of the two halves of the stator winding. Sometimes the shorted coil may be located by the fact that one coil feels much hotter than the other. An increase over normal in the magnetic noise (hum) may also indicate a shorted stator. |
| (c) Rotor rubbing stator | (c) Extraneous matter may be between the rotor and the stator, or the bearings may be badly worn. |
| (d) Excessive load | (d) Be sure proper pulleys are on the motor and the machine. Driving the load at higher speeds requires more horsepower. Take an ammeter reading. If current draw exceeds the nameplate amperes for full load, the answer is evident. |
| (e) Low voltage | (e) Measure the voltage at the motor terminals with the switch closed. The reading should not vary more than 10% from the value stamped on the nameplate. |
| (f) High voltage | (f) See (e) above. |
| (g) Incorrect line connection to the motor | (g) Check the connection diagram sent with the motor. |
| **Motor burns out** | |
| (a) Frozen bearing | (a) See Paragraphs (a), (b), and (c) under *Excessive Bearing Wear*. |
| (b) Some condition of prolonged excessive overload | (b) Before replacing the burned-out motor, locate and remove the cause of the overload. Certain jobs that present a heavy load will, under unusual conditions of operation, apply prolonged overloads that may destroy a motor and be difficult to locate unless examined carefully. On jobs where it is assumed somewhat intermittent service will normally prevail, and that consequently are closely motored, the load cycle should be especially checked, as a change in this feature will easily produce excessive overload on the motor. |
| **Motor is noisy** | |
| (a) Unbalanced rotor | (a) When transportation handling has been so rough as to damage the heavy shipping case, it is well to test the motor for unbalanced conditions at once. It is even possible (though it rarely happens) that a shaft may be bent. In any event, the rotor should be rebalanced dynamically. |
| (b) Worn bearings | (b) See Paragraphs (a), (b), and (c) under *Excessive Bearing Wear*. |
| (c) Rough commutator, or brushes not seating properly | (c) Noise from this cause occurs only during the starting period, but conditions should be corrected to avoid consequent trouble. |
| (d) Excessive end play | (d) Proper end play is as follows: $\frac{1}{3}$ hp (248.7 W) and smaller—0.127 to 0.762 mm; $\frac{1}{2}$ (373 W) to 1 hp (0.746 kW)—0.254 to 1.905 mm. Washers supplied by the factory should be used. Be sure to tell factory all figures involved. Remember that too little end play is as bad as too much. |
| (e) Motor not properly aligned with the driven machine | (e) Correct the mechanical condition. |
| (f) Motor not firmly fastened to mounting base | (f) All small motors have steel bases so they can be firmly bolted to their mounting without fear of breaking. It is, of course, not to be expected that the base should be strained out of shape in order to make up for roughness in the mounting base. |
| (g) Loose accessories in motor | (g) Such parts as oil covers, guards (if any), end plates, etc., should be checked, especially |

| Symptom and Possible Cause | Possible Remedy |
|---|---|
| | if they have been removed for inspection of any sort. The conduit box should be tightened when the top is fitted after the connections are made. |
| (h) Air gap not uniform | (h) This results from a bent shaft or an unbalanced rotor. See Paragraph (a). |
| (i) Amplified motor noises | (i) When this condition is suspected, set the motor on a firm floor. If the motor is now quiet, then the mounting is acting as an amplifier to bring about certain noises in the motor. Frequently, the correction of slight details in the mounting will eliminate this, but rubber mounts are the surest cure. |

**Excessive brush wear**

| | |
|---|---|
| (a) Dirty commutator | (a) Clean with fine sandpaper (never use emery). |
| (b) Poor contact with commutator | (b) See that the brushes are long enough to reach the commutator, that they move freely in the slots, and that the spring tension gives firm but not excessive pressure. |
| (c) Excessive load | (c) If the brush wear is due to overload, it can usually be checked by noting the time required for the brushes to lift from the commutator. The proper time is less than 10 seconds. |
| (d) Failure to lift promptly and stay off during the running period | (d) Examine for conditions listed under *Motor Operates Without Lifting Brushes*. |
| (e) High mica | (e) Examination will show this condition. Take a very light cut off the commutator face and polish with fine sandpaper. Undercut the mica. |
| (f) Rough commutator | (f) True up on lathe. |

**Brush-holder or rocker-arm wear**

| | |
|---|---|
| (a) Failure to lift properly and stay off during the running period | (a) No noticeable wear of this part should occur during the life of the motor. Troublesome wear indicates faulty operation. See under *Motor Operates Without Lifting Brushes*. |

**Radio interference**

| | |
|---|---|
| (a) Faulty ground | (a) Check for poor ground connections, and repair. Static electricity generated by the belts may cause radio noises if the motor frame is not thoroughly grounded. Check for loose connections or contacts in the switch, fuses, or starter. |

**Capacitor-Start Induction Motors**

**Failure to start**

| | |
|---|---|
| (a) Blown fuses or overload device tripped | (a) Examine motor bearings. Be sure that they are in good condition and properly lubricated. Be sure the motor and driven machine both turn freely. Check the circuit voltage at the motor terminals against the voltage stamped on the motor nameplate. Examine the overload protection of the motor. Overload relays operating on either magnetic or thermal principles (or a combination of the two) offer adequate protection to the motor. Ordinary fuses of sufficient size to permit the motor to start do not protect against burnout. A combination fuse and thermal relay, such as *Buss Fusetron*, protects the motor and is inexpensive. If the motor does not have overload protection, the fuses should be replaced with overload relays or *Buss Fusetrons*. After installing suitable fuses and resetting the overload relays, allow the machine to go through its operating cycle. If the protective devices again operate, check the load. If the motor is excessively overloaded, take the matter up with the manufacturer. |
| (b) No voltage or low voltage | (b) Measure the voltage at the motor terminals with the switch closed. See that it is within 10% of the voltage stamped on the motor nameplate. |
| (c) Open-circuited field | (c) Indicated by a humming sound when the switch is closed. Examine for broken wires and connections. |
| (d) Incorrect voltage or frequency | (d) Requires motor built for operation on power supply available. AC motors will not operate on dc circuit, or vice versa. |
| (e) Cutout switch faulty | (e) The operation of the cutout switch may be observed by removing the inspection plate in the end bracket. If the governor disk does not hold the switch closed, the motor cannot start. This may call for additional end-play washers between the shaft shoulder and the bearing. Dirty or corroded contact points may also keep the motor from starting. See that the contacts are clean. With load disconnected from the motor, close the starting switch. If the motor does not start, start it by hand and listen for the characteristic click of the governor as the motor speeds up and also when the switch has been opened and the motor slows down. Absence of this click may indicate that the |

| Symptom and Possible Cause | Possible Remedy |
|---|---|
| | governor weights have become clogged, or that the spring is too strong. Continued operation under this condition may cause the phase winding to burn up. Remove the motor to the service shop for adjustment. |
| (f) Open field | (f) These motors have a main and phase winding in the stator. With the leads disconnected from the capacitor, apply current to the motor. If the main winding is all right, the motor will hum. If the main winding tests satisfactorily, connect a test lamp between the phase lead (the black lead) from the capacitor and the other capacitor lead. Close the starting switch. If the phase winding is all right, the lamp will glow and the motor may attempt to start. If either winding is open, remove the motor to the service shop for repairs. |
| (g) Faulty capacitor | (g) If the starting capacitor (electrolytic) is faulty, the motor starting torque will be weak and the motor may not start at all, but may run if started by hand. A capacitor can be tested for open circuit or short circuit as follows: Charge it with dc (if available), preferably through a resistance or test lamp. If no discharge is evident on immediate short circuit, an open or a short is indicated. If no dc is available, charge with ac. Try charging on ac several times to make certain that the capacitor has had a chance to become charged. If the capacitor is open, short-circuited, or weak, replace it. Replacement capacitors should not be of a lower capacity or voltage than the original. In soldering the connections, *do not use acid flux*.<br>***Note 1***—Electrolytic capacitors, if exposed to temperatures of 20°F (-6.7°C) and lower, may temporarily lose enough capacity so that the motor will not start and may cause the windings to burn up. The temperature of the capacitor should be raised by running the motor idle, or by other means. Capacitors should not be operated in temperatures exceeding 165°F. (74°C).<br>***Note 2***—The frequency of operation of electrolytic capacitors should not exceed two starts per minute of 3 seconds' acceleration each, or three to four starts per minute at less than 2 seconds' acceleration, provided that the total accelerating time (i.e., the time before the switch opens) does not exceed 1 to 2 minutes per hour. This may be approximately determined by checking the ampere input with the nameplate marking. Excessive load may prevent the motor from accelerating to the speed at which the governor acts, and thus cause the phase winding to burn up. |
| **Radio interference** | |
| (a) Faulty ground | (a) Check for poor ground connections. Static electricity generated by the belts may cause radio noises if the motor frame is not thoroughly grounded. |
| (b) Loose connections | (b) Check for loose connections or contacts in the switch, fuses, or starter. Capacitor motors ordinarily will not cause radio interference. Sometimes vibration may cause the capacitor to move so that it touches the metal container. This may cause radio interference. Open the container, move the capacitor, and replace the paper packing so that the capacitor cannot shift. |

*Source:* Courtesy of Lincoln Electric Co.

# IV

# Selection of Dual-Element Fuses for Motor-Running Overload Protection

**Motors Marked With Not Less Than 1.15 S.F. Or Temp. Rise Not Over 40°C**

| Ampere Ratings | | Ampere Ratings | | Ampere Ratings | |
|---|---|---|---|---|---|
| Motor | Fuse* | Motor | Fuse* | Motor | Fuse* |
| 1.00 to 1.11 | 1¼ | 6.40 to 7.19 | 8 | 72.0 to 79.9 | 90 |
| 1.12 to 1.27 | 1⁶/₁₀ | 7.20 to 7.99 | 9 | †80.0 to 87.9 | 100 |
| 1.28 to 1.43 | 1⁸/₁₀ | 8.00 to 9.59 | 10 | 88.0 to 99.9 | 110 |
| 1.44 to 1.59 | 1⁸/₁₀ | 9.60 to 11.9 | 12 | 100 to 119 | 125 |
| 1.60 to 1.79 | 2 | 12.0 to 13.9 | 15 | 120 to 139 | 150 |
| 1.80 to 1.99 | 2¼ | 14.0 to 15.9 | 17½ | 140 to 159 | 175 |
| 2.00 to 2.23 | 2½ | 16.0 to 19.9 | 20 | †160 to 179 | 200 |
| 2.24 to 2.55 | 2⁸/₁₀ | 20.0 to 23.9 | 25 | 180 to 199 | 225 |
| 2.56 to 2.79 | 3²/₁₀ | †24.0 to 27.9 | 30 | 200 to 239 | 250 |
| 2.80 to 3.19 | 3½ | 28.0 to 31.9 | 35 | 240 to 279 | 300 |
| 3.20 to 3.59 | 4 | 32.0 to 35.9 | 40 | 280 to 319 | 350 |
| 3.60 to 3.99 | 4¼ | 36.0 to 39.9 | 45 | †320 to 359 | 400 |
| 4.00 to 4.47 | 5 | 40.0 to 47.9 | 50 | 360 to 399 | 450 |
| 4.48 to 4.99 | 5⁶/₁₀ | 48.0 to 55.9 | 60 | 400 to 480 | 500 |
| 5.00 to 5.59 | 6¼ | 56.0 to 63.9 | 70 | 480 to 521 | 600 |
| 5.60 to 6.39 | 7 | 64.0 to 71.9 | 80 | | |

*Use FUSETRON Fuses, FRN-R (250V) or FRS-R (600V); or LOW-PEAK fuses LPN-RK (250V) or LPS-RK (600V).

†Disconnect switch must have an ampere rating at least 115% of motor ampere rating (430-110a). Next larger size switch with fuse reducers may be required.

Abnormal Installations—May require FUSETRON or LOW-PEAK Fuses of a larger size than shown; will provide only short-circuit protection. These applications include:

(Courtesy of Bussmann)

**All Other Motors (Less Than 1.15 S.F. Or Temp. Rise Greater Than 40°C)**

| Ampere Ratings | | Ampere Ratings | | Ampere Ratings | |
|---|---|---|---|---|---|
| Motor | Fuse* | Motor | Fuse* | Motor | Fuse* |
| 1.00 to 1.08 | 1⅛ | 6.09 to 6.95 | 7 | 69.6 to 78.2 | 80 |
| 1.09 to 1.21 | 1¼ | 6.96 to 7.82 | 8 | 78.3 to 86.9 | 90 |
| 1.22 to 1.39 | 1⁴/₁₀ | 7.83 to 8.69 | 9 | †87.0 to 95.6 | 100 |
| 1.40 to 1.56 | 1⁶/₁₀ | 8.70 to 10.0 | 10 | 95.7 to 108 | 110 |
| 1.57 to 1.73 | 1⁸/₁₀ | 10.5 to 12.0 | 12 | 109 to 125 | 125 |
| 1.74 to 1.95 | 2 | 13.1 to 15.0 | 15 | 131 to 150 | 150 |
| 1.96 to 2.17 | 2¼ | 15.3 to 17.3 | 17½ | 153 to 173 | 175 |
| 2.18 to 2.43 | 2½ | 17.4 to 20.0 | 20 | †174 to 195 | 200 |
| 2.44 to 2.78 | 2⁸/₁₀ | 21.8 to 25.0 | 25 | 196 to 217 | 225 |
| 2.79 to 3.04 | 3²/₁₀ | †26.1 to 30.0 | 30 | 218 to 250 | 250 |
| 3.05 to 3.47 | 3½ | 30.5 to 34.7 | 35 | 261 to 300 | 300 |
| 3.48 to 3.91 | 4 | 34.8 to 39.1 | 40 | 305 to 347 | 350 |
| 3.92 to 4.34 | 4½ | 39.2 to 43.4 | 45 | †348 to 391 | 400 |
| 4.35 to 4.86 | 5 | 43.5 to 50.0 | 50 | 392 to 434 | 450 |
| 4.87 to 5.43 | 5⁶/₁₀ | 52.2 to 60.0 | 60 | 435 to 480 | 500 |
| 5.44 to 6.08 | 6¼ | 60.9 to 69.5 | 70 | | |

(a) FUSETRON Fuses or LOW-PEAK Fuses in high ambient temperatures environments.
(b) A motor started frequently or rapidly reversed.
Motor is directly connected to a machine that cannot be brought up to full speed quickly (large fans, centrifugal machines such as extractors and pulverizers. Machines having large fly wheels such as large punch presses).
(c) Motor has a high Code Letter (or possibly no Code Letter) with full voltage start.

# Tables and Formulas

The data contained in this section are provided for reference only.* Many formulas are for estimating purposes only because they cannot consider all factors in every machine application. Many formulas can assist the reader by demonstrating basic physical or electrical relationships needed to understand a more abstract concept in control or motor operation. Other data, such as conversion factors, are included for your convenience to provide a more comprehensive resource when working in an international design environment.

**NOTE:**
The following equations for calculating horsepower are meant to be used for estimating purposes only. These equations do not include any allowance for machine friction, windage, or other factors. These factors must be considered when selecting a drive for a machine application.

## HORSEPOWER FORMULAS

### Rotating Objects

$$hp = \frac{T \times N}{63,000}$$

where $T$ = torque (lb – in.)
$N$ = speed (rpm)

$$hp = \frac{T \times N}{5252}$$

where $T$ = torque (lb – ft)
$N$ = speed (rpm)

### Objects in Linear Motion

$$hp = \frac{F \times V}{33,000}$$

where $F$ = force (lb)
$V$ = velocity (ft/min)

$$hp = \frac{F \times V}{396,000}$$

where $F$ = force (lb)
$V$ = velocity (in./min)

### Pumps

$$hp = \frac{gpm \times head \times specific\ gravity}{3960 \times efficiency\ of\ pump}$$

$$hp = \frac{gpm \times psi \times specific\ gravity}{1713 \times efficiency\ of\ pump}$$

where gpm = gallons per minute
head = height of water (ft)
efficiency of pump = %/100
psi = pounds per square inch

Specific gravity of water = 1.0
1 cu ft per sec = 448 gpm
1 psi = a head of 2.309 ft — for water weighing 62.36 lb per cu ft at 62°F

---

*This information on formulas and tables is courtesy of Allen-Bradley Co.

## Fans and Blowers

$$\text{hp} = \frac{\text{cfm} \times \text{psf}}{33{,}000 \times \text{efficiency of fan}}$$

$$\text{hp} = \frac{\text{cfm} \times \text{piw}}{6356 \times \text{efficiency of fan}}$$

$$\text{hp} = \frac{\text{cfm} \times \text{psi}}{229 \times \text{efficiency of fan}}$$

where cfm = cubic feet per minute
 psf = pounds per square foot
 piw = inches of water gage
 psi = pounds per square inch
 efficiency of fan = %/100

## Conveyors

$$\text{hp (vertical)} = \frac{F \times V}{33{,}000}$$

$$\text{hp (horizontal)} = \frac{F \times V \times \text{coefficient of friction}}{33{,}000}$$

where $F$ = force (lb)
 $V$ = velocity (ft/min)

Coefficient of friction:

 Ball or roller slide = 0.02
 Dovetail slide = 0.20
 Hydrostatic ways = 0.01
 Rectangle ways with gib = 0.1–0.25

## TORQUE FORMULAS

$$T = \frac{\text{hp} \times 5252}{N}$$

where $T$ = torque (lb-ft)
 hp = horsepower
 $N$ = speed (rpm)

$$T = F \times R$$

where $T$ = torque (lb−ft)
 $F$ = force (lb)
 $R$ = radius (ft)

$$T(\text{accelerating}) = \frac{\omega k^2 \times \Delta\text{rpm}}{308 \times t}$$

where $T$ = torque (lb−ft)
 $\omega k^2$ = inertia reflected to the motor shaft (lb−ft$^2$)
 $\Delta$rpm = change in speed
 $t$ = time to accelerate (sec)

**Note:**
To change lb-ft$^2$ to in.-lb-sec$^2$: Divide by 2.68
To change in.-lb-sec$^2$ to lb-ft$^2$: Multiply by 2.68

## AC MOTOR FORMULAS

$$\text{sync speed} = \frac{\text{frequency} \times 120}{\text{number of poles}}$$

where sync speed = synchronous speed (rpm)
 frequency = frequency (Hz)

$$\% \text{ slip} = \frac{(\text{sync speed} - \text{FL speed}) \times 100}{\text{sync speed}}$$

where FL speed = full-load speed (rpm)
 sync speed = synchronous speed (rpm)

$$\text{reflected } \omega k^2 = \frac{\omega k^2 \text{ of load}}{(\text{reduction ratio})^2}$$

where $\omega k^2$ = inertia (lb-ft$^2$)

## ELECTRICAL FORMULAS

### Ohm's Law

$$I = \frac{E}{R} \qquad R = \frac{E}{I} \qquad E = I \times R$$

where $I$ = current (amperes)
 $E$ = EMF or voltage (volts)
 $R$ = resistance (ohms)

### Power in DC Circuits

$$P = I \times E \qquad \text{hp} = \frac{I \times E}{746}$$

$$\text{kW} = \frac{I \times E}{1000} \qquad \text{kWh} = \frac{I \times E \times \text{hours}}{1000}$$

where $P$ = power (watts)
 $I$ = current (amperes)
 $E$ = EMF or voltage (volts)
 kW = kilowatts
 kWh = kilowatthours

### Power in AC Circuits

$$\text{kVA (one-phase)} = \frac{I \times E}{1000}$$

$$\text{kVA (three-phase)} = \frac{I \times E \times 1.73}{1000}$$

where kVA = kilovolt-amperes
 $I$ = current (amperes)
 $E$ = EMF or voltage (volts)

$$\text{kW (one-phase)} = \frac{I \times E \times \text{PF}}{1000}$$

$$\text{kW (two-phase)} = \frac{I \times E \times \text{PF} \times 1.42}{1000}$$

$$\text{kW (three-phase)} = \frac{I \times E \times PF \times 1.73}{1000}$$

$$PF = \frac{W}{V \times I} = \frac{kW}{kVA}$$

where kW = kilowatts
$I$ = current (amperes)
$E$ = EMF or voltage (volts)
PF = power factor
W = watts
V = volts
kVA = kilovolt-amperes

## Calculating Motor Amperes

$$\text{motor amperes} = \frac{hp \times 746}{E \times 1.732 \times Eff \times PF}$$

$$\text{motor amperes} = \frac{kVA \times 1000}{1.73 \times E}$$

$$\text{motor amperes} = \frac{kW \times 1000}{1.73 \times E \times PF}$$

where hp = horsepower
$E$ = EMF or voltage (volts)
Eff = efficiency of motor (%/100)
kVA = kilovolt-amperes
kW = kilowatts
PF = power factor

## Calculating AC Motor Locked-Rotor Amperes

$$LRA = \frac{hp \times \left(\frac{\text{start kVA}}{hp}\right) \times 1000}{E \times 1.73}$$

where LRA = locked-rotor amperes
hp = horsepower
kVA = kilovolt-amperes
$E$ = voltage (volts)

$$\text{LRA at freq. } X = \frac{\text{60-Hz LRA}}{\sqrt{\frac{60}{\text{freq. } X}}}$$

where 60 Hz LRA = locked-rotor amperes
freq. $X$ = desired frequency (Hz)

## ER FORMULAS

### Calculating Accelerating Force for Linear Motion

$$F\text{(acceleration)} = \frac{W \times \Delta V}{1933 \times t}$$

where $F$ = force (lb)
$W$ = weight (lb)
$\Delta V$ = change of velocity (fpm)
$t$ = time to accelerate weight (sec)

### Calculating Minimum Accelerating Time of a Drive

$$t = \frac{\omega k^2 \times \Delta N}{308 \times T}$$

where $t$ = Time required to accelerate load (sec)
$\omega k^2$ = total inertia that the motor must accelerate (lb-ft$^2$; includes motor rotor, gear reducer, and load
$\Delta N$ = change in speed required (rpm)
$T$ = accelerating torque (lb-ft)

**Note:**
To change lb-ft$^2$ to in.-lb-sec$^2$: Divide by 2.68
To change in.-lb-sec$^2$ to lb-ft$^2$: Multiply by 2.68

$$rpm = \frac{fpm}{0.262 \times D}$$

where rpm = revolutions per minute
fpm = feet per minute
$D$ = diameter (ft)

$$\omega k^2 \text{ reflected to motor} = \text{load } \omega k^2 \times \left(\frac{\text{load rpm}}{\text{motor rpm}}\right)^2$$

where $\omega k^2$ = inertia (lb-ft$^2$)
rpm = revolutions per minute

## ENGINEERING CONSTANTS

### Temperature

0°C = freezing point of water
32°F = freezing point of water = 0°C
100°C = boiling point of water at atmospheric pressure
212°F = boiling point of water at atmospheric pressure
1.8°F change = 1°C
0.252 kilocalorie = 1 Btu
−270°C = absolute zero
−459.6°F = absolute zero

### Length

1760 yd = 1 mile
25.4 mm = 2.54 cm = 1 in.
3 ft = 1 yd
3.2808 ft = 1 m
39.37 in. = 1 m = 100 cm = 1000 mm
5280 ft = 1 mile
0.62137 mile = 1 km

## Weight

16 oz = 1 lb
2.2046 lb = 1 kg
2.309 ft water at 62°F = 1 psi
28.35 g = 1 oz
59.76 lb = weight of 1 cu ft of water at 212°F
0.062428 lb per cu ft = 1 kg/cu m
62.355 lb = weight of 1 cu ft water at 62°F
8⅓ (8.32675) lb = weight 1 gal water at 62°F

## Power

1.3410 hp = 1 kW
2.545 Btu per hr = 1 hp
33,000 ft-lb per min = 1 hp
550-ft-lb per sec = 1 hp
745.7 W = 1 hp

## Area

10.764 sq ft = 1 m$^2$
1,273,239 circular mils = 1 sq in.
144 sq in. = 1 sq ft
645 mm$^2$ = 1 sq in.
9 sq ft = 1 sq yd
0.0929 m$^2$ = 1 sq ft

## Mathematic

1.4142 = square root of 2
1.7321 = square root of 3
3.1416 = $\pi$ = ratio of circumference of circle to diameter = ratio of area of a circle to square of radius
57.296 degrees = 1 rad (angle)
0.7854 × diameter squared = area of a circle

## Pressure

14.223 psi = 1 kg per cm$^2$ = 1 "metric atmosphere"
2.0355 in. Hg at 32°F = 1 psi
2.0416 in. Hg at 62°F = 1 psi
2116.3 psf = atmospheric pressure
27.71 in. water at 62°F = 1 psi
29.921 in. Hg at 32°F = atmospheric pressure
30 in. Hg at 62°F = atmospheric pressure (approximate)
33.974 ft water at 62°F = atmospheric pressure
0.433 psi = 1 ft of water at 62°F
5196 psf = 1 in. water at 62°F
760 mm Hg = atmospheric pressure at 0°C
0.07608 lb = weight 1 cu ft air at 62°F and 14.7 psi

## Volume

1728 cu in. = 1 cu ft
231 cu in = 1 gal (U.S.)
277.274 cu in = 1 gal (British)
27 cu ft = 1 cu yd
31 gal (31.5 U.S. gal) = 1 barrel
35.314 cu ft = 1 m$^3$
3.785 liters = 1 gal
61.023 cu in. = 1 liter
7.4805 gal = 1 cu ft

## Units of Pressure

kg per cm$^2$ = kilograms per square centimeter
Hg = symbol for mercury
psi = pounds per square inch
psf = pounds per square foot

## Units of Volume

cu in. = cubic inch
gal = gallon
cu ft = cubic feet
ml = milliliter
fl oz = fluid ounce (U.S.)

# CONVERSION FACTORS

## Length

| To Convert: | To: | Multiply by: |
|---|---|---|
| meters | feet | 3.281 |
| meters | inches | 39.37 |
| inches | meters | 0.0254 |
| feet | meters | 0.3048 |
| millimeters | inches | 0.0394 |
| inches | millimeters | 25.4 |
| threads/inch | millimeter pitch | Divide into 25.4 |
| yards | meters | 0.914 |

*Example:* 10 m × 3.281 = 32.81 ft

## Area

| To Convert: | To: | Multiply by: |
|---|---|---|
| circular mil | meter$^2$ | 0.50 × 10$^{-9}$ |
| yard$^2$ | meter$^2$ | 0.8361 |

*Example:* 100 circular mils × 0.5 × 10$^{-9}$ = 0.5 × 10$^{-7}$ m$^2$ (0.5 × 10$^{-7}$ circular mil = 0.00000005 m$^2$)

## Power

| To Convert: | To: | Multiply by: |
|---|---|---|
| watts | hp | 0.00134 |
| ft-lb/min | hp | 0.0000303 |
| hp | kW | 0.746 |

*Example:* 1500 W × 0.00134 = 2.01 hp

## Rotation/Rate

| To Convert: | To: | Multiply by: |
|---|---|---|
| rpm | deg/sec | 6.00 |
| rpm | rad/sec | 0.1047 |
| deg/sec | rpm | 0.1667 |
| rad/sec | rpm | 9.549 |
| fpm | m/s | 0.00508 |
| fps | m/s | 0.3048 |
| gal/min | cm³/s | 63.09 |
| in./sec | m/s | 0.0254 |
| km/h | m/s | 0.2778 |
| mph | m/s | 0.447 |
| mph | km/h | 1.609 |
| rpm | rad/s | 0.1047 |
| yd³/min | m³/s | 0.01274 |

*Example:* 1800 rpm × 6.00 = 10800 deg/sec

## Moment of Inertia

| To Convert: | To: | Multiply by: |
|---|---|---|
| newton-meters² | lb-ft² | 2.42 |
| oz-in.² | lb-ft² | 0.000434 |
| lb-in.² | lb-ft² | 0.00694 |
| slug-ft² | lb-ft² | 32.17 |
| oz-in.-sec² | lb-ft² | 0.1675 |
| in.-lb-sec² | lb-ft² | 2.68 |

*Example:* 25 newton-meters² × 2.42 = 60.5 lb-ft²

## Mass/Weight

| To Convert: | To: | Multiply by: |
|---|---|---|
| oz | g | 31.1 |
| kg | lb | 2.205 |
| lb | kg | 0.4536 |
| newtons | lb | 0.2248 |

*Example:* 50 oz × 31.1 = 1555 g

## Torque

| To Convert: | To: | Multiply by: |
|---|---|---|
| newton-meters | lb-ft | 0.7376 |
| lb-ft | newton-meters | 1.3558 |
| lb-in. | lb-ft | 0.0833 |
| lb-ft | lb-in. | 12.00 |

*Example:* 30 Newton-Meters × 0.7376 = 22.13 lb-ft

## Volume

| To Convert: | To: | Multiply by: |
|---|---|---|
| cm³ (ml) | m³ | 0.000001 |
| fl oz | cm³ | 29.57 |
| ft³ of water (39.2°F) | kg (or liter) | 28.32 |
| cfm | m³/s | 0.000472 |
| liters | m³ | 0.001 |
| yd³ | m³ | 0.7646 |

*Example:* 250 cm³ × .000001 = .00025 m³

## Temperature

| To Convert: | To: | Use the Formula: |
|---|---|---|
| °F | °C | $°C = \dfrac{°F - 32}{1.8}$ |
| °C | °F | $°F = (°C \times 1.8) + 32$ |

*Example:* $68°F = \dfrac{68 - 32}{1.8} = 20°C$

$20°C = (20 \times 1.8) + 32 = 68°F$

## Fractional Inch to Equivalent Millimeters and Decimals

| Inch | Equivalent mm | Decimal | Inch | Equivalent mm | Decimal | Inch | Equivalent mm | Decimal | Inch | Equivalent mm | Decimal |
|---|---|---|---|---|---|---|---|---|---|---|---|
| 1/64 | 0.3969 | 0.0156 | 17/64 | 6.7469 | 0.2656 | 33/64 | 13.0969 | 0.5156 | 49/64 | 19.4469 | 0.7656 |
| 1/32 | 0.7938 | 0.0313 | 9/32 | 7.1438 | 0.2813 | 17/32 | 13.4938 | 0.5313 | 25/32 | 19.8438 | 0.7813 |
| 3/64 | 1.1906 | 0.0469 | 19/64 | 7.5406 | 0.2969 | 35/64 | 13.8906 | 0.5469 | 51/64 | 20.2406 | 0.7969 |
| 1/16 | 1.5875 | 0.0625 | 5/16 | 7.9375 | 0.3125 | 9/16 | 14.2875 | 0.5625 | 13/16 | 20.6375 | 0.8125 |
| 5/64 | 1.9844 | 0.0781 | 21/64 | 8.3344 | 0.3181 | 37/64 | 14.6844 | 0.5781 | 53/64 | 21.0344 | 0.8281 |
| 3/32 | 2.3813 | 0.0938 | 11/32 | 8.7313 | 0.3438 | 19/32 | 15.0813 | 0.5938 | 27/32 | 21.4313 | 0.8438 |

| Inch | Equivalent mm | Decimal | Inch | Equivalent mm | Decimal | Inch | Equivalent mm | Decimal | Inch | Equivalent mm | Decimal |
|---|---|---|---|---|---|---|---|---|---|---|---|
| 7/64 | 2.7781 | 0.1094 | 23/64 | 9.1281 | 0.3594 | 39/64 | 15.4781 | 0.6094 | 55/64 | 21.8281 | 0.8594 |
| 1/8 | 3.1750 | 0.1250 | 3/8 | 9.5250 | 0.3750 | 5/8 | 15.8750 | 0.6250 | 7/8 | 22.2250 | 0.8750 |
| 9/64 | 3.5719 | 0.1406 | 25/64 | 9.9219 | 0.3906 | 41/64 | 16.2719 | 0.6406 | 57/64 | 22.6219 | 0.8906 |
| 5/32 | 3.9688 | 0.1563 | 13/32 | 10.3188 | 0.4063 | 21/32 | 16.6688 | 0.6563 | 29/32 | 23.0188 | 0.9063 |
| 11/64 | 4.3656 | 0.1719 | 27/64 | 10.7156 | 0.4219 | 43/64 | 17.0656 | 0.6719 | 59/64 | 23.4156 | 0.9219 |
| 3/16 | 4.7625 | 0.1875 | 7/16 | 11.1125 | 0.4375 | 11/16 | 17.4625 | 0.6875 | 15/16 | 23.8125 | 0.9375 |
| 13/64 | 5.1594 | 0.2031 | 29/64 | 11.5094 | 0.4531 | 45/64 | 17.8594 | 0.7031 | 61/64 | 24.2094 | 0.9531 |
| 7/32 | 5.5563 | 0.2188 | 15/32 | 11.9063 | 0.4688 | 23/32 | 18.2563 | 0.7188 | 31/32 | 24.6063 | 0.9688 |
| 15/64 | 5.9531 | 0.2344 | 31/64 | 12.3031 | 0.4844 | 47/64 | 18.6531 | 0.7344 | 63/64 | 25.0031 | 0.9844 |
| 1/4 | 6.3500 | 0.2500 | 1/2 | 12.700 | 0.5000 | 3/4 | 19.0500 | 0.7500 | | | |

## Inertia of a Solid Steel Shaft (lb-ft² per inch of length)

| Diameter (in.) | $\omega k^2$ | Diameter (in.) | $\omega k^2$ | Diameter (in.) | $\omega k^2$ | Diameter (in.) | $\omega k^2$ | Diameter (in.) | $\omega k^2$ | Diameter (in.) | $\omega k^2$ |
|---|---|---|---|---|---|---|---|---|---|---|---|
| 3/4 | 0.00006 | 8 | 0.791 | 14 1/4 | 7.97 | 37 | 360.70 | 62 | 2844.3 | 87 | 11,028 |
| 1 | 0.00007 | 8 1/4 | 0.895 | 14 1/2 | 8.54 | 38 | 401.30 | 63 | 3032.3 | 88 | 11,544 |
| 1 1/4 | 0.0005 | 8 1/2 | 1.000 | 14 3/4 | 9.15 | 39 | 445.30 | 64 | 3229.5 | 89 | 12,077 |
| 1 1/2 | 0.001 | 8 3/4 | 1.130 | 15 | 9.75 | 40 | 492.78 | 65 | 3436.1 | 90 | 12,629 |
| 1 3/4 | 0.002 | 9 | 1.270 | 16 | 12.59 | 41 | 543.90 | 66 | 3652.5 | 91 | 13,200 |
| 2 | 0.003 | 9 1/4 | 1.410 | 17 | 16.04 | 42 | 598.80 | 67 | 3879.0 | 92 | 13,790 |
| 2 1/4 | 0.005 | 9 1/2 | 1.550 | 18 | 20.16 | 43 | 658.10 | 68 | 4115.7 | 93 | 14,399 |
| 2 1/2 | 0.008 | 9 3/4 | 1.750 | 19 | 25.03 | 44 | 721.40 | 69 | 4363.2 | 94 | 15,029 |
| 2 3/4 | 0.011 | 10 | 1.930 | 20 | 30.72 | 45 | 789.30 | 70 | 4621.7 | 95 | 15,679 |
| 3 | 0.016 | 10 1/4 | 2.130 | 21 | 37.35 | 46 | 861.80 | 71 | 4891.5 | 96 | 16,349 |
| 3 1/2 | 0.029 | 10 1/2 | 2.350 | 22 | 44.99 | 47 | 939.30 | 72 | 5172 | 97 | 17,041 |
| 3 3/4 | 0.038 | 10 3/4 | 2.580 | 23 | 53.74 | 48 | 1021.80 | 73 | 5466 | 98 | 17,755 |
| 4 | 0.049 | 11 | 2.830 | 24 | 63.71 | 49 | 1109.60 | 74 | 5774 | 99 | 18,490 |
| 4 1/4 | 0.063 | 11 1/4 | 3.090 | 25 | 75.02 | 50 | 1203.07 | 75 | 6090 | 100 | 19,249 |
| 4 1/2 | 0.079 | 11 1/2 | 3.380 | 26 | 87.76 | 51 | 1302.2 | 76 | 6422 | | |
| 5 | 0.120 | 11 3/4 | 3.680 | 27 | 102.06 | 52 | 1407.4 | 77 | 6767 | 110 | 28,183 |
| 5 1/2 | 0.177 | 12 | 4.000 | 28 | 118.04 | 53 | 1518.8 | 78 | 7125 | 120 | 39,914 |
| 6 | 0.250 | 12 1/4 | 4.350 | 29 | 135.83 | 54 | 1636.7 | 79 | 7498 | 130 | 54,978 |
| 6 1/4 | 0.296 | 12 1/2 | 4.720 | 30 | 155.55 | 55 | 1761.4 | 80 | 7885 | 140 | 73,948 |
| 6 1/2 | 0.345 | 12 3/4 | 5.110 | 31 | 177.77 | 56 | 1893.1 | 81 | 8286 | 150 | 97,449 |
| 6 3/4 | 0.402 | 13 | 5.58 | 32 | 201.80 | 57 | 2031.9 | 82 | 8703 | 160 | 126,152 |
| 7 | 0.464 | 13 1/4 | 5.96 | 33 | 228.20 | 58 | 2178.3 | 83 | 9135 | 170 | 160,772 |
| 7 1/4 | 0.535 | 13 1/2 | 6.42 | 34 | 257.20 | 59 | 2332.5 | 84 | 9584 | 180 | 202,071 |
| 7 1/2 | 0.611 | 13 3/4 | 6.91 | 35 | 386.80 | 60 | 2494.7 | 85 | 10,048 | 190 | 250,858 |
| 7 3/4 | 0.699 | 14 | 7.42 | 36 | 323.20 | 61 | 2665.2 | 86 | 10,529 | 200 | 307,988 |

# VI

# Full-Load Currents of AC and DC Motors

## THREE-PHASE, 60-Hz AC INDUCTION MOTORS[a]

| HP | RPM | Full Load Current | | | | | |
|---|---|---|---|---|---|---|---|
| | | 200 v | 230 v | 460 v | 575 v | 2200 v | 4000 v |
| 1/4 | 1800 | 1.10 | 0.96 | 0.48 | 0.38 | .... | .... |
| | 1200 | 1.33 | 1.16 | 0.58 | 0.46 | .... | .... |
| | 900 | 1.67 | 1.45 | 0.73 | 0.58 | .... | .... |
| 1/3 | 1800 | 1.33 | 1.16 | 0.58 | 0.47 | .... | .... |
| | 1200 | 1.64 | 1.43 | 0.72 | 0.58 | .... | .... |
| | 900 | 2.01 | 1.75 | 0.88 | 0.71 | .... | .... |
| 1/2 | 1800 | 1.93 | 1.68 | 0.84 | 0.67 | .... | .... |
| | 1200 | 2.38 | 2.07 | 1.04 | 0.83 | .... | .... |
| | 900 | 3.34 | 2.90 | 1.45 | 1.16 | .... | .... |
| 3/4 | 1800 | 2.68 | 2.33 | 1.17 | 0.93 | .... | .... |
| | 1200 | 3.28 | 2.85 | 1.43 | 1.14 | .... | .... |
| | 900 | 3.97 | 3.45 | 1.73 | 1.38 | .... | .... |
| 1 | 3600 | 3.16 | 2.75 | 1.38 | 1.10 | .... | .... |
| | 1800 | 3.51 | 3.05 | 1.53 | 1.22 | .... | .... |
| | 1200 | 4.07 | 3.54 | 1.77 | 1.42 | .... | .... |
| | 900 | 4.30 | 3.74 | 1.87 | 1.50 | .... | .... |
| 1 1/2 | 3600 | 4.80 | 4.17 | 2.09 | 1.67 | .... | .... |
| | 1800 | 4.92 | 4.28 | 2.14 | 1.71 | .... | .... |
| | 1200 | 5.58 | 4.85 | 2.43 | 1.94 | .... | .... |
| | 900 | 6.68 | 5.81 | 2.91 | 2.32 | .... | .... |
| 2 | 3600 | 6.39 | 5.56 | 2.78 | 2.22 | .... | .... |
| | 1800 | 6.62 | 5.76 | 2.88 | 2.30 | .... | .... |
| | 1200 | 7.30 | 6.35 | 3.18 | 2.54 | .... | .... |
| | 900 | 8.29 | 7.21 | 3.61 | 2.88 | .... | .... |
| 3 | 3600 | 9.05 | 7.87 | 3.94 | 3.14 | .... | .... |
| | 1800 | 9.53 | 8.29 | 4.14 | 3.32 | .... | .... |
| | 1200 | 10.3 | 8.92 | 4.46 | 3.56 | .... | .... |
| | 900 | 11.7 | 10.20 | 5.09 | 4.08 | .... | .... |
| 5 | 3600 | 14.6 | 12.7 | 6.34 | 5.08 | .... | .... |
| | 1800 | 15.2 | 13.2 | 6.60 | 5.28 | .... | .... |
| | 1200 | 16.2 | 14.1 | 7.05 | 5.64 | .... | .... |
| | 900 | 17.9 | 15.6 | 7.80 | 6.24 | .... | .... |
| 7 1/2 | 3600 | 22.1 | 19.2 | 9.6 | 7.68 | .... | .... |
| | 1800 | 22.2 | 19.3 | 9.7 | 7.72 | .... | .... |
| | 1200 | 23.3 | 20.3 | 10.2 | 8.12 | .... | .... |
| | 900 | 27.4 | 23.8 | 11.9 | 9.51 | .... | .... |
| 10 | 3600 | 28.2 | 24.5 | 12.3 | 9.8 | .... | .... |
| | 1800 | 29.0 | 25.2 | 12.6 | 10.1 | .... | .... |
| | 1200 | 30.6 | 26.6 | 13.3 | 10.6 | .... | .... |
| | 900 | 33.2 | 28.9 | 14.5 | 11.6 | .... | .... |
| | 600 | 38.9 | 33.8 | 16.9 | 13.5 | .... | .... |
| 15 | 3600 | 42.2 | 36.7 | 18.4 | 14.7 | .... | .... |
| | 1800 | 43.8 | 38.1 | 19.1 | 15.2 | .... | .... |
| | 1200 | 45.9 | 39.9 | 20.0 | 16.0 | .... | .... |
| | 900 | 48.2 | 41.9 | 21.0 | 16.8 | .... | .... |
| | 600 | 55.5 | 48.3 | 24.2 | 19.3 | .... | .... |
| 20 | 3600 | 56.4 | 49.0 | 24.5 | 19.6 | 5.2 | 2.9 |
| | 1800 | 58.1 | 50.5 | 25.3 | 20.2 | 5.3 | 3.0 |
| | 1200 | 59.5 | 51.7 | 25.9 | 20.6 | 5.4 | 3.1 |
| | 900 | 62.8 | 54.6 | 27.3 | 21.8 | 5.8 | 3.2 |
| | 600 | 70.7 | 61.5 | 30.8 | 24.6 | 6.4 | 3.5 |
| 25 | 3600 | 68.1 | 59.2 | 29.6 | 23.6 | 6.3 | 3.4 |
| | 1800 | 72.1 | 62.7 | 31.3 | 25.0 | 6.5 | 3.6 |
| | 1200 | 74.4 | 64.7 | 32.3 | 25.8 | 6.7 | 3.7 |
| | 900 | 77.5 | 67.4 | 33.7 | 27.0 | 6.9 | 3.8 |
| | 600 | 82.7 | 71.9 | 35.9 | 28.8 | 8.1 | 4.4 |

*Source:* Courtesy of Allen-Bradley.

[a]The full-load currents listed are "average values" for horsepower-rated motors of several manufacturers at the more common rated voltages and speeds. These "average values," along with the similar values listed in the NEC should be used only as a guide for selecting suitable components for the motor branch circuit. The rated full-load current, shown on the motor nameplate, may vary considerably from the value listed, depending on the specific motor design. The nameplate full-load current should always be used in determining the rating of the devices used for motor running overcurrent protection.

| HP | RPM | Full Load Current | | | | | |
|---|---|---|---|---|---|---|---|
| | | 200 v | 230 v | 460 v | 575 v | 2200 v | 4000 v |
| 30 | 1800 | 83.7 | 72.8 | 36.4 | 29.2 | 7.8 | 4.3 |
| | 1200 | 88.7 | 77.1 | 38.6 | 30.8 | 8.0 | 4.4 |
| | 900 | 91.3 | 79.4 | 39.7 | 31.8 | 8.2 | 4.5 |
| | 600 | 101.1 | 87.9 | 43.9 | 35.2 | 9.3 | 5.0 |
| 40 | 1800 | 112.7 | 98 | 49.0 | 39.2 | 10.0 | 5.5 |
| | 1200 | 113.9 | 99 | 49.5 | 39.6 | 10.3 | 5.7 |
| | 900 | 119.6 | 104 | 52.0 | 41.6 | 10.6 | 5.8 |
| | 600 | 130.0 | 113 | 56.5 | 45.2 | 11.5 | 6.3 |
| 50 | 1800 | 139.2 | 121 | 60.5 | 48.4 | 12.3 | 6.8 |
| | 1200 | 140.3 | 122 | 61.0 | 48.8 | 12.4 | 6.8 |
| | 900 | 146.1 | 127 | 63.5 | 50.8 | 13.1 | 7.2 |
| | 600 | 158.7 | 138 | 69.0 | 55.2 | 14.2 | 7.8 |
| 60 | 1800 | 164.5 | 143 | 71.5 | 57.2 | 14.6 | 8.0 |
| | 1200 | 170.2 | 148 | 74.0 | 59.2 | 14.9 | 8.2 |
| | 900 | 173.7 | 151 | 75.5 | 60.4 | 15.4 | 8.5 |
| | 600 | 186.3 | 162 | 81.0 | 64.8 | 16.7 | 9.2 |
| 75 | 1800 | 204.7 | 178 | 89.0 | 71.2 | 18.0 | 9.9 |
| | 1200 | 208.2 | 181 | 90.5 | 72.4 | 18.2 | 10.0 |
| | 900 | 215.1 | 187 | 93.5 | 74.8 | 19.0 | 10.5 |
| | 600 | 228.9 | 199 | 99.5 | 79.6 | 21.0 | 11.6 |
| 100 | 1800 | 268.0 | 233 | 116 | 93.2 | 23.6 | 13.0 |
| | 1200 | 274.9 | 239 | 120 | 95.6 | 24.2 | 13.3 |
| | 900 | 281.8 | 245 | 123 | 98.0 | 24.8 | 13.6 |
| | 600 | 295.6 | 257 | 128 | 103.0 | 26.4 | 14.5 |
| | 450 | 333.5 | 290 | 145 | 116.0 | 29.8 | 16.4 |
| 125 | 1800 | 332.4 | 289 | 144 | 115 | 29.2 | 16.1 |
| | 1200 | 342.7 | 298 | 149 | 119 | 29.9 | 16.4 |
| | 900 | 350.8 | 305 | 153 | 122 | 30.9 | 17.0 |
| | 720 | 361.1 | 314 | 157 | 126 | 31.3 | 17.2 |
| | 600 | 368.0 | 320 | 160 | 128 | 32.8 | 18.0 |
| | 450 | 403.7 | 351 | 175 | 140 | 36.0 | 19.8 |
| 150 | 1800 | 397.9 | 346 | 173 | 138 | 34.8 | 19.1 |
| | 1200 | 402.5 | 350 | 175 | 140 | 35.5 | 19.5 |
| | 900 | 417.5 | 363 | 182 | 145 | 37.0 | 20.4 |
| | 720 | 432.4 | 376 | 188 | 150 | 37.0 | 20.4 |
| | 600 | 434.7 | 378 | 189 | 151 | 38.8 | 21.3 |
| | 450 | 480.7 | 418 | 209 | 166 | 42.0 | 23.1 |
| 200 | 1800 | 529.0 | 460 | 230 | 184 | 46.7 | 25.7 |
| | 1200 | 535.9 | 466 | 233 | 186 | 47.0 | 25.9 |
| | 900 | 563.5 | 490 | 245 | 196 | 49.4 | 27.2 |
| | 720 | 568.1 | 494 | 247 | 197 | 49.0 | 27.0 |
| | 600 | 572.7 | 498 | 249 | 199 | 50.9 | 28.0 |
| | 450 | 607.2 | 528 | 264 | 211 | 53.7 | 29.5 |
| 250 | 1800 | 657.8 | 572 | 286 | 229 | 57.5 | 31.6 |
| | 1200 | 667.0 | 580 | 290 | 232 | 58.5 | 32.2 |
| | 900 | 694.6 | 604 | 302 | 242 | 61.5 | 33.8 |
| | 720 | 718.8 | 625 | 312 | 250 | 61.5 | 33.8 |
| | 600 | 724.5 | 630 | 315 | 252 | 61.0 | 33.6 |
| | 450 | 724.5 | 630 | 315 | 252 | 65.3 | 35.9 |
| | 360 | 777.4 | 676 | 338 | 270 | 70.0 | 38.5 |
| 300 | 1800 | 787.8 | 685 | 342 | 274 | 69.0 | 38.0 |
| | 1200 | 800.4 | 696 | 348 | 278 | 70.0 | 38.5 |
| | 900 | 830.3 | 722 | 361 | 289 | 73.5 | 40.4 |
| | 600 | 830.3 | 722 | 361 | 289 | 72.3 | 39.8 |
| | 450 | 874.0 | 760 | 380 | 304 | 76.0 | 41.8 |
| | 360 | 954.5 | 830 | 415 | 332 | 82.8 | 45.5 |

# VII
# Power Factor Correcting Capacitors

## SUGGESTED MAXIMUM CAPACITOR RATING WHEN MOTOR AND CAPACITOR ARE SWITCHED AS UNIT

| Induction Motor Horse-Power Rating | Nominal Motor Speed (rpm): | | | | | | | | | | | |
|---|---|---|---|---|---|---|---|---|---|---|---|---|
| | 3600 | | 1800 | | 1200 | | 900 | | 720 | | 600 | |
| | Capacitor Rating (kVAR) | Line Current Reduction (%) | Capacitor Rating (kVAR) | Line Current Reduction (%) | Capacitor Rating (kVAR) | Line Current Reduction (%) | Capacitor Rating (kVAR) | Line Current Reduction (%) | Capacitor Rating (kVAR) | Line Current Reduction (%) | Capacitor Rating (kVAR) | Line Current Reduction (%) |
| 3 | 1.5 | 14 | 1.5 | 15 | 1.5 | 20 | 2 | 27 | 2.5 | 35 | 3.5 | 41 |
| 5 | 2 | 12 | 2 | 13 | 2 | 17 | 3 | 25 | 4 | 32 | 4.5 | 37 |
| 7½ | 2.5 | 11 | 2.5 | 12 | 3 | 15 | 4 | 22 | 5.5 | 30 | 6 | 34 |
| 10 | 3 | 10 | 3 | 11 | 3.5 | 14 | 5 | 21 | 6.5 | 27 | 7.5 | 31 |
| 15 | 4 | 9 | 4 | 10 | 5 | 13 | 6.5 | 18 | 8 | 23 | 9.5 | 27 |
| 20 | 5 | 9 | 5 | 10 | 6.5 | 12 | 7.5 | 16 | 9 | 21 | 12 | 25 |
| 25 | 6 | 9 | 6 | 10 | 7.5 | 11 | 9 | 15 | 11 | 20 | 14 | 23 |
| 30 | 7 | 8 | 7 | 9 | 9 | 11 | 10 | 14 | 12 | 18 | 16 | 22 |
| 40 | 9 | 8 | 9 | 9 | 11 | 10 | 12 | 13 | 15 | 16 | 20 | 20 |
| 50 | 12 | 8 | 11 | 9 | 13 | 10 | 15 | 12 | 19 | 15 | 24 | 19 |
| 60 | 14 | 8 | 14 | 8 | 15 | 10 | 18 | 11 | 22 | 15 | 27 | 19 |
| 75 | 17 | 8 | 16 | 8 | 18 | 10 | 21 | 10 | 26 | 14 | 32.5 | 18 |
| 100 | 22 | 8 | 21 | 8 | 25 | 9 | 27 | 10 | 32.5 | 13 | 40 | 17 |
| 125 | 27 | 8 | 26 | 8 | 30 | 9 | 32.5 | 10 | 40 | 13 | 47.5 | 16 |
| 150 | 32.5 | 8 | 30 | 8 | 35 | 9 | 37.5 | 10 | 47.5 | 12 | 52.5 | 15 |
| 200 | 40 | 8 | 37.5 | 8 | 42.5 | 9 | 47.5 | 10 | 60 | 12 | 65 | 14 |

| Induction Motor Horse-Power Rating | Nominal Motor Speed (rpm): | | | | | | | | | | | |
|---|---|---|---|---|---|---|---|---|---|---|---|---|
| | 3600 | | 1800 | | 1200 | | 900 | | 720 | | 600 | |
| | Capacitor Rating (kVAR) | Line Current Reduction (%) | Capacitor Rating (kVAR) | Line Current Reduction (%) | Capacitor Rating (kVAR) | Line Current Reduction (%) | Capacitor Rating (kVAR) | Line Current Reduction (%) | Capacitor Rating (kVAR) | Line Current Reduction (%) | Capacitor Rating (kVAR) | Line Current Reduction (%) |
| 250 | 50 | 8 | 45 | 7 | 52.5 | 8 | 57.5 | 9 | 70 | 11 | 77.5 | 13 |
| 300 | 57.5 | 8 | 52.5 | 7 | 60 | 8 | 65 | 9 | 80 | 11 | 87.5 | 12 |
| 350 | 65 | 8 | 60 | 7 | 67.5 | 8 | 75 | 9 | 87.5 | 10 | 95 | 11 |
| 400 | 70 | 8 | 65 | 6 | 75 | 8 | 85 | 9 | 95 | 10 | 105 | 11 |
| 450 | 75 | 8 | 67.5 | 6 | 80 | 8 | 92.5 | | 100 | 9 | 110 | 11 |
| 500 | 77.5 | 8 | 72.5 | 6 | 82.5 | 8 | 97.5 | 9 | 107.5 | 9 | 115 | 10 |

Note: For use with three-phase, 60-Hz NEMA class B motors to raise full-load power factor to approximately 95%.

# KILOWATT MULTIPLIERS TO DETERMINE CAPACITOR KILOVAR REQUIRED FOR POWER FACTOR CORRECTION

| Original Power Factor | Corrected Power Factor | | | | | | | | | | | | | | | | | | | |
|---|---|---|---|---|---|---|---|---|---|---|---|---|---|---|---|---|---|---|---|---|
| | 0.80 | 0.81 | 0.82 | 0.83 | 0.84 | 0.85 | 0.86 | 0.87 | 0.88 | 0.89 | 0.90 | 0.91 | 0.92 | 0.93 | 0.94 | 0.95 | 0.96 | 0.97 | 0.98 | 0.99 | 1.0 |
| 0.50 | 0.982 | 1.008 | 1.034 | 1.060 | 1.086 | 1.112 | 1.139 | 1.165 | 1.192 | 1.220 | 1.248 | 1.276 | 1.306 | 1.337 | 1.369 | 1.403 | 1.440 | 1.481 | 1.529 | 1.589 | 1.732 |
| 0.51 | 0.937 | 0.962 | 0.989 | 1.015 | 1.041 | 1.067 | 1.094 | 1.120 | 1.147 | 1.175 | 1.203 | 1.231 | 1.261 | 1.292 | 1.324 | 1.358 | 1.395 | 1.436 | 1.484 | 1.544 | 1.687 |
| 0.52 | 0.893 | 0.919 | 0.945 | 0.971 | 0.997 | 1.023 | 1.050 | 1.076 | 1.103 | 1.131 | 1.159 | 1.187 | 1.217 | 1.248 | 1.280 | 1.314 | 1.351 | 1.392 | 1.440 | 1.500 | 1.643 |
| 0.53 | 0.850 | 0.876 | 0.902 | 0.928 | 0.954 | 0.980 | 1.007 | 1.033 | 1.060 | 1.088 | 1.116 | 1.144 | 1.174 | 1.205 | 1.237 | 1.271 | 1.308 | 1.349 | 1.397 | 1.457 | 1.600 |
| 0.54 | 0.809 | 0.835 | 0.861 | 0.887 | 0.913 | 0.939 | 0.966 | 0.992 | 1.019 | 1.047 | 1.075 | 1.103 | 1.133 | 1.164 | 1.196 | 1.230 | 1.267 | 1.308 | 1.356 | 1.416 | 1.559 |
| 0.55 | 0.769 | 0.795 | 0.821 | 0.847 | 0.873 | 0.899 | 0.926 | 0.952 | 0.979 | 1.007 | 1.035 | 1.063 | 1.093 | 1.124 | 1.156 | 1.190 | 1.227 | 1.268 | 1.316 | 1.376 | 1.519 |
| 0.56 | 0.730 | 0.756 | 0.782 | 0.808 | 0.834 | 0.860 | 0.887 | 0.913 | 0.940 | 0.968 | 0.996 | 1.024 | 1.054 | 1.085 | 1.117 | 1.151 | 1.188 | 1.229 | 1.277 | 1.337 | 1.480 |
| 0.57 | 0.692 | 0.718 | 0.744 | 0.770 | 0.796 | 0.822 | 0.849 | 0.875 | 0.902 | 0.930 | 0.958 | 0.986 | 1.016 | 1.047 | 1.079 | 1.113 | 1.150 | 1.191 | 1.239 | 1.299 | 1.442 |
| 0.58 | 0.655 | 0.681 | 0.707 | 0.733 | 0.759 | 0.785 | 0.812 | 0.838 | 0.865 | 0.893 | 0.921 | 0.949 | 0.979 | 1.010 | 1.042 | 1.076 | 1.113 | 1.154 | 1.202 | 1.262 | 1.405 |
| 0.59 | 0.619 | 0.645 | 0.671 | 0.697 | 0.723 | 0.749 | 0.776 | 0.802 | 0.829 | 0.857 | 0.885 | 0.913 | 0.943 | 0.974 | 1.006 | 1.040 | 1.077 | 1.118 | 1.166 | 1.226 | 1.369 |
| 0.60 | 0.583 | 0.609 | 0.635 | 0.661 | 0.687 | 0.713 | 0.740 | 0.766 | 0.793 | 0.821 | 0.849 | 0.877 | 0.907 | 0.938 | 0.970 | 1.004 | 1.041 | 1.082 | 1.130 | 1.190 | 1.333 |
| 0.61 | 0.549 | 0.575 | 0.601 | 0.627 | 0.653 | 0.679 | 0.706 | 0.732 | 0.759 | 0.787 | 0.815 | 0.843 | 0.873 | 0.904 | 0.936 | 0.970 | 1.007 | 1.048 | 1.096 | 1.156 | 1.299 |
| 0.62 | 0.516 | 0.542 | 0.568 | 0.594 | 0.620 | 0.646 | 0.673 | 0.699 | 0.726 | 0.754 | 0.782 | 0.810 | 0.840 | 0.871 | 0.903 | 0.937 | 0.974 | 1.015 | 1.063 | 1.123 | 1.266 |
| 0.63 | 0.483 | 0.509 | 0.535 | 0.561 | 0.587 | 0.613 | 0.640 | 0.666 | 0.693 | 0.721 | 0.749 | 0.777 | 0.807 | 0.838 | 0.870 | 0.904 | 0.941 | 0.982 | 1.030 | 1.090 | 1.233 |
| 0.64 | 0.451 | 0.474 | 0.503 | 0.529 | 0.555 | 0.581 | 0.608 | 0.634 | 0.661 | 0.689 | 0.717 | 0.745 | 0.775 | 0.806 | 0.838 | 0.872 | 0.909 | 0.950 | 0.998 | 1.068 | 1.201 |
| 0.65 | 0.419 | 0.445 | 0.471 | 0.497 | 0.523 | 0.549 | 0.576 | 0.602 | 0.629 | 0.657 | 0.685 | 0.713 | 0.743 | 0.774 | 0.806 | 0.840 | 0.877 | 0.918 | 0.966 | 1.026 | 1.169 |
| 0.66 | 0.388 | 0.414 | 0.440 | 0.466 | 0.492 | 0.518 | 0.545 | 0.571 | 0.598 | 0.626 | 0.654 | 0.682 | 0.712 | 0.743 | 0.775 | 0.809 | 0.846 | 0.887 | 0.935 | 0.995 | 1.138 |
| 0.67 | 0.358 | 0.384 | 0.410 | 0.436 | 0.462 | 0.488 | 0.515 | 0.541 | 0.568 | 0.596 | 0.624 | 0.652 | 0.682 | 0.713 | 0.745 | 0.779 | 0.816 | 0.857 | 0.905 | 0.965 | 1.108 |
| 0.68 | 0.328 | 0.354 | 0.380 | 0.406 | 0.432 | 0.458 | 0.485 | 0.511 | 0.538 | 0.566 | 0.594 | 0.622 | 0.652 | 0.683 | 0.715 | 0.749 | 0.786 | 0.827 | 0.875 | 0.935 | 1.078 |
| 0.69 | 0.299 | 0.325 | 0.351 | 0.377 | 0.403 | 0.429 | 0.456 | 0.482 | 0.509 | 0.537 | 0.565 | 0.593 | 0.623 | 0.654 | 0.686 | 0.720 | 0.757 | 0.798 | 0.846 | 0.906 | 1.049 |
| 0.70 | 0.270 | 0.296 | 0.322 | 0.348 | 0.374 | 0.400 | 0.427 | 0.453 | 0.480 | 0.508 | 0.536 | 0.564 | 0.594 | 0.625 | 0.657 | 0.691 | 0.728 | 0.769 | 0.817 | 0.877 | 1.020 |
| 0.71 | 0.242 | 0.268 | 0.294 | 0.320 | 0.346 | 0.372 | 0.399 | 0.425 | 0.452 | 0.480 | 0.508 | 0.536 | 0.566 | 0.597 | 0.629 | 0.663 | 0.700 | 0.741 | 0.789 | 0.849 | 0.992 |
| 0.72 | 0.214 | 0.240 | 0.266 | 0.292 | 0.318 | 0.344 | 0.371 | 0.397 | 0.424 | 0.452 | 0.480 | 0.508 | 0.538 | 0.569 | 0.601 | 0.635 | 0.672 | 0.713 | 0.761 | 0.821 | 0.964 |
| 0.73 | 0.186 | 0.212 | 0.238 | 0.264 | 0.290 | 0.316 | 0.343 | 0.369 | 0.396 | 0.424 | 0.452 | 0.480 | 0.510 | 0.541 | 0.573 | 0.607 | 0.644 | 0.685 | 0.733 | 0.793 | 0.936 |
| 0.74 | 0.159 | 0.185 | 0.211 | 0.237 | 0.263 | 0.289 | 0.316 | 0.342 | 0.369 | 0.397 | 0.425 | 0.453 | 0.483 | 0.514 | 0.546 | 0.580 | 0.617 | 0.658 | 0.706 | 0.766 | 0.909 |
| 0.75 | 0.132 | 0.158 | 0.184 | 0.210 | 0.236 | 0.262 | 0.289 | 0.315 | 0.342 | 0.370 | 0.398 | 0.426 | 0.456 | 0.487 | 0.519 | 0.553 | 0.590 | 0.631 | 0.679 | 0.739 | 0.882 |

| Original Power Factor | Corrected Power Factor | | | | | | | | | | | | | | | | | | | | |
|---|---|---|---|---|---|---|---|---|---|---|---|---|---|---|---|---|---|---|---|---|---|
| | 0.80 | 0.81 | 0.82 | 0.83 | 0.84 | 0.85 | 0.86 | 0.87 | 0.88 | 0.89 | 0.90 | 0.91 | 0.92 | 0.93 | 0.94 | 0.95 | 0.96 | 0.97 | 0.98 | 0.99 | 1.0 |
| 0.76 | 0.105 | 0.131 | 0.157 | 0.183 | 0.209 | 0.235 | 0.262 | 0.288 | 0.315 | 0.343 | 0.371 | 0.399 | 0.429 | 0.460 | 0.492 | 0.526 | 0.563 | 0.604 | 0.652 | 0.712 | 0.855 |
| 0.77 | 0.079 | 0.105 | 0.131 | 0.157 | 0.183 | 0.209 | 0.236 | 0.262 | 0.289 | 0.317 | 0.345 | 0.373 | 0.403 | 0.434 | 0.466 | 0.500 | 0.537 | 0.578 | 0.626 | 0.686 | 0.829 |
| 0.78 | 0.052 | 0.078 | 0.104 | 0.130 | 0.156 | 0.182 | 0.209 | 0.235 | 0.262 | 0.290 | 0.318 | 0.346 | 0.376 | 0.407 | 0.439 | 0.473 | 0.510 | 0.551 | 0.599 | 0.659 | 0.802 |
| 0.79 | 0.026 | 0.052 | 0.078 | 0.104 | 0.130 | 0.156 | 0.183 | 0.209 | 0.236 | 0.264 | 0.292 | 0.320 | 0.350 | 0.381 | 0.413 | 0.447 | 0.484 | 0.525 | 0.573 | 0.633 | 0.776 |
| 0.80 | 0.000 | 0.026 | 0.052 | 0.078 | 0.104 | 0.130 | 0.157 | 0.183 | 0.210 | 0.238 | 0.266 | 0.294 | 0.324 | 0.355 | 0.387 | 0.421 | 0.458 | 0.499 | 0.547 | 0.609 | 0.750 |
| 0.81 | | 0.000 | 0.026 | 0.052 | 0.078 | 0.104 | 0.131 | 0.157 | 0.184 | 0.212 | 0.240 | 0.268 | 0.298 | 0.329 | 0.361 | 0.395 | 0.432 | 0.473 | 0.521 | 0.581 | 0.724 |
| 0.82 | | | 0.000 | 0.026 | 0.052 | 0.078 | 0.105 | 0.131 | 0.158 | 0.186 | 0.214 | 0.242 | 0.272 | 0.303 | 0.335 | 0.369 | 0.406 | 0.447 | 0.495 | 0.555 | 0.698 |
| 0.83 | | | | 0.000 | 0.026 | 0.052 | 0.079 | 0.105 | 0.132 | 0.160 | 0.188 | 0.216 | 0.246 | 0.277 | 0.309 | 0.343 | 0.380 | 0.421 | 0.469 | 0.529 | 0.672 |
| 0.84 | | | | | 0.000 | 0.026 | 0.053 | 0.079 | 0.106 | 0.134 | 0.162 | 0.190 | 0.220 | 0.251 | 0.283 | 0.317 | 0.354 | 0.395 | 0.443 | 0.503 | 0.646 |
| 0.85 | | | | | | 0.000 | 0.027 | 0.053 | 0.080 | 0.108 | 0.136 | 0.164 | 0.194 | 0.225 | 0.257 | 0.291 | 0.328 | 0.369 | 0.417 | 0.477 | 0.620 |
| 0.86 | | | | | | | 0.000 | 0.026 | 0.053 | 0.081 | 0.109 | 0.137 | 0.167 | 0.198 | 0.230 | 0.264 | 0.301 | 0.342 | 0.390 | 0.450 | 0.593 |
| 0.87 | | | | | | | | 0.000 | 0.027 | 0.055 | 0.083 | 0.111 | 0.141 | 0.172 | 0.204 | 0.238 | 0.275 | 0.316 | 0.364 | 0.424 | 0.567 |
| 0.88 | | | | | | | | | 0.000 | 0.028 | 0.056 | 0.084 | 0.114 | 0.145 | 0.177 | 0.211 | 0.248 | 0.289 | 0.337 | 0.397 | 0.540 |
| 0.89 | | | | | | | | | | 0.000 | 0.028 | 0.056 | 0.086 | 0.117 | 0.149 | 0.183 | 0.220 | 0.261 | 0.309 | 0.369 | 0.512 |
| 0.90 | | | | | | | | | | | 0.000 | 0.028 | 0.058 | 0.089 | 0.121 | 0.155 | 0.192 | 0.233 | 0.281 | 0.341 | 0.484 |
| 0.91 | | | | | | | | | | | | 0.000 | 0.030 | 0.061 | 0.093 | 0.127 | 0.164 | 0.205 | 0.253 | 0.313 | 0.456 |
| 0.92 | | | | | | | | | | | | | 0.000 | 0.031 | 0.063 | 0.097 | 0.134 | 0.175 | 0.223 | 0.283 | 0.426 |
| 0.93 | | | | | | | | | | | | | | 0.000 | 0.032 | 0.066 | 0.103 | 0.144 | 0.192 | 0.252 | 0.395 |
| 0.94 | | | | | | | | | | | | | | | 0.000 | 0.034 | 0.071 | 0.112 | 0.160 | 0.220 | 0.363 |
| 0.95 | | | | | | | | | | | | | | | | 0.000 | 0.037 | 0.079 | 0.126 | 0.186 | 0.329 |
| 0.96 | | | | | | | | | | | | | | | | | 0.000 | 0.041 | 0.089 | 0.149 | 0.292 |
| 0.97 | | | | | | | | | | | | | | | | | | 0.000 | 0.048 | 0.108 | 0.251 |
| 0.98 | | | | | | | | | | | | | | | | | | | 0.000 | 0.060 | 0.203 |
| 0.99 | | | | | | | | | | | | | | | | | | | | 0.000 | 0.143 |
| | | | | | | | | | | | | | | | | | | | | | 0.000 |

# Switch Symbols

**COMMON SWITCH SYMBOLS**

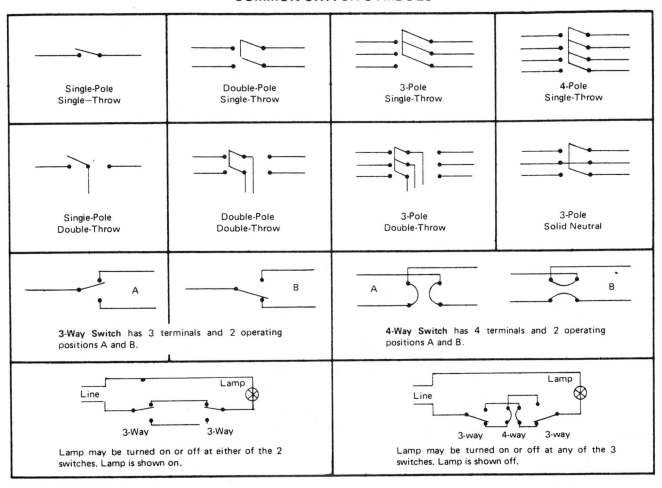

## OTHER SWITCHES

| Disconnect | Circuit Interrupter | Circuit Breaker W/Thermal O.L. | Circuit Breaker W/Magnetic O.L. | Circuit Breaker W/Thermal and Magnetic O.L. | Limit Switches | | Foot Switches |
|---|---|---|---|---|---|---|---|
| | | | | | Normally Open / Normally Closed | Held Closed / Held Open | N.O. / N.C. |

| Pressure & Vacuum Switches | | Liquid Level Switch | | Temperature Actuated Switch | | Flow Switch (Air, Water, Etc.) | |
|---|---|---|---|---|---|---|---|
| N.O. | N.C. | N.O. | N.C. | N.O. | N.C. | N.O. | N.C. |

# Wiring Diagram Symbols

## STANDARD ELEMENTARY WIRING DIAGRAM SYMBOLS

## CONTACTS

| Instant Operating | | | | Timed Contacts — Contact Action Retarded After Coil Is | | | |
|---|---|---|---|---|---|---|---|
| With Blowout | | Without Blowout | | Energized | | De-Energized | |
| N.O. | N.C. | N.O. | N.C. | N.O.T.C. | N.C.T.O. | N.O.T.O. | N.C.T.C. |

| INDUCTORS | TRANSFORMERS | | | | |
|---|---|---|---|---|---|
| Iron Core / Air Core | Auto | Iron Core | Air Core | Current | Dual Voltage |

| AC MOTORS | | | | DC MOTORS | | | |
|---|---|---|---|---|---|---|---|
| Single Phase | 3 Phase Squirrel Cage | 2 Phase 4 Wire | Wound Rotor | Armature | Shunt Field (Show 4 Loops) | Series Field (Show 3 Loops) | Comm. or Compens. Field (Show 2 Loops) |

**432** Wiring Diagram Symbols

# Unit Prefixes

| Prefix | Symbol | Power of 10 | Value | Name |
|---|---|---|---|---|
| tera | T | $10^{12}$ | 1,000,000,000,000 | trillion |
|  |  | $10^{11}$ | 100,000,000,000 | hundred-billion |
|  |  | $10^{10}$ | 10,000,000,000 | ten-billion |
| giga | G | $10^{9}$ | 1,000,000,000 | billion |
|  |  | $10^{8}$ | 100,000,000 | hundred-million |
|  |  | $10^{7}$ | 10,000,000 | ten-million |
| mega | M | $10^{6}$ | 1,000,000 | million |
|  |  | $10^{5}$ | 100,000 | hundred-thousand |
| myria | my | $10^{4}$ | 10,000 | ten-thousand |
| kilo | k | $10^{3}$ | 1,000 | thousand |
| hecto | h | $10^{2}$ | 100 | hundred |
| deka | da | $10^{1}$ | 10 | ten |
|  |  | $10^{0}$ | 1 | unit |
| deci | d | $10^{-1}$ | .1 | tenth |
| centi | c | $10^{-2}$ | .01 | hundredth |
| milli | m | $10^{-3}$ | .001 | thousandth |
|  |  | $10^{-4}$ | .000 1 | ten-thousandth |
|  |  | $10^{-5}$ | .000 01 | hundred-thousandth |
| micro | $\mu$ | $10^{-6}$ | .000 001 | millionth |
|  |  | $10^{-7}$ | .000 000 1 | ten-millionth |
|  |  | $10^{-8}$ | .000 000 01 | hundred-millionth |
| nano | n | $10^{-9}$ | .000 000 001 | billionth |
|  |  | $10^{-10}$ | .000 000 000 1 | ten-billionth |
|  |  | $10^{-11}$ | .000 000 000 01 | hundred-billionth |
| pico | p | $10^{-12}$ | .000 000 000 001 | trillionth |
|  |  | $10^{-13}$ | .000 000 000 000 1 | ten-trillionth |
|  |  | $10^{-14}$ | .000 000 000 000 01 | hundred-trillionth |
| femto | f | $10^{-15}$ | .000 000 000 000 001 | quadrillionth |
|  |  | $10^{-16}$ | .000 000 000 000 000 1 | ten-quadrillionth |
|  |  | $10^{-17}$ | .000 000 000 000 000 01 | hundred-quadrillionth |
| atto | a | $10^{-18}$ | .000 000 000 000 000 001 | quintillionth |

# XI
# Conversion Factors

| Multiply: | By: | To Obtain: |
|---|---|---|
| British thermal units | 778.3 | foot-pounds |
| British thermal units | $3.931 \times 10^{-4}$ | horsepower-hours |
| British thermal units | 1055 | joules |
| centimeters | $3.281 \times 10^{-2}$ | feet |
| centimeters | 0.3937 | inches |
| centimeters per second | 1.969 | feet per minute |
| circular mils | $5.067 \times 10^{-6}$ | square centimeters |
| circular mils | $7.854 \times 10^{-7}$ | square inches |
| cubic centimeters | $3.531 \times 10^{-5}$ | cubic feet |
| cubic centimeters | $6.102 \times 10^{-2}$ | cubic inches |
| cubic feet | 1728 | cubic inches |
| cubic feet | 0.02832 | cubic meters |
| feet | 30.48 | centimeters |
| grams | 0.3527 | ounces |
| horsepower | 550 | foot pounds per second |
| horsepower | 0.7457 | kilowatts |
| inches | 2.540 | centimeters |
| kilometers | 3281 | feet |
| liters | 0.2642 | gallons |
| lumens per sq ft | 1.0 | foot-candles |
| meters | 39.37 | inches |
| miles per hour | 88 | feet per minute |
| pounds per sq ft | 0.06804 | atmospheres |
| radians | 57.30 | degrees |
| yards | 91.44 | centimeters |

# XII Decibel Table

| dB | Power Ratio | Voltage or Current Ratio |
|---|---|---|
| 0 | 1.00 | 1.00 |
| 0.5 | 1.12 | 1.06 |
| 1.0 | 1.26 | 1.12 |
| 1.5 | 1.41 | 1.19 |
| 2.0 | 1.58 | 1.26 |
| 3.0 | 2.00 | 1.41 |
| 4.0 | 2.51 | 1.58 |
| 5.0 | 3.16 | 1.78 |
| 6.0 | 3.98 | 2.00 |
| 7.0 | 5.01 | 2.24 |
| 8.0 | 6.31 | 2.51 |
| 9.0 | 7.94 | 2.82 |
| 10 | 10.0 | 3.16 |
| 15 | 31.6 | 5.62 |
| 20 | 100 | 10 |
| 25 | 316 | 17.8 |
| 30 | 1,000 | 31.6 |
| 40 | 10,000 | 100 |
| 50 | $10^5$ | 316 |
| 60 | $10^6$ | 1,000 |
| 70 | $10^7$ | 3,162 |
| 80 | $10^8$ | 10,000 |
| 90 | $10^9$ | 31,620 |
| 100 | $10^{10}$ | $10^5$ |

# Index

## A

Acceleration, solid-state
    step-less, 226
Adapter, phase sequence, 43
Additional information, 399
Air gap, 108
Alphabet, Greek, 9
Alternator, 309
    automotive, 8
    single-phase, 311
    three-phase, 312
    two-phase, 311
Alternating current
    induction motor, types, 146
    motor, operation curve, 279
    motors, magnetic field, 142
    motors, series, 142
    motors, types, 142
Aluminum wire, 9
Ambient temperature
    relays, 128
    temperature, 384
Ammeter, 39
    ac, 14
Ampere rating, 50
Amperes, 6
ANSI, 57
Antenna, 188
Anti-plugging diagrams, 256
Anti-plugging, 254
Armature, 108
    exciter, 310
    reactance, 314
    reaction, 140, 141, 314
    resistance, 314
    rotation, 137
    types, 139
ASCII Code, 353
    seven-bit digital code, 353
Assemblies, 111
Atomic structure, 2
Atoms, 2
Automatic sequence, 214
    control, 250
Automatic transfer switching, 344
Autotransformer starting,
    209–211, 236
Autotransformers, 289, 293, 294
    single-phase, 292
    three-phase, 292

## B

Backspin, 259
Backup operations, 355
Balance analyzer, 43
Bar code, 188, 189
BCD code to Decimal Digit, 182
Bearing tool, 35
Brake
    electric motor, 255–257
    hydraulic, 259
    ratings, 131
Braking, 255
    electromechanical, 257
    mechanical, 257
Bridge circuit, 168
Brush
    lifting, 201
    terminology, 389
Brushless synchronous, 273
Buck and boost transformers, 304
Bushing tool set, 35

## C

Cable
    benders, 44
    cutter, 38
    stripper, 37, 38
    tray, 346, 347
        trough type, 348
Calculating single-phase, kVAs, 295
Calculating three-phase, kVAs, 296
Capacitor
    motors, uses, 203
    start motor, 203
    testing, 378
Capacity of a capacitor, 278
Careers, 394
    general information, 396
Cell
    controllers, 360
    photoelectric, 6
    secondary, 7
CEMF, 137
Centers, motor control, 275, 276
Centrifugal switch, testing, 377
Chemical action, 6
CIM, 361
Circuit
    board puller, 43, 44

Circuit (*Cont.*):
  breaker units, 345
  breaker, magnetic, 24
  breakers, 79
  bridge, 168
  complete, 9
  door chime, 21
  live, 16
  open, 11
  parallel, 11
  protection, 49
  relay, 25
  series, 10, 11
  seriesparallel, 11
  short, 11, 12
  simple, 11
  time delay, 163
Circuits, 9
  isolating, 302
  short, 50
  subdividing, 303
Circular mils, 38
Classifications, motor, 136
Clean commutators, 370
Codes, electrical, 51
Coil
  relay, 132
  servicing, 196
  voltage, 110
Color code
  OSHA, 53
  resistor, 27
Common switch symbols, 64
Commutators, 386, 387
  damaged, 387, 388
  worn, 387
Compelling relay, 214
Conductors, 3
Conduit
  bender, 44
  reamers, 44
Configurations, screw-head, 31
Connection diagrams, 372
Construction electrician, 395
  number in USA, 395
Contract construction, 131
Contact
  life, 131
  relay, 115
  sensors, 172
Continuity tester, 40
Control
  circuits, 80
  motor speed, 234
Controller
  consequence-pole motor, 215

feeder disconnect, 270
four-speed, 219, 220
selection, 214
smart motor, 248
solid-state, 248, 274
squirrel-cage, 270–272
synchronous, non-reversing, 272
Controllers, 356
  duplex motor, 269
  full voltage, 217, 221
  medium voltage, 269
  programmable, 40, 42
  solid-state adjustable speed, 241
Controlling electricity, 84
Controls
  closed loop, 284
  electronic features, 352
  open loop, 282,
  pressure, 176
  solid-state digital, 285
Conversion, 85
Cooling, transformer, 297
Copper
  losses, 290
  wire, 10
Core construction, 290
Corner grounded delta, 59
Coulomb, 106
Counter electromotive force, 137
Crimper, 36
Cross-sectional area, 8
CSA, 52
Current, 6
  flow 3
  limit module, 227, 228
  monitoring, 120
  ramp module, 227, 228
  rating, 110
  relay, 121

## D

D'Arsonval meter movement, 13, 14
Danger signals, 370
DC drives, 282, 284
DC motors, 127, 136
  applications, 141
  characteristics, 141
  compound wound, 139
  series wound, 138
  shunt wound, 138
  starting resistance, 141
  troubleshooting, 141
  types, 138
DC switching, 86
Decoders, 188

Definitions, brush, 388, 389
Delta connection, 312
De-rating a motor, 385
Devices, safety, 49
Diac, 229
   symbols, 229
Diagnosis, problems, 372
Diagrams, sequence control, 249
Dielectric constants, 166
Digital
   logic probe, 40, 41
   meters, 17
   timers, 159, 160
Diodes, 25
Direct current motor circuit, 73
Direction of rotation, 139
Disconnects, 56
Display systems, 359
Distribution
   electricity, 328
   systems, 328
Disturbances, power supply, 375
Double pole switch, 72
Double wiring, 301
DPST switch, 21
Drives
   adjustable frequency ac, 278
   adjustable speed, 278
   variable pitch, 281
   wound rotor ac motor, 282
Drum
   controllers, 240
   switch, 84, 85
Dry type transformer, 296
Dual-element fuses, 50

## E

Earnings, electricians, 398
Eddy current, 381
   losses, 289
   current, 281, 108
EIA, 354
Electric motors, maintenance, 370
Electric shock, 48
Electrical
   engineer, 395
   noise, 354, 376
   symbols, 62
   symbols, architectural plans, 63
Electrician
   employment, 394
   industrial, 394
   knives, 38
   troubleshooting, 366
Electricity, 2
   controlling, 20
   generating, 3
   measuring, 13
Electrodynamometer, 19
Electromagnet, 109, 195
   devices, 231
   relays, 128
Electromechanical
   devices, 224
   relays, 130
Electronic
   loads, 317
   speed switch, 242
   symbols, 62, 66, 67, 68, 69
Elements, 2
EMF, 140, 167
Employment
   electricians, 398
   outlook, 395
Encoders, 179
   programmable, 179
Energy, 3
Environmental effects, 354
EPA highlights, 292
EPROM 357, 358
Equipment
   grounding, 58
   standards, 53
Excitation, 235
Exciters, 310
Exotic generators, 6, 8
Eye protection, 56

## F

Fans, blowers, and pumps, 278
Fault currents, 334
Fiber optics, 359
Fire extinguishers, 54
   how to select, 55
Fish tape, 37
Float switches, 86, 88, 89, 179
   automatic, 180, 181
Floor trench duct, 346
Flow meter, strain gage, 170
Flow switches, 86, 89
Foot switches, 89
Formulas, 153
Forward-reverse-stop, 267
Fossil fuel power generators, 324, 326
Four wire-to-three wire, 302
Frame and shaft, 310
Franklin, Benjamin, 106
Free electrons, 3
Frequency, 313
   control, 241

Friction, 4
Full line voltage starting, 236
Fuse puller, 36, 37
Fuses, 50

## G

Gage-type thermostat, 75
Gas furnace, 108
Gear pullers, 35
Generation, basic principle, 308
Generator
    diesel powered, 316
    exotic, 6
    falling water, 324
    MHD, 6, 8
    multiple-engine, 317
    nuclear power, 324, 326
    standby, 315
    tractor driven, 315
GFI, 48, 49
Glands, packing, 177
Glossary, 401–404
Governor, 206
Ground fault
    interrupter, 48, 49
    receptacles, 367, 369
Ground monitor, 367
Grounded phase, 380
Grounding, 294, 355
Grounds, looking for, 380
Guide, troubleshooting, 380–384

## H

Hacksaw, 33
Hammer, 32
    ball peen, 32
    using properly, 56
Hardware, system, 356
Harmonics, third, 293
Heat, 4
Heating effect, 386
High pressure cut-out switch, 75
Highlights, EPA, 292
Horsepower, 13
Hum, ac, 111
Hydraulic brakes, 259

## I

Identification, automatic, 186
Immersion well, 177
Impedances, 293
Inclined coil iron vane meter, 20
Index, 439–450

Induction motor, 146, 201
    capacitor-start, 148
    resistance-start, 149
Industrial electrician, 394
Input-output, 353
    modules, 358
Installation practices, 355
Insulation systems, classification, 152
Insulators, 3
Intelligent I/O modules, 358
Interfacing, 353
Interlock switches, 90
Interlocking
    auxiliary contact, 92
    double-circuit, 92
    mechanical, 90
    pushbutton, 91
Inter-poles, 141
Inverters
    CSI, 279
    PWM, 280
    variable-voltage, 279
Iron core transformers, 288
Iron, soldering, 36

## J

Jack knife, 38
Jogging, 252
    forward and reverse, 253
Joystick, 90

## K

Kilowatthour meter, 18
    computer checking, 19
Kilowatts, 13
    to horsepower, 305
Kirchoff's law of voltages, 11
Knife switch, 20, 21
Knockout punch, 37

## L

Ladder diagram, 157, 158
Leadership, developing, 398
LED, 40, 185, 186, 228, 391
Level sensor, solid-state, 165
Light, 6
    emitting diodes, 231
    sources, 185
Lightning protection, 232
Limit switch, 93
    circuits, 96
    coil-tight, 89
    operating levers, 173
    types of, 93, 95
Limited energy electrician, 396

Line diagram or ladder diagram, 74
Line voltage monitor, 261
Load, 137
   changes, 314
   condition, 370
   converter relays, 123
   converter, 125
   detector, 123–125
      rock crusher belt, 124
   transfer systems, 318
Lock box, 58
Lodestone, 106
Logic rack, 226
Losses,
   copper, 290
   eddy current, 289
   hysteresis, 290
   transformer, 289
Low-voltage protection, 76, 218
Lubrication, 370

## M

Magnet, permanent, 109
Magnetic
   brakes, 258
   field, rotating, 144
   field, two-phase, 143
   starter, 209, 212
   starters, 240
Magnetism, 6, 106
Magnetite, 106
Maintenance electrician, 396
Make ratings, 131
Mallet, 32
Manual starters, 246
Manually operated three-phase starter, 73
MCM, 38, 39
Megger, 39
Megohmmeter, 39
Melting alloy overload relay, 128
Meters, 13
   clamp-on, 17, 18
   digital, 17
   inclined coil-iron core, 20
   kilowatthour, 18
   other types, 19
   vibration, 43
MHD, 6, 8
Micro-cell controllers, 360
Microphone, crystal, 5
Milliamperes, 6
Milliwatt, 13
Mini-cell controller, 361
Module, remote temperature
      detector, 262
Modules, 226, 275

Moisture in panel board, 366
Molecular structure, 2
Monitor
   ground, 367, 368
   load side, 119, 120
Monograms, 52
Motor
   and generator safety, 57
   applications, two- and three-phase, 147
   ball bearing, 370
   capacitor-start, 202
   characteristics, two- and three-phase, 147
   classifications, 136
   commutator type, 372
   connections, 373, 374
   control circuitry, 352
      control, ac motor, solid-state,
         241, 246
   electric, 200
   life, 384
   problems, 371
   poly-phase, 206
   problems, squirrel cage, 372
      programmable, 262
   protection, 259, 386
   reversibility, 201
   single-phase induction, 148
   sleeve bearing, 371
   speed, 139, 150
   split-phase, 200, 205
   squirrel cage, 145, 146, 234
   starting, 200
   stator, two-phase, 143
   synchronous, 234
   three-phase, 266
   troubleshooting, 279
   uses, 201
   wound rotor, 238
Mu, 8
Multi-meter, 14–16
Multiplexing, 182, 189
Multi-speed starters, 242

## N

National Electrical Code, 48, 51
National Electrical Manufacturers
      Association, 53
National Joint Apprenticeship and Training
      Committee, 397
NEC, 386
NEMA, 53, 131, 151, 254, 352, 375
Networks, 331, 332
   local area, 361
Nonflammable, 297
Number of poles, 150
Nut driver set, 35

## O

Oersted, 106
Ohm, 9
  Georg 106
  law, 12
    formula, 12
Ohmmeter, 16
Oil
  selection, 371
  standard, 271
  transformer, 291
Omega, 9
Open phase
  finding, 380
  looking for, 380
Oscilloscope, 42, 43
OSHA, 53
Overcurrent protection, control circuits, 78, 79
Overcurrents, 49
Overheated, 386
Overload protection, 260
Overload relays, 79
  automatic reset, 260
Overload trip times, 227
Overlubrication, 371
Overloads, 49
  bimetallic, 126

## P

Panel boards, 345
Parallel ports, 354
Paralleling, 318
Part-winding starting, 211
PCB, 292
Peak load sharing, 320
Performance characteristics, 385
Performance objectives, 2, 30, 48, 62, 72, 84, 106, 114, 136, 156, 172, 192, 200, 224, 234, 246, 266, 278, 288, 308, 324, 352, 366, 394
Permanent split-capacitor motor, 204
Phase
  failure relays, interfacing, 119, 120
  reversal, 119
Photoelectric switches,
  miniature, 186
  operation, 184
PID loop control station, 357
PLC, 353
  future, 361
Pliers, 31
Plugging, 175, 254
  diagrams, 255
PMP card, 263
Polychlorinated biphenyls, 292
Pouch, electrician's, 30

Power, 13
  balanced delta, 313
  balanced wye, 313
  failures, 343
  management systems, 320
  measurement, 335
  outages, 377
  pole, 226
  sources, 314
  switching, automatic, 342
  three-phase, 334
  uninterruptible, 317
Precipitator, 5
Pressure, 4
  controls, 98, 176, 178
  switches, 96, 97
Preventive maintenance, 366
Primary resistor starting, motors, 208
Prime mover systems, 319
Probe, digital logic, 40, 41
Problems
  DC motor, 386
  motor, 370
  using meters to locate, 279
Processors, 356
Programmable controller, 353
Protection, surge, 116
Protective module, 262
Pushbutton
  switches, 97
  stations, 98
PVC cutters, 44
PWM
  drive, 280
  wave forms, 280

## R

Raceways, 345
Radial systems, 329–331
Radio frequency
  identification systems, 187
RAM, 357, 358
Reactance starting, 237
Readers, bar code, 188
Recorder, 42
Reduced voltage
  starting, 206, 224
Related occupations, 398
Relays, 25, 114
  and altitude, 132
  and shock, 132
  and vibrations, 132
  armature, 114
  characteristics, 131
  contact amplifier, 123
  contact symbols, 62

current type, 129
electromechanical, 130
general purpose timing, 158
inverse time current, 261
line current monitoring, 261
motor winding, 129
over/under, 122
phase-failure, 119
pneumatic timing, 161, 162
potential type, 129
solenoid state, 115
solenoids, 114
solid-state monitoring, 121
time delay, 157
users for, 114
Reliability, solid-state, 355
Removing a ground, 367
Reset timer, 163
Resistance, 4, 8
    speed regulation, 238
    starting, 237
Resistive loads, 122
Resistor, 25, 240
    carbon composition, 26, 27
    color code, 62
    ranges, 209
    wattage rating, 27
    wire-wound, 26, 208
Reversibility, 202, 205
Reversing starter, 268
Revolving field synchronous motor, 145
RF tag, 188
Rheostat, speed control, 140
Right hand rule, 136
Rochelle salts, 4
Rotating-cam, 174, 175
Rotation, 150
Rotors, 309

## S

Safety
    electrical, 48
    notice, 299
    switches, 98, 99
Scanners
    hand-held, 189
    moving, 189
SCR, 117, 224, 225
    ac operation, 118
    biasing, 118
    construction, 118
    operation of, 118
    shorted, 228
Screwdrivers, 30

Screw-head configurations, 31
Seebeck effect, 167
Selector switches, 99
Semiconductors, 3
Sensing
    presence, 172
    temperature, 166
Sensors, 165
    capacitor types, 166
    classes, 172
    contact, 172
    mark, 185
    noncontact, 172
    pressure, 169
    temperature, 169
Sequence control, 164, 249
    diagrams, 250, 251
Serial ports, 354
Series circuit, 10
Service rating, 370
Shaded pole
    induction motors, 149
    motor, 204
    starting, 149
Shock, prevention of, 367
Short circuits, 50
    currents, 334
    looking for, 377
    testing for, 378
Shunt motor, 283
Shunts, 14
Signal generator, 43
Silicon-controlled rectifier, 117, 224
Sine-wave, 309
Single phasing, 385
Skinning knife, 38
Slip, 147
Snap switches, 95
Socket wrench, 34
Soldering
    gun, 36
    iron, 36
Solenoids, 23, 24, 106, 192
    applications, 195
    classes, 195
    coils, 109, 195
    construction of, 108
    frames, 192
    in circuits, 196
    industrial, 192
    lockout, 175
    movement, 110
    tubular, 193, 194
    valves, 107, 196
Solid-state
    electronics, 352

Solid-state
   equipment, 389
   relay switching, 122
Sound levels, 294
Sources, information, 397
Special transformers, 297
Speed, 235
   control, types, 239
   monitoring, 242
   ranges, 243
   switch, operation, 175
Split-phase induction
      motors, 148
Squirrel cage motors
   motor, 145
   rotors, 377
   slip and speed, 150
   starting, 152
Standard wiring diagrams, 62
   symbols
Standards, other countries, 52
Start circuits, 72
Starters, 266
   delta connections, 213
   full voltage reversing, 267
   two-speed, 268
   manual, 248
   multi-speed, 214, 242
   multi-switch, 239
   single-phase, 247
   three-phase, 247
   wye-delta, 211–213
Starting
   characteristics, motors, 208
   methods, 236
   motor, 236
   other methods, 238
   sequence, 217
   your own business, 397
Start-stop-job, 267
Static exciters, 310
Stators, 309
   construction, 146
Steam generator, 327
Step down
   transformer, 81
   voltage, 302
Stepping-up voltage, 302
Strain gage, 169
Strip chart, 42
Stripper, cable, 37
Suppressor
   surge, 231, 232
   transient, 116
Switches, 20, 23
   applications, 186, 187
   boards, 344
   centrifugal, 372
   dip, 161
   DPST, 21
   float, 179
   four-way, 23
   home and office, 23
   inductive cylindrical, 182, 183
   internal, 370
   knife, 20, 21
   limit, 172
   photoelectric, 183, 184
   programmable, 174
   proximity, 182, 183
   rocker, 22
   single pole, 100
   solid-state, 182
   SPDT, 21
   speed, 174
   start-stop, 100
   temperature, 100, 167
   three-way, 23
   thumbwheel, 160
   toggle, 21, 22
   transistor, 100
   vacuum, 102
Switching, 341
   relay loads, 122
Switchyard, 324
Symbols, common switches, 64
Synchronous motors, 144
   applications, 146
   characteristics, 146
   speed, 150
   starting, 145
   uses, 238
System grounding, 58
System, vision, 189
Systems analysis, 333
Systems, RFI, 187

## T

Tachometer, 37
   generator, 243
Taps, transformer, 297
Temperature, 9
   compensation, 126
   controls, 177, 178
   switches, 101
Tester
   continuity, 42
   polarized plug, 40
   polarized receptacle, 42
Thermal
   overload relays, 125

protectors, 79
  units, mounting, 128
Thermistor, 167, 168
  relay, 123
Thermocouple, 5, 167
Thermostat
  control, 75
  relay, 75
Third harmonics, 293
Three-phase motor, 146
  ampere rating, 375
  buck and boost transformer, 296
  diagrams, 207
    rating data, 372
  rotating field, 144
Three wire
  controls, 77, 78
  control circuits, 76
Thruster brakes, 258
Time-delay protection, 218
Time/count modes, 159
Time-delay relays, 157
Timers, 156
  dashpot, 156
  motor driven, 163
  programmable, 158, 159, 165
  solid-state, 156
  synchronous clock, 156
  time delay, 156,157
Timing modes, 159
Toggle switches, 101
  residential, 22
Tools
  electrical, 38
  hand, 30
  portable toolbox, 30
  safety, 53
  special, 43
Torque
  breakdown, 151
  horsepower, 151
  locked-rotor, 151, 152
Transducers
  piezo-resistance, 169
  strain gage, 169
Transfer switch, 314
  automatic, 316
Transformers, 288, 336
  applications, 301
  banks, 339
  buck and boost, 294
  calculating the load, 294
  connections, 338, 339
  construction, 336
  cooling, 297
  customer wiring, 295
  cutaway, 291
  dry type, 296
  high voltages, 301
  in control circuits, 79
  iron core, 288
  kVA rating, 304
  liquid filled, 299
  maintenance, 299
  motor, 302
  oil, 291
  operating equipment, 301
  pad mounted, 299, 300
  power, 300
  sound levels, 294
  special, 297
  submersible, 299
  three-phase, cast epoxy, 299
  three-phase connections, 337
  three-phase, boost and buck, 296
  two-winding, 294
  wiring diagrams, 298
Transients, 132, 376
Transistor, 115
Triac, 116, 230
  applications, 117
  symbols, 230
Trip current, 127
Troubleshooting, 389
  aids, 372
Turns ratio, 288, 289
Two speed manual motor starter, 73
Two-coil circuit, 130
Two-speed motor, 216
Two-wire control circuits, 74
Types of thermal overload relays, 125

## U

UL, 51, 52
Under-voltage release, 74
Under-voltage protection, 76
Underwriters Laboratories, 51, 52
Universal motor, 136
Utility-isolated peaking transfer, 320
Utility-powered parallel peaking, 321

## V

Valve, refrigeration, 197
Vectors, 338
Ventilation, 384
Vibration meter, 43
Vision system, 189
  programmable, 190
Voltage, 6
  fluctuations, 376

Voltage (*Cont.*):
   levels, 305
   module, 227, 228
   rating, 50
   recorder, 42
   regulation, 314
   relay, 121
   spikes, 376
   tester, 39
   unbalance, 385
   variations, 110, 200
Volt-ammeter, clamp-on, 379
Voltmeter, 15, 39
   DC, 15
Volt-ohm-milliameter, 40
Volts, 6
VOM, 38, 40
   analog, 40

## W

Wall duct, 346
Wattage rating, 27
Wattmeter, 18
Waveforms, two-phase, 143
Wells, 177
Windings, compensating, 140, 141
Wire
   gages, 36
   size, 9
   wrapper, 44
   wound resistors, 208
Wiring
   adequate, 370
   devices, 367
   diagram, three-phase motor, 72
   solid-state, 228
Working
   clothes, 56
   practices, safe, 53
Worn brushes, 370
Wound rotor motors, 238
Wrench, 33
   adjustable, 33
   Allen, 34
   box-end, 33
   torque, 34, 35
Wye
   connected stator, 144
   connection, 312

## Z

Zero-current turn off
   relay, 122
Zero-voltage turn on
   relay, 122
Zig-zag transformer, 59